Tactics of Interfacing

Leonardo

Roger F. Malina, Executive Editor
Sean Cubitt, Editor-in-Chief

Synthetics: Aspects of Art and Technology in Australia, 1956–1975, Stephen Jones, 2011

Hybrid Cultures: Japanese Media Arts in Dialogue with the West, Yvonne Spielmann, 2012

Walking and Mapping: Artists as Cartographers, Karen O'Rourke, 2013

The Fourth Dimension and Non-Euclidean Geometry in Modern Art, revised edition, Linda Dalrymple Henderson, 2013

Illusions in Motion: Media Archaeology of the Moving Panorama and Related Spectacles, Erkki Huhtamo, 2013

Relive: Media Art Histories, edited by Sean Cubitt and Paul Thomas, 2013

Re-collection: Art, New Media, and Social Memory, Richard Rinehart and Jon Ippolito, 2014

Biopolitical Screens: Image, Power, and the Neoliberal Brain, Pasi Väliaho, 2014

The Practice of Light: A Genealogy of Visual Technologies from Prints to Pixels, Sean Cubitt, 2014

The Tone of Our Times: Sound, Sense, Economy, and Ecology, Frances Dyson, 2014

The Experience Machine: Stan VanDerBeek's Movie-Drome and Expanded Cinema, Gloria Sutton, 2014

Hanan al-Cinema: Affections for the Moving Image, Laura U. Marks, 2015

Writing and Unwriting (Media) Art History: Erkki Kurenniemi in 2048, edited by Joasia Krysa and Jussi Parikka, 2015

Control: Digitality as Cultural Logic, Seb Franklin, 2015

New Tendencies: Art at the Threshold of the Information Revolution (1961–1978), Armin Medosch, 2016

Screen Ecologies: Art, Media, and the Environment in the Asia-Pacific Region, Larissa Hjorth, Sarah Pink, Kristen Sharp, and Linda Williams, 2016

Pirate Philosophy: For a Digital Posthumanities, Gary Hall, 2016

Social Media Archeology and Poetics, edited by Judy Malloy, 2016

Practicable: From Participation to Interaction in Contemporary Art, edited by Samuel Bianchini and Erik Verhagen, 2016

Machine Art in the Twentieth Century, Andreas Broeckmann, 2016

Here/There: Telepresence, Touch, and Art at the Interface, Kris Paulsen, 2017

Voicetracks: Attuning to Voice in Media and the Arts, Norie Neumark, 2017

Ecstatic Worlds: Media, Utopias, Ecologies, Janine Marchessault, 2017

Interface as Utopia: The Media Art and Activism of Fred Forest, Michael F. Leruth, 2017

Making Sense: Art, Computing, Cognition, and Embodiment, Simon Penny, 2017

Weather as Medium: Toward a Meteorological Art, Janine Randerson, 2018

Laboratory Lifestyles: The Construction of Scientific Fictions, edited by Sandra Kaji-O'Grady, Chris L. Smith, and Russell Hughes, 2018

The Invisible Colors: The Arts of the Atomic Age, Gabrielle Decamous, 2018

Virtual Menageries: A Mediology of Animal Representations, Jody Berland, 2019

From Fingers to Digits: An Artificial Aesthetic, Ernest Edmonds and Margaret A. Boden, 2019

Material Witness: Media, Forensics, Evidence, Susan Schuppli, 2020

Tactics of Interfacing: Encoding Affect in Art and Technology, Ksenia Fedorova, 2020

See http://mitpress.mit.edu for a complete list of titles in this series.

Tactics of Interfacing

Encoding Affect in Art and Technology

Ksenia Fedorova

The MIT Press
Cambridge, Massachusetts
London, England

This book was set in ITC Stone Serif Std and ITC Stone Sans Std by New Best-set Typesetters Ltd. Printed and bound in the United States of America.

Library of Congress Cataloging-in-Publication Data

Names: Fedorova, K. (Kseniiā) author.
Title: Tactics of interfacing : encoding affect in art and technology / Ksenia Fedorova.
Description: Cambridge, Massachusetts : The MIT Press, [2020] | Series: Leonardo | Includes bibliographical references and index.
Identifiers: LCCN 2019047637 | ISBN 9780262044158 (hardcover)
Subjects: LCSH: Human-machine systems—Psychological aspects. | Digital media—Psychological aspects.
Classification: LCC T14 .F35 2020 | DDC 620.8/2019—dc23
LC record available at https://lccn.loc.gov/2019047637

10 9 8 7 6 5 4 3 2 1

To my parents

Contents

Series Foreword

Leonardo, the International Society for the Arts, Sciences, and Technology, and the affiliated French organization Association Leonardo have some very simple goals:

1. To advocate, document, and make known the work of artists, researchers, and scholars developing the new ways in which the contemporary arts interact with science, technology, and society.
2. To create a forum and meeting places where artists, scientists, and engineers can meet, exchange ideas, and, when appropriate, collaborate.
3. To contribute, through the interaction of the arts and sciences, to the creation of the new culture that will be needed to transition to a sustainable planetary society.

When the journal *Leonardo* was started some fifty years ago, these creative disciplines usually existed in segregated institutional and social networks, a situation dramatized at that time by the "Two Cultures" debates initiated by C. P. Snow. Today we live in a different time of cross-disciplinary ferment, collaboration, and intellectual confrontation enabled by new hybrid organizations, new funding sponsors, and the shared tools of computers and the Internet. Sometimes captured in the "STEM to STEAM" movement, new forms of collaboration seem to integrate the arts, humanities, and design with science and engineering practices. Above all, new generations of artist-researchers and researcher-artists are now at work individually and collaboratively bridging the art, science, and technology disciplines. For some of the hard problems in our society, we have no choice but to find new ways to couple the arts and sciences. Perhaps in our lifetime we will see the emergence of "new Leonardos," hybrid creative individuals or teams that will not only develop a meaningful art for our times but also drive new agendas in science and stimulate technological innovation that addresses today's human needs.

For more information on the activities of the Leonardo organizations and networks, please visit our websites at http://www.leonardo.info/ and http://www.olats.org/. The Leonardo Book Series and journals are also available on our ARTECA art science technology aggregator: http://arteca.mit.edu/.

Roger F. Malina
Executive Editor, Leonardo Publications

Acknowledgments

This work has grown out of my long-term interest in philosophy and in the intersection of art and technologies. It crystallized in the present form during my time at the University of California–Davis for my second PhD and was finalized thanks to the Alexander von Humboldt Foundation fellowship in Berlin.

First and foremost I would like to express deep gratitude to Kriss Ravetto-Biagioli, my dissertation chair, who has rigorously read multiple versions of this text at its different stages. Her generous theoretical insights, lessons in critical discernment, commitment to high-level thinking, and genuine care about the subject matter at hand will stay with me. I would like to also acknowledge the inspiring support of my committee members, Joseph Dumit and Colin Milburn. Joe's many comments and questions have taught me to be open to potential surprises at any corner of the maze of intellectual work, and our many discussions of the issues of embodiment, plasticity, feedback, and scientific imagery were always a joy and introduced me to an array of new concepts and approaches. Colin Milburn helped to contextualize my work within the broader field of media studies and encouraged me to set high standards for thinking about its future. I was fortunate to have the steady support of Kris Fallon, who became a trusted sounding board for processing many ideas and a collaborator on writing and practice-as-research projects.

Special thanks go to the faculty at UC Davis (as well as at UC Berkeley and MIT) whose seminars have expanded my views on various issues and methodologies of the studies of science and technology, critical theory, and aesthetics—Mario Biagioli, Timothy Lenoir, David Bates, Christina Cogdell. Thanks to Colin Johnson, Jonathan Ducette, Emelie Mahdavian, Mayee

Wong, Xan Chacko, Mikhail Kurtov, Nina Sosna, Margarete Pratschke, and Asko Lehmuskallio for valuable informal feedback on different parts of this text. Claire Farago, Inge Hinterwaldner, Oliver Grau, and Ed Shanken, among others, nourished me with encouragement and motivating collegial conversations.

This project would have been impossible without the insightful and inspiring work of all the artists, many of whom I got to know in person. I am grateful for their additional feedback on my writing about their projects, as well as for kindly providing the images. Over the years, I have appreciated opportunities to exchange ideas with members of communities at the International Symposium on Electronic Art; Leonardo/ISAST; Media Art Histories; the Society for Literature, Science, and the Arts; and other conferences. In addition to UC Davis, the Berkeley Center for New Media was a reliable source of stimulating lectures, symposia, and informal conversations. I am grateful to my current colleagues at the Institute for Art and Visual History at Humboldt University and at other Berlin institutions for the chance to present and discuss parts of this work.

I thank the Alexander von Humboldt Foundation for their generosity in making it possible for me to finish this work. Grants from the UC Davis Humanities Institute, the UC Davis Program in the Humanities, the Mellon Research Initiative in Digital Cultures, as well as the Danish Agency for International Education allowed me to conduct field research by visiting artists' studios and having conversations with researchers at institutions such as the Royal College of Art, Goldsmiths (University of London), MIT, and Aarhus University.

I thank the anonymous reviewers of the manuscript for their insights and the MIT team for supporting this project.

The roots of this work go back to my formative studies and exchanges in Russia—with my dear teachers at the philosophy department of the Ural Federal University, who nurtured and helped to develop my initial interests in thinking traditions and particularly in contemporary philosophy and aesthetics; my collaborator Nina Sosna, who encouraged me to delve deeper into media theory, setting high standards of intellectual work; and the friendly and energetic community of media art practitioners in Russia, whose inner freedom and determination continue to be inspiring. I am grateful also to my colleagues at the Ural branch of the National Center for Contemporary Arts for giving me an opportunity to run the Art Science

Technology program. My friends in my hometown of Ekaterinburg became imaginary (and sometimes actual) conversational partners who helped to sharpen ideas and to sustain my intellectual interests. I can never repay the debt I feel to my parents and grandparents for respecting and supporting my choice to pursue my scholarly path so far away from home.

The ways in which I am grateful to my soul-partner Marc, who supported me on this path most closely, are too multiple to list. I feel honored and lucky for his presence in my life, and the example of selflessness, courage, enthusiasm, "thinking big," life assertiveness, and love he has demonstrated to me every day. The profound impact his way of thinking has had on me carries throughout this work as well.

Introduction

"How am I looking tonight?" a young man asks into the air while fixing his hair in front of his bathroom mirror. The question is not just rhetorical, but has an addressee; for a number of days, the man has been receiving very specific advice on what kind of appearance and behavior might help his self-esteem. In our era of smart home agents, such interactions are hardly novel. Amazon's Alexa, for example, can learn details about your diet, or your home heating and lighting preferences, or access personal style tips. Yet in this case, the digital assistant is not Alexa but "Lauren" (figure 0.1). Like Alexa, "Lauren" operates through electronic devices, such as cameras, microphones, and various sensors distributed throughout the home. "Lauren" is there not only to assist with immediate tasks but to "surprise [you] and to make you feel something." She is there to give a "human" kind of support—because, in fact, she *is* a human. In this performative art project, artist Lauren McCarthy comes to people's homes and installs custom-made devices to monitor their lives for several days, in order to anticipate and satisfy their needs through remote actions and advice (which appears as a text on a specially designed screen).[1] She puts herself in the position of a machine, testing people's expectations, but also giving them attention, the kind of attention a machine would not (yet) be able to provide. "Lauren" helps its users realize how intimately they allow technology to infiltrate their lives and how subtle the boundaries between human and nonhuman may become.

The history of human cohabitation with technologies often shows how quickly people can become accustomed to such new conveniences. The ease of interaction, increasingly in the form of natural language-like dialogue, is becoming commonplace, yet the implications are hard to overlook.

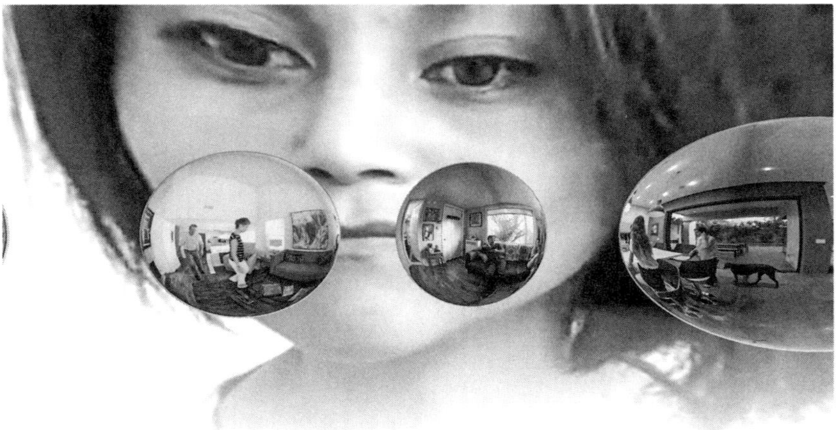

Figure 0.1
Lauren McCarthy, *LAUREN* (2017). Screen capture from video. Courtesy of Lauren McCarthy.

Children growing up in a world permeated by technology perceive actions as inputs and expect outputs—tangible changes in their material environment. They swipe picture books with their fingers, as if those are tablet devices, expecting the images to respond; they embrace home assistants, like Alexa, as family members. More and more everyday life choices and decisions are now delegated to algorithms. Technologies invented to be controlled establish more and more control over our behavior.

This book explores how new digital technologies affect the way we conceive of a human "self" and its relations to the world. With the introduction of ubiquitous computing in all facets of daily life, the "self" increasingly becomes an object of technological application. Information technologies, such as biofeedback, machine vision, and remote sensing, influence how we define ourselves based on the new kinds of data we receive about ourselves. These sensing data are principally different from the results of traditional techniques of self-observation, such as diaries, which are mostly descriptive. In contrast, technological monitoring is automatic and pervasive. Machines today can detect much more than what we can realize by ourselves. They can track our emotional states by analyzing bodily characteristics and facial microexpressions, and read our communications for mood predispositions. They assist us in daily tasks, like giving nearly omniscient driving directions or regulating temperatures in our homes, but they become more than mere technology, serving also as our companions, psychotherapists, and whole environments that can stimulate or calm us down, help us to reorient our focus, remind us of something important from our personal history, and so on. The issue at stake is that tracking technologies enhance and augment, but also simulate the sense and image of who we are. Subtly, they encourage versions of ourselves that are easier to interpret computationally. Allowing deeper, more granular, and often quite intimate connection with ourselves, they simultaneously become a means for manipulation and control.

At the center of many of these dramatic changes is communication with and via technologies, a process defined by the phenomenon of an interface. In a conventional sense, or in relation to computational technologies, the interface is a place of connection between a human and a digital system that allows them to communicate with one another in order to generate and exchange information. Besides the so-called WIMP (windows, menus, mouse, and pointer) of the classical graphical user interface that originated in the 1980s, interfaces today include virtual and augmented reality tools; biochip implants and wearable biofeedback sensing devices; forms of interaction within intelligent environments; and auditory, tangible, gesture-based, and robotic interfaces. All interfaces offer ways to navigate through data—be it images, numbers, text, sound, or even bodily sensations like vibrations. They also open diverse avenues for embodied interaction and affective experience, thus generating new constellations of meaning.

I approach the interface not so much as a technology, but as a condition that brings to the fore and gives structure to the relational nature of being human. The implied relations may include a relation toward oneself (self-reflection) or to other human and nonhuman beings, or to a particular situatedness in life—the list can be continued. Whether the attention is turned toward one's own needs or toward another human, these relations are always mediated—if not by technologies, then by our own perceptions and attitudes. Unlike well-known approaches to relationality in contemporary art and culture, such as Nicolas Bourriaud's relational aesthetics with its focus on formal configurations of social interactions, I consider specifically forms of mediation via digital technology. Yet to define this project in categories of interaction aesthetics, such as the ones introduced by Katja

Kwastek in *Aesthetics of Interaction in Digital Art*, would also not be completely correct.[2] The term "interfacing" in this book does imply both relational and interactive scenarios, but it calls for a shift of attention toward the very textures and operational grammar that enable these scenarios within the digitally determined world. "To interface" means to activate the condition of an interface: to enter a dynamic framework of relations organized into a certain procedural structure. I argue that the logical organization of computer-mediated communication and experience, within a given interfacial condition, can open up room for new experiences of the self and a renewed conception of what may be constitutive to a human self.

Projects like *LAUREN* highlight the psychological effects produced within interactive experiences. They bring to the foreground expectations and projections, enchantments and disenchantments, and, most importantly, the shifts in perspective that may be subtle but profoundly transforming. *LAUREN*'s participants ("users") report that, like an intimate companion, she helps them find their car keys and reminds them to take their medications, but they also realize that there are zones which they prefer to keep private, and that there is a line between her being "in support" and "in control." The interface that she sets up comprises elements such as placement of the camera, the form of communication with her (e.g., by simply calling her name), and things that she can control remotely, such as light, heat, or music. This interface structures the participants' relations both to her as a human and to their own daily routines—to how they express themselves in private and what those expressions (seen from the outside by another human) may tell them about themselves. She is present through observing them while staying invisible and giving only written comments, yet her presence permeates their lives as she exerts control over the ambient conditions of their homes, reminds them of things only they themselves would know, and comments on their feelings. The very idea of a relation is thus highlighted and problematized by the interface: it is the technological interface that stands between Lauren as a human and her projects' participants, evoking new realizations about what conditions behavior and interactions.[3]

The specific focus of this book is contemporary media art practices placed in dialogue with the broader theories of the "self," mediation, and interactivity. I argue here that art is distinctly suited to exploring the impact of ever-increasing digital interventions on our self-perception. Art offers a space from which we can reflect on and expand the conceptual dimensions of the kinds of experience and knowledge received in the form of biometric and other types of sensing data. Because of its more speculative and experimental status, art promises to be a "safer" way to try out new experiences and explore the hidden facets of one's self. (What may be unacceptable in other circumstances is here encouraged.) Through immersing its participant in a relational feedback loop, art reveals how sensing one's self can generate its own sense and how awareness of the constructedness of the offered experience can give a sense of potentiality, and even feel liberating. When Lauren McCarthy puts herself in place of a machine (an Alexa-like home assistant), she disrupts the unspoken agreement that delineates the zone of machinic competence and purview. By entering that zone with her human sensibility and drawing attention to operations that usually are taken for granted, she also foregrounds the capacities and power of human relations, making clear what in these relations cannot (or maybe can) be replaced by a machine. Yet this awareness comes exactly from the experience of being enclosed in the system. It is specifically through active participation that one acquires a more self-reflexive position and a different perspective on how the system as a whole operates.

Despite the centrality of the art material in this analysis, I would not position it as an art historical investigation (at least not in a classical sense). Rather, art serves here as a lens that brings together diverse philosophical and media theoretical conceptions of the aesthetic and ethical aspects of technologies' impacts (as in work by Mark B. N. Hansen, Brian Massumi, Luciana Parisi, Alexander Galloway, and Wendy Chun, as well as Friedrich Kittler, Gilbert Simondon, and others). While some of these authors (e.g., Hansen and Galloway, but also Anna Munster and Kriss Ravetto-Biagioli) appeal to art as a ground for developing their theoretical findings, the constellation of topics and examples discussed here is different, and so is the focus—modalities of experiences of relationality under the condition of their technological encoding.

I feature projects such as the interactive installations *Unstable Empathy* by Mattia Casalegno and *Telephone Rewired* by Sean Montgomery, where artists track levels of brainwave synchronization between different participants, thus placing them in an intersubjective space of unarticulated dialogue. The same topic of sharing intimate body expressions with a complete stranger has also been explored in better-known pieces such as Christa

Sommerer and Laurent Mignonneau's *Mobile Feelings*, in which handheld objects communicate over a distance the holder's heartbeat (transmitted in the form of a subtle rhythmic vibration) as well as breath, which—held next to a cheek—feels like a kiss on the other end, creating a sense of telematic intimacy. Another well-known example is Raphael Lozano-Hemmer's *Pulse Park*, which filled Madison Square Park in New York (as well as other public spaces in other locations) with green light beams pulsating with the rhythm of the heartbeats of passersby, thus serving as a public interface for the collective organism of the city.

The interaction in these installations is still digitally mediated, but instead of data, the emphasis is more on what we feel their transmission means to us. The meaning assigned may be a purely private one, not easily shareable and convertible. The form of communication that gets established goes beyond quantification and rather occurs through images, fictionalized associations, and metaphors, while also being an acutely sensory experience.

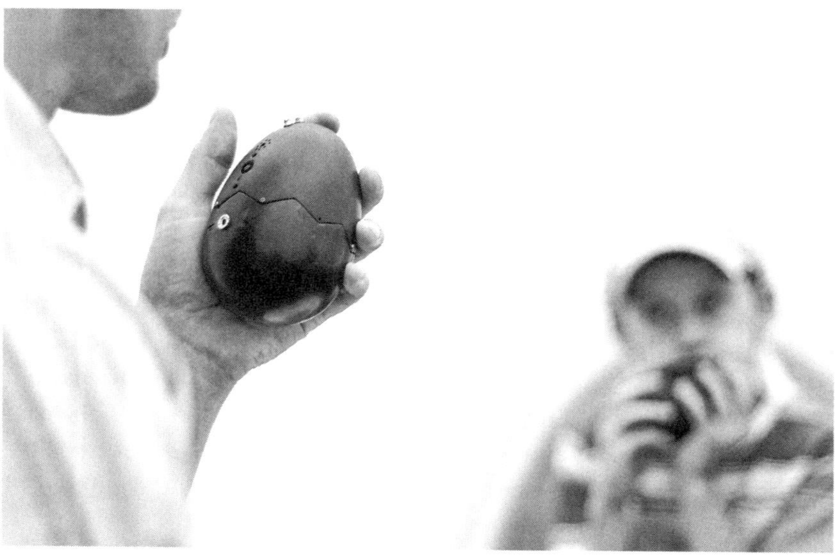

Figure 0.2
Christa Sommerer and Laurent Mignonneau, *Mobile Feelings* (2002–2003). © Christa Sommerer and Laurent Mignonneau, in collaboration with France Telecom Studio Creatif, Paris, France. Photo by Victor Voropaev. Courtesy of the Ural Branch of the National Center for Contemporary Art, Russia.

The aesthetic dimension of this "being in relation"—with one's self, the other, and the world—calls for a highly subjective, time- and place-specific sense of presence. The embodied and psychological situatedness of the experiencing subject is crucial in the selection of works presented here and distinguishes this analysis from other writings on interface and art, for example by Christian Ulrik Andersen and Søren Bro Pold, who focus on the social and political aspects of the interface as a cultural construct that defines today's infrastructures of cultural content creation and consumption, labor organization, and perception of a globalized world.[4] Rather than the broader cultural impacts of platforms and global digital economies, at stake here are coded operations that concern an individual experience (for example, of commonly used technologies such as location tracking or biofeedback-based ones). The kinds of embodied and narrativized accounts discussed in this book are often open-ended and even troubling, inviting further questions and complex emotional responses. For instance, works using locative media, like *Indeterminate Hikes+* by the EcoArtTech collective, instead of orienting one in space, emphasize the unpredictable nature of one's immediate surroundings, encouraging its participants to stay open to the emergent and unprescribed expressions of the location, such as its sounds, colors, and animate and inanimate elements. They shift attention toward the contingencies of personal experience, the zones that perhaps still cannot be fully precoded and rather should be discerned and made sense of on the terms of the individual herself. I thus explore these spaces of negotiation between possible personal interpretations of given situations and the codes that frame these situations.

Placed in the larger context of media design and technocultural practices, art reveals a distinctive critical capacity. Works like *LAUREN* intervene in this dominant context from within, using the same media and vocabulary of justification, but with a twist. The custom-made surveillance and control devices that McCarthy distributes in the homes of the project's "subscribers" are deliberately modeled on commercial ones, yet they are not preprogrammed and respond only to the artist's own commands. Indeed, by "camouflaging" themselves as certain recognizable forms of media (smartphone applications, virtual assistants, etc.), many media art projects intentionally confuse their audiences. They feed into contemporary sensibilities that have already been affected by technologies, such as people's reliance on smartphones, biofeedback tracking, and other devices.

However, their strategy is to subvert the assumptions we have about the smooth functionality, and often the presumed neutrality and objectivity, of these devices, as well as the human qualities we have projected onto them, such as "loyalty" and "friendliness." The ultimate goal of these artworks is to discover and bring to light the various "gray" zones of technological culture.

This type of interventionist approach and call for alternatives to mainstream media structures is precisely what characterized the idea of the "tactical" within the digital and software activist practices of the 1990s, also known as "tactical media."[5] Characterized famously by Matthew Fuller as "not just art" in 1998, it also became the motto of net art and "artware"—the art of software platforms, developed as an alternative, community-driven means of exchange, disclosing regimes of control, autonomy, and political agendas in software.[6] Both usages were inspired by the concept of "tactics" in Michel de Certeau's *The Practice of Everyday Life* where he defines a "tactic" in opposition to "strategies" of technocratic institutions, as a "calculus which cannot count on a 'proper' (spatial or institutional localization)" and thus "insinuates itself into the other's place, fragmentarily, without taking it over in its entirety, without being able to keep it at a distance."[7] De Certeau associates tactics with "ways of operating" and "styles of action," like "ways of walking, reading, producing, speaking, etc." These "traverses" of the dominant systems, as he also calls them, "circulate, come and go, overflow and drift over an imposed terrain, like the snowy waves of the sea slipping in among the rocks and defiles of an established order."[8] While tactical media, as well as other activist cultural traditions, take as their guiding line the resistance mode of de Certeau's text, I invite readers to explore some other possible aspects of the concept of "tactics."

One useful characteristic of "tacticity" in this expanded reading is that it speaks to concrete and situated practices, and to responses to particular sets of challenges. What is relevant to the present study is both the specificity and individuality of such a response, the "degree of plurality and creativity"[9] that it instantiates, and the fact that it is indeed a reaction exposing an attitude toward something; that is, it does not exist in a vacuum, but also requires a position, a stake—be it a conscious or an intuitive choice. This, then, suggests a question: Whose position? Who or what is the "actor" of "interfacing" that performs the "tactics"? What does being an "agent" capable of producing change involve? But there is another question also: Is the actor indeed only an individual subject and always only a human? The context of interfacial relations makes this system dynamic much more tangled. Interfacing itself should be seen as a tactic; for example, as a means of diverting and "tricking" the "dominant system" of the habitual or socially constructed "self." Another aspect that I try to bring to the foreground is the affective dimension of the attitude behind a "tactic": What kinds of emotions are involved ("encoded") in the *dérives* modeled by the artists in response to the challenges of today's digital culture? How are ethics and aesthetics intertwined in them?

Among the key features of digital processing that require particular critical reflection when applied to human matters are automaticity and programmability. If more and more decisions are delegated to machines, how does this trend constrain the expression of individual human will? How well justified are the claims by scientists, such as by those engaged in bioinformatics about the computability of bodily processes—but even more importantly for the present context, by the representatives of the field known as affective computing about the de/codability of emotions (e.g., "reading" emotions by analyzing facial microexpressions or biofeedback data, such as heart rate variability or galvanic skin response)? The aesthetic regime of engagement helps to reveal the complexities of human response to these machinic characteristics. Interfacing can therefore be approached as a tactic of the body itself as it resists such codifications and computable structures by extending itself across and beyond them in its creative expressivity (e.g., through gesture and dance, or an affective engagement with a physical location in a feeling of proprioception).

Of course, the application of algorithms to human processes has its history, which goes back to the cybernetic understanding of a human as a system operating through feedback. Indeed, the behavior of a human as a living organism is structured through its interaction with the environment. Perception of the environment (such as identifying danger) both precedes and concludes action by comparing the images it creates before and after the act. These feedback loops are foundational for the existence of any natural being since they are the basis for learning and adaptation (which in turn is key to the survival of a species). Interfacing also fundamentally presupposes feedback.

The principle of feedback was initially applied to descriptions of self-regulatory processes in organic and mechanical systems. Generally, it implies that the responses of a system to the changes that provoke those responses are fed back to the source of the initial changes, thus forming a loop; think, for example, of how a thermostat regulates room temperature by constantly measuring it against the set number, or how an organism releases sweat to keep its own inner temperature stable. Feedback is foundational for the adaptive behavior of not only organic and mechanical systems, but also information-based systems and artificial intelligence. It is crucial to remember that the primary interest of the mathematicians and physicists associated with cybernetic theory (during and right after World War II) was the issue of control and the possibility of calculating "restrictions of outputs" for the missile trajectories directed toward a moving target. Such origins necessarily informed the instrumentalist, management-oriented logic of most of the later adaptations of those scientific insights, including in the field of affective computing. The interfacing scenarios that I discuss are all structured through negotiations with such control mechanisms and, principally, with their predictive and *anticipatory* nature.

As has been recognized by cybernetics theorists (particularly Arturo Rosenblueth, Norbert Wiener, and Julian Bigelow, in their classical paper on self-regulatory behavior), a "margin of error" is needed to correct behavior in real time, or as they put it, "in the course of the behavior": "the signals from the goal are used to restrict outputs which would otherwise go beyond the goal."[10] The principle of feedback loops is thus the basis of algorithmic processes as time-based decision-making mechanisms. Initially, cyberneticians, such as Ross Ashby and Wiener, described the desired state for both natural and machinic systems as a state of homeostasis, or a state of balanced stability, equilibrium.[11] Yet feedback itself, even as a loop, is a dynamic process. It stems from and generates differences and discontinuities, and thus keeps the system in a state of constant suspense.[12] Checking the state of a system against the external conditions, it establishes a relation, but the level of difference between the system and those conditions is not immediately known. That moment of delay (between measurement and reaction) is a source of its own effects, which are, as I argue, "affective" in their nature.

In this book, I analyze works that provide insights into the human experience of how it *feels* to be locked in a feedback loop. Such is, for instance, the experience of the dancer Ekaterina Zharinova in her media performance *Inside the Chain* (2016). In this performance, Zharinova moves together with her own screen doubles that are created via motion capture technique and are slightly delayed. The work involves the casting of doubles and her response to her own moves as they are presented to her through the figures on the screen. A different and yet similar interaction is modeled in the interactive installation *Alter Ego* by Alexa Wright, where the viewer is faced with a 3D-rendered copy of oneself that can effectively respond to its "original's" expressions of surprise, concern, or bedazzlement by simulating its own facial expressions. The affective reaction that is born within the experience of this relational dynamic is far from the state of equilibrium or homeostasis. Such a state should rather be called *metastable*, since the source of its stability is provided not by the individual entities but by the relations between them. Within such a model, relations matter more than what they connect.[13]

Metastability is known as a "false" or unstable equilibrium that can persist for an unbounded time before being broken by even the slightest modification of the system's parameters. It implies a state of continuous oscillation. Metastability characterizes the system's potential of being in relation (in system theory, it is literally the potential energy needed to balance the excess or lack of other parameters). Gilbert Simondon relates it to the state of potentiality more generally and to the process of "becoming" out of that state, characteristic to both technical objects and humans.[14] The artworks that I discuss place the beholder in the midst of such a state, where responses are continuously (and, potentially, endlessly) built upon responses, and the borders of individual identities are put in flux.[15] What holds the responses together and organizes them into a particular experience is the condition of the interface. Across the works of diverse media and varying conceptual agendas, interfacing plays a role not only as a meeting point between the audience and a piece but as a relational scenario for living through that state of confused boundaries and in constant (suspenseful) readiness to respond to the changing conditions. Throughout the book I show how awareness of the interface can make us rethink our position in the world as a stable one and helps us to become more open to experimenting with our emotional responses to the unavoidable conditions of change.

In my focus on these situations of the "in-between," I partly build upon the heterogeneous tradition of critical writing associated with the so-called affective turn, united by the attention of its authors to the nonsymbolizable, nonrepresentational, presubjective, and embodied modes of being and meaning production.[16] Yet the aim here is not to theorize the affect and emotion per se. Rather, my contribution to these debates is grounded on comparative analysis of particular modalities of experience with technologies of prediction, tracking, and simulation. I attempt to understand what happens in interaction with digital "doubles"—in their visual forms but also based on other types of expression of the self, such as that which is registered in pools of data stemming from personal online search histories and other traceable behavioral features. This inquiry connects to recent studies of new, technologically conditioned affective phenomena, such as the conception of the digital uncanny developed by Kriss Ravetto-Biagioli. She writes: "We are now confronted with a new type of uncanny experience, an uncanny evoked by parallel processing, aggregate data, and cloud computing. The digital uncanny does not refer to human anxieties projected onto nonhuman devices (automata, dolls, avatars, body doubles) but rather to how nonhuman devices (surveillance technologies, algorithms, feedback, and data flows) anticipate human gestures, emotions, actions, and interactions, thus intimating that we are machines and that our behavior may be predictable precisely because we are machinic."[17] Given the results of the application of data collection and analysis (including algorithms of AI), such unsettling concerns are indeed well justified. Hence, the need for a counterargument becomes ever more urgent. Without promising any particularly hopeful outcomes, I suggest that at least one way to counteract the dystopian prognoses could be the cultivation of zones where we as humans could still resist such machinization of our own views on ourselves: for example, through creativity and irony, through the tactic of refusing to agree with the solidifying approximation-based behavioral "patterns" and "best match" options routinely offered to us. With their emphasis on interfaciality, on the process of relating, the artistic scenarios I feature here attempt to disturb these analogy- and reproducibility-oriented schemas.

The interface is the place where feedback happens and various potentials play out. In fact, my use of the term is closer to the original definition that dates back to the nineteenth century. The term "interface" was coined by the engineer James Thompson in relation to the behavior of fluids in thermodynamic theory. It was defined as a *boundary condition* that produces fluidity from the differences in energy pressure within parts of a liquid. Thus, from the beginning an interface was understood not as a dividing surface, but as a location and a state of unequal distribution of qualities (e.g., energy) that creates movement within a substance (e.g., a liquid). Its main meaning is in its *effects* (such as fluidity); in fact, it reveals itself only in effects. In other words, it is defined through the *potentiality* of these effects. While the effects of the interface can be multiple, it is important that they are describable, and it is so because they take a certain form, even if it is a dynamic form. As Branden Hookway describes, "dynamic form is less a form than a *forming*, a process active across space and time, and elusive to formal analysis unless captured in some way."[18] The forming occurs out of potentiality. (One way to capture dynamic form is also through the concept of *work*, which is similar to "effect," e.g., in the sense of thermodynamics—the production of heat or mechanical work.) Many pieces I discuss present images emerging out of other images, such as the morphing of the contours of participants' facial features as their brainwaves synchronize in the above-mentioned installation *Unstable Empathy*. The resulting "hybrid" image literally *forms* itself as a projection on a glass surface placed between two people, facing one another, while they try to "tune in" to each other's state of mind. Another example is the well-known video work *Three Transitions* by Peter Campus, where the features of the face are gradually erased, only to reveal the same face left behind. The interface here is not only the screen with a projection, but the whole setup of cameras, sensors (in case of the former), and videotapes (in case of the latter) and their temporal relations. But interfacing as a procedure and a tactic also implies what happens with the subject position of the human, whose face reemerges after being erased.

What counts as an effect produced by an interface is a real-time transformation of a system that happens in response to manipulation of the individual components of the system. Just as in thermodynamic systems, changes in particular characteristics, such as heat level, affect the state of the system as a whole. The same part-whole relation takes place in computer interfaces we use every day. What we know today as icons within graphical user interface (GUI) not only signify an action (as its "symbols") but are direct triggers of that action: they can switch it on or off (if it is reversible). The icons are indexical signs of specific material or immaterial effects: they represent visually what happens if this or that button gets pressed (for example, an icon of a photo camera indicates that the button is used to take pictures). But the effect of an interface is more than just the changes triggered by manipulation of the icons or other input controllers. At its deeper level, its effect involves a vision of what constitutes the "whole" of a system—be it technical or human. It places agency in the very logical and material connections between the controller parts (icons) and the whole (the action they trigger). The aim of this project is to foreground our relations to these connections and what their latent potential can make us feel. Yet first there should be an awareness of that relation itself, an awareness of being affected by the machinic operations. This does not come automatically and often requires its own situational scenarios, like the one *LAUREN* offers.

We expect to use interfaces in order to have control, but we do not usually realize how our very actions and thinking can be controlled by the protocols embedded in the devices (just as a child might swipe paper pages, or an iPhone user strengthens intonation and articulation for Siri's understanding). Recent critical studies on the interface have approached it as a politically and ethically charged place of concealment.[19] One of the biggest changes that has occurred since the first computer user interfaces were developed in the 1960s through 1980s, aside from their spread and commercialization, has been the expansion of the principle of invisible, organic, and imperceptible interfaces of ubiquitous computing—everything from multitouch displays and Apple black-box products to voice-controlled smart environments, like the Alexa system. These interfaces prioritize seamlessness and function above the user's knowledge of how exactly they function. As Lori Emerson points out, "all these interfaces share a common goal underlying their designs: to efface the interface altogether and so also efface our ability to read, let alone write, the interface, definitely turning us into consumers rather than producers of content."[20] Behind user-friendliness and ease of use is hidden a complex set of algorithmic decisions and predetermined choices. With pervasive tracking of multiple aspects of our daily lives (our diet, our information and product consumption, our friendships, and our political opinions), the notion of choice itself may soon recede to the background. Ultimately, what may be endangered is individual human agency itself. In his *Interface Effect,* Alexander Galloway shifts the center of the critique from the interfaces per se to the kinds of political order they represent. He suggests understanding a computer in terms of its ethics, implying precisely aspects of control embedded already within representational structures that can be either coherent or incoherent and, with that, allowing different strategies of choice.[21]

The problem is not only (lack of) digital literacy, but the ability to understand the work of complex sets of protocols and code routines that underlie the perceptible layer and specify the rules of communication between the input signs and software components responsible for triggering action (the so-called application program interface, or API). It may be enough to remember the general principle of the *mise en abyme* of computer architecture, the layeredness of its interfaces, including an input (a mechanical signal, like touch, or an object, like a USB plug), code layer (rules of translation between data formats), as well as data representation as an output. This book calls attention to an *aesthetic* type of awareness about the qualities of the interface: what touches you in its work and how exactly you perceive it. As Christian Ulrik Andersen and Søren Pold put it, "If it is not possible to fully 'unveil' the '*mise en abyme*' of the interface, we can, however, elucidate how interfaces can embed choices, conduct, languages, and ultimately values, worldviews and aesthetics into technical infrastructures."[22] Such analysis does not require understanding the technical nuances, but trains one to remember the potential implications of one's own actions, feelings, and even thoughts.

"Interface" (at least in English) relates to "face"; it is a zone in between two faces. Extension of this definition to human-computer interaction leads to the assumption that both a human *and* a computer would have "facial" features. The model of a face has long been one of the key instrumental features of human-computer interaction design as it strived to attain "user-friendliness" (think of the famous Mac smiley face, or the many digital assistants represented by animated faces). The "friendliness" of a machinic face does not have to be literal—it can be expressed through verbal approval (or disapproval) of an action, beeps, flashes, and other kinds of reaction. The key is that the response should be one that a human should be able to intuitively understand—just as the input from a human should be understandable by a computer. Machines did not always have a face in the sense of a screen interface. Before the more familiar screen interfaces, computing machines had tangible but cumbersome forms of interfacing with them, including punch cards, scrolls of paper, or a mesh of wires on a wall. Still, what unites these tools with today's screens is that all of them are systems that perform a form of faciality. They do that through presenting and perceiving information and by establishing a feedback relation. The components of this relation include operations of sensing, recognition, and interpretation, but also mutual encoding—transferring onto the other the operational codes needed for understanding each other. One of the goals of this work is to demonstrate how exactly that happens and what the equivalents of a code as language can be at the level of a human.[23]

Within the technical milieu, the interface is often described in the context of so-called object-oriented programming, where "objects" may be any element in the system: software, hardware, as well as people (users). The interface mediates their relations. Taken within a longer cultural history of material forms for recording and reproducing information (such as scrolls and books, but also optical media from camera obscura to cinematic screens), the interface can be seen as almost synonymous with the concept of a medium more generally. In this sense, it can indeed play the role of an "active metaphor" and a "translator" assigned to media by Marshall McLuhan ("all media are active metaphors in their power to translate experiences into new forms").[24] The interface as a "translator" enables the representation of "objects" that do not have a phenomenal or material existence—such as imaginary and virtual objects, but also fantasies and desires.[25] As Galloway points out, we can also compare "interface" with the concept of "intr*a*face"—interface within an interface, a "zone of indecision" between the inside and the outside.[26] "Intraface" may, then, remind us that while the interface can be thought of as a threshold, a point of connection to something beyond itself (like a window or a door), in the end it works more as a frame for an image, or, perhaps, as a frame entirely subsumed by an image, which, ultimately, is still (only) an image. And yet the interactive possibility of the interface makes this image *feel* real and allows one to use this zone of suspension and indeterminacy to play out in it an imaginative scenario (as if it were an "objectively" given reality). The interface is thus a point of entry of the imaginary and the unconscious.

My project takes inspiration from an understanding of the mediating capacity of the interface as "forming" and as the space of "becoming" of a relation—regardless of the distinction between the real and the imaginary, the "objective" and the "subjective." The "system" at stake here is the "self." What in thermodynamics is "work" and in a graphical user interface is a response of a software system, in the case of a human self is transformation of its position as a relational being. I will use the word "self" alongside concepts of the "subject" and "subjectivity." Whereas the "subject" position can sometimes be related to a machine as an entity with agency (in contrast to an object), the "self" remains a category of the human (in contrast to the "other").

The self is seen here less as an entity than as an expression of its relational qualities. The idea of an identity and the self is inscribed in an individual already through language—through the personal pronoun "I" that a person needs to communicate with others and to distinguish his or her position from that of others. This perspective was deepened in Jacques Lacan's psychoanalytic theory of the formation of the sense of self in relation to a mirror image of the self, as an adaptation of the external view that others have of one's self. A similar thought can be traced back to the concept of the "reflected" or "looking-glass self" developed by the American sociologist Charles Cooley in 1902. Specifically, for Cooley it meant a knowledge of one's self formed through internalization of other people's judgments that function as one's mirror image (I am what I think you think I am). Although neither he nor most of his followers in social psychology went into the intricacies of the unconscious and its relation to the real and the symbolic, his looking-glass self, analogously to Lacan's, is also based on the perceived, or *imagined* idea of one's self and imagined relations between the ego selves. My emphasis here, though, is on how an aesthetic experience—that is, a tuning of one's perceptory field to the outside world—can become a special condition for configuring these relations. I show that they do not always function as identification or introjection, and instead often result in confusion and disorientation of the very schematic "givens," such as those established through language and culture. The interface, then, plays the role of a structure for this experience, a condition, rather than a thing or even a process. By "tactics of interfacing," I imply particular practical scenarios of activating that structure.

The examples of artworks discussed here include projects in a variety of media forms, ranging from video, light, and sound installations to wearable devices, smartphone applications, and mixed-reality and locative media performances. All these forms of media appeal to different modalities of perception and engage both the body and the cognitive apparatus. Although many projects attempt to break through (or away from) the dominance of the visual sense, the idea of the image reoccurs in different forms. One of them is the image of the self that can include both an inner view of one's self and an external perspective, given through a mirror image or through data (biometric, social media, etc.). Such an image of one's self corresponds to two very different, but paradoxically overlapping spheres—the imaginary, intuitively given, and the externally observable and calculable. "Modalities of perception" primarily implies human sensory abilities, but as machines increasingly participate in constructing their own "versions" of us, the spectrum of capacities through which something can be sensed or "known" grows wider. The aim here is to explore the implications that these new types of perspectives enabled by sensing technology have for our own sensing and making sense of ourselves.

Each chapter addresses various tactics of interfacing via feedback used in media art and how they reflect the issue of relationality, or being-in-relation: with one's self, with the other, and with the environment. In the course of this study, I consider several modes of relationality that involve embodied and disembodied modalities, each adding different aspects to what can (and cannot) be given to us in feedback through digital systems.

The first chapter, "Face to Interface," develops the issues of interfaciality through facial perception and interaction. "Facing" one another, humans and machines need to quickly decode each other's respective actions and respond accordingly. A face indicates agency, an ability to act and respond to actions; it represents a certain individual standing and a position—however preprogrammed it may be. What is critical is that in such an exchange the human also acquires a position—that of an interlocutor who has to adjust to the understanding capabilities of its machinic partner. I begin with distinguishing characteristics of the face as a medium, then analyze a number of interactive installations that engage techniques of image morphing, facial expression recognition, and computer vision algorithms to explore issues of fluid and networked identity and the simulation of emotions. The participants are emotionally immersed in interactions with the digital versions of themselves or with representations of others, yet the very interface of the interaction keeps them wondering about the mechanisms of its effects. The face gives way to an interface.

In the second chapter, "Body Image and the Algorithmic Organic," I focus on the ideas of self-image and the boundaries of the self, understood through the perspective of technological mediation of an embodied experience. Body data-scanning technologies allow more and more accurate tracking of the inner bodily processes, including those related to affective states that lie below the threshold of conscious control. The proliferation of "smart" wearable devices and software applications that help you track yourself is a signifier of a cultural trend (exemplified, for instance, by the Quantified Self movement). It is oriented toward embracing the potential of technologies to give us a more nuanced "data image" of ourselves that can help us discover otherwise invisible patterns in our behavior and make more informed personal decisions based on that knowledge. This strategy, though, implies subjectification of the self to the algorithms mediating the work of the sensing devices. The resulting "image" of the self is a "data proxy," and there are still questions about the spectrum of its effects on the idea of self-knowledge.[27] Such a technologically generated "picture" of one's self can be analyzed within the long philosophical tradition of self-perception and of the role of the image in configuring the self. One useful perspective is Maurice Merleau-Ponty's distinction between a body image (where a body is an object of intentional consciousness) and a body schema (a nonconscious configuration stemming from the experience of the body itself). I use examples of the video, time-based installation, and performative works created via motion capture technique to address the issue of doubling the body through an image and in other ways. Other works, using wearable biofeedback technologies, are also discussed to challenge the dominant narrative of self-optimization, offering alternative creative paths of self-exploration. The concept of the image is understood here broadly: as a sensible or perceptible entity (not necessarily only visual). It continues the analysis started in the first chapter about communication through an image and the complications for human natural perception arising from the attempts to codify that image. Here, the image is interpreted not only in terms of its potential codification (as when facial expressions are seen as informational units) but as engaging certain types of feedback mechanisms at various levels. I argue that images can be seen as interfacial, while the body, with its grounding in the organic and the biological, provides an important contrast and represents a logic of its own that points beyond the principle of the self and enables experience of relationality on its own terms.

In the third chapter, "Mixed-Reality Interfaces and the ELIZA Effect," I address the question of how technologies reshape the relation between the self and the other at a cognitive level. The cases I feature in this chapter are positioned in between the actual physical reality in which the viewer's body is present and the "virtual" datascape reality—the space of self-representation through self-description, as well as through various forms of activity tracking (for example, the kind that Lauren can observe and analyze). This implies expansion of the field of intimacy, raising questions of trust. Would people trust Lauren more if she were an algorithm or a real person? How exactly does the illusion of reciprocity work, and what are the ethical and political implications of its simulation? To explore these issues I consider the effect of the psychological mechanism of projection. In the context of digital culture, it is exemplified by the "ELIZA effect," named after the "psychotherapist" chatbot ELIZA created by Joseph Weizenbaum in the 1960s, a phenomenon where people assume that a machine has more human qualities than it actually has. I look at the interfaces and algorithms behind psychotherapeutic relations with artificial intelligence agents, and the problems this type of communication entails. An illustrative and provocative example is a smartphone application "Karen," by the British art collective Blast Theory, that tricks you into a trusting relationship with a real-looking but preprogrammed character. Played by an actress, "Karen" is an algorithm, and in that sense she is the exact opposite of Lauren, who is invisible but is a human and remotely present with you in real time.

The fourth chapter, "Interfaces of Spatial Relationality," focuses on interfaces for relations with the surrounding environment and the ways they augment and alter natural or unmediated spatial perception. Geotagging, augmented reality, and other technologies are creating new locative sense perceptions and new feelings of agency at a distance in the extended environment. At question is the sense of spatial and temporal situatedness of the embodied self. The types of interfacing I consider imply combinations of sensory modalities, including vision, movement (kinesthetics), touch, and the auditory sense. Proprioception—our sense of spatial positioning—can be seen as key in exploring these interactions as embodied experiences of relationality. Embodied perception of space works differently with and without map-based navigation, while sound adds yet another perspective and alters the sense of situatedness and orientation in a different way that is more subtle and immersive. Technologies (most significantly GPS, as well as other positioning sensors like gyroscopes and accelerometers) treat space as a database and allow new types of control within it. I observe how the shifts in perception and effects of disorientation reveal the zones for enacting the latent dimensions of space, and with that, the feeling of the potentiality and expansion of the bodily self.

Together, these chapters demonstrate that art and aesthetic experience offer especially worthwhile ways to reflect on the effects of contemporary technological culture. Feedback encounters with digitally mediated versions of ourselves place us in an intersubjective space of indeterminacy and suspended expectations, shifting our perspectives on the boundaries of the "self," challenging and augmenting the very capacity of feeling human. The interfaces for such encounters are places of negotiation and the experience of being-in-relation. They provide this experience with a frame or structure, which itself is metastable. Art creatively enriches the operational capabilities of interactive digital interfaces to address the current cultural conditions in its own ways. With the open-ended nature of its inquiries, media art refrains from making specific diagnoses and instead gives its audiences room to engage with the posed questions in a profoundly subjective and individual manner. Grounded in an embodied and intimate sense of one's self, such engagement allows more acute and deeper insights into the extents of the transformation of our sensibilities and how we understand ourselves affectively via technologies. It challenges us to not settle for convenient answers but to keep reflecting further on what exactly may be at stake—and what exactly makes us human.

1 Face to Interface

To attain user-friendliness, human-computer interaction design relies on the anthropomorphization of computer operations. This is particularly exemplified by the model of a face. "Facing" one another, humans and machines need to quickly decode each other's respective actions and respond accordingly. To facilitate this interaction, computer interfaces have been equipped with all kinds of face images—from the Mac operating system's face indicators to friendly looking helpers (including Microsoft's reviled and soon-retired animated paperclip, Clippy) and—most famously—emoticons. To fulfill the main function of a face in such an interaction—to make it feel more natural and engaging to humans—these figures are often reduced to the key recognizable facial features, namely, the eyes and mouth. Thus the relationship that is formed is iconographic in nature: one is invited to interact with an image, an icon. Facial iconographies of digital systems reflect their treatment in terms of agency: a face invites the expectation of an active response. Even in cases without direct visual "facing," as in distributed systems and "smart" environments, the mode of control, such as voice commands, is still human-centered and anthropomorphic—the machines are still agents of action that one can literally talk to. Along with agency, the figure of a face brings with it a sense of subjectivity: the feeling that the actions produced by software in response to human commands are governed by a certain type of subjective "will." The machinic behavior is preprogrammed to appropriate parts of human expressivity and to create this impression of subjectivity. In such an exchange, the human also takes over a very particular role—not only as a "user," but as a subject whose expressions, exactly because they are being mimicked, become limited by how the machinic "interlocutor" can understand them.[1] To receive an acceptable response from the machine, the human has to adopt its language, which is not purely "machinic" anymore but also does not reflect the complexities of human psychology, particularly the nuances of emotional reasoning.

The face-to-face model predominant since the late 1980s replaced the initial model of human-computer interaction (HCI) as primarily *body*-machine interaction (e.g., when control happens through hand gestures), which followed with the rise of usability discourse (specifically in web design) throughout the 1990s.[2] As the computer was acquiring the more subject-like position of a conversationalist, its relations with the human required a new form of organization and signification, a role for which the face, with its implication of communication between subjects, became an ultimate candidate. The desire to eliminate interface altogether arose as a consequence of the machine's new subject status as a space of differentiation between human and computer. As Anna Munster describes it in her analysis of the role of the face in the evolution of paradigms of interfaciality, "It has become common HCI parlance to desire the erasure of the interface but to nevertheless reinscribe the power of the face as a single form of expression. . . . The need for the interface to act as a surface for translation is apparently lessened if the human can be translated into a complex type of 'informational processing unit.' Both the computer system and the human can then be represented as internal to a virtual world: as information."[3] The model of a face, or the face-to-face as a shared informational and codifiable continuum, came to replace the original idea of an interface as a place of translation between the human and the computer levels. This replacement was able to happen, and the principle of faciality continues to be vital in present-day interfaces, due to the special qualities of the face—particularly its connection to codification.

This chapter does not go into a full analysis of the nature of faciality adopted in HCI design, but rather focuses on how machinic operations affect and transform the human experience of a face and facial communication. I will discuss not only the *logic* of the face but also how it lines up with the *sign* of the face, that is, the visual representation of faciality and the relations it implies. This distinction becomes clearer when the two diverge, when the face gets reduced to the function of communicability, to interchangeable "vacuous" animated icons and codifiable functions. To explain this, I will use the optics of a number of artworks. As I argue throughout this book, art is uniquely positioned to address the impacts of technological development on human perception: art focuses on experiences, and yet it is also responsible for their construction. This double capacity highlights the dense intertwining of the aesthetic dimension (and its related phenomenological and ethical aspects), on the one hand, and the political dimension (the forces and motivations that go into the construction of something), on the other. Particular characteristics of faciality, such as facial expressions, as well as the sense of subjectivity and identity embedded in the face, become a special battleground where simulation technologies are concerned. Here art comes to play its special role of being capable of both framing the questions and providing sensible personal experiences of them.

So what exactly is the face?

The Face as a Medium

There are many ways to approach the concept of a face. We usually think of a face as a carrier of unique features, a surface onto which they are drawn and sculpted, an expressive exterior through which an individual presents and communicates their being. The peculiar nature of a (human) face is that these features are both individual and generic, that is, universally shared. To understand the specifics of an individual face, we need to see a general composition and structure of facial features (e.g., the elemental schema of two eyes and a mouth, so successfully adopted by emoticons). These features are usually identifiable as at once general and subject-specific, or unique. This doubleness becomes apparent through an experiment considered by Georges Bataille: a large number of photographs of faces, shot with the heads positioned similarly, are superimposed upon one another to produce an image of a generic face, consisting of a set of necessary recognizable features. According to Bataille, the composite image would illustrate proof of a geometrical "common measure" that he treats as a fundamentally Platonic characteristic.[4] Indeed, understanding the face in a Platonic sense is also supported by the traditional supposition that the face *represents* something—an implicit thought, emotion, or attitude. Yet the form of representation and what is being represented remain separate. The "common measure" of the face excludes anything radically singular and can only be the idealized generic. Both of these suppositions—that distinguishing the individual is impossible without its being mapped onto a general schema, and the representational nature of the face—are crucial for our context. The model of a composite face is the basis of a number of artworks featured later in this chapter. The pieces (namely Megan Daalder's *The Mirrorbox*, Karen Lancel and Hermen Maat's *Saving Face*, and Mattia Casalegno's *Unstable Empathy*) make the discrepancy between the singular and the generic felt through the physical experience of having an image of one's face superimposed on the image of another person's face. These pieces show how universalities may work only at the level of machinic vision (when it is able to recognize and categorize the needed features accordingly), while at the level of human experience they are problematic: one cannot convincingly merge one's face into an abstract generic image, as the resulting composite image will unavoidably have its own specificity and identity, however fictional.

If we trace the etymology of the word "face," we find two related origins. The Latin *facia* (or *facies*) means "appearance"; the French *visage* derives from *visus*, "sight"; the Greek *prosopon* is literally "toward-look"; and similarly, the root *lik/lice* in Slavic languages also refers to the visual aspect—"icon," "image." But the other meaning of *facia* is "form, shape, figure," which is also supported by the related Latin verb *facere*, "to make," "to put something into form."[5] Both meanings fit well with the Platonic reading of the face as representational, as a visible outer layer and a physical shape of something that corresponds to it at another order of things (e.g., the realm of ideas) and cannot be accessed otherwise. The face as an image and as something that has been "shaped" implies a certain degree of unity. It is perceived as a whole and is governed by the principles of visual perception: it is—first—*seen*, and only then can it be interpreted, analyzed, *read*. (These characteristics will be important for our later discussion of the transformation of facial—and interfacial—perception in the digital era).

One of the primary functions of a human face is communicative. Hence, it can be treated as a type of media. Again, there are at least two aspects to this that are worth stressing. One has to do with the social and the related ethical and political dimensions, and the other with the structure of the face as an apparatus, for example for signal translation and transformation. I am interested in tracking how one aspect informs the other and how those interconnections instigate other possible aspects of the work of faciality.

Faces serve as a ground for commonality, as an indicator and a producer of the fundamental connectedness of human beings. Indeed, we come into the world knowing what a face is; the face is what infants respond to the most.[6] Not only does distinguishing faces from one another enable us to recognize them as different from ourselves—as the "other"—but faces allow us to *negotiate* the differences within a communally shared space, by communicating our desires, expectations, frustrations, and beliefs and trying to understand the emotions coming from others—all without having to name what is being communicated. Faces thus constitute what political philosophy calls a language of diplomacy that goes beyond linguistic practices. For instance, Hannah Arendt and Giorgio Agamben connect this characteristic to the foundation of the *polis*, a space where social life is organized. For Arendt, the face is a medium of appearance, which—as public appearance—is always a political act; and it is so not only through the visibility of a face but also through maintaining the idea of public space as a space for sharing and exchange, where anyone has a right to appear and have a place.[7] Similarly, Agamben interprets the face as "the only location of community, the only possible city"[8] because the face provides an opening toward the space of pure communicability. He emphasizes the nonlinguistic or extralinguistic potential of the face: "what the face expresses and reveals is not *something* that could be formulated as a signifying proposition," but "a revelation of language itself." He adds: "Such a revelation, therefore, does not have any real content and does not tell the truth about this or that state of being, about this or that aspect of human beings and of the world: it is *only* opening, *only* communicability. To walk in the light of the face means to be this opening—and to suffer it, and to endure it."[9] What Arendt calls appearance, or the space that makes negotiation possible, Agamben calls exposition, which adds slightly more explicitly an element of subjective will and control over who or what exposes. Like appearance, exposition is political ("Exposition is the location of politics").[10] The face allows one *to take possession of the exposition* of oneself, and more precisely—possession of not only *what* is being exposed, but the fact of the exposition, or appearance itself.[11] In other words, the face has to face itself as the means of exposition, and this is where it works as an indicator of agency—something that made faciality crucial for simulating a sense of reciprocity in computer-mediated communication.

The era of new media makes this shift from a face as an image to a face as a medium ever more relevant and brings to the fore political and ethical dimensions of the act of exposition.

The act of exposition to the other through the face matters not only politically, but also ethically. Media scholar Amit Pinchevski identifies three functions of the face as a medium: "the surface by which we appear, the interface through which we interact, and the face-to-face in which we care for others."[12] Pinchevski points out that in Hebrew *panim* (face) shares a common root with *pniya*, "to turn" and "to address," which reminds him of what Emmanuel Levinas had to say about the face: "The face of the Other at each moment destroys and overflows the plastic image it leaves me."[13] The reference to Levinas should not be surprising, since in some ways the principles of his ethical philosophy were influential for contemporary political discourse on the face, as demonstrated by Arendt and Agamben. The face of the other goes beyond mere surface image, in order to *address* one directly. As Pinchevsky explains, it "exceeds the idea of the Other in me. . . . The face resists full visibility. . . . So for the face to maintain its ethical import it must somehow appear beyond what is seen, beyond representation; and in order to do so, it must somehow issue a call, an expression that fractures the *sur*face."[14] The "beyond representation," despite its obvious religious connotations in Levinasian context, is still productive: it means going beyond mere visibility or exposition *of something*, beyond appearance in the sense of a visage; and yet the reference, the link to this "beyond," is contained nowhere else but *within* the image itself.[15] Such a reading thus undermines the (Platonic) representational model of a face, since what matters is not only *what* is being communicated, but the act of opening, exposing one's self (the image of oneself) to the other as an invitation to go beyond such distinctions (self and other), entering the space of exchange where there is nothing beyond this face of exposition, beyond representation of representation (face showing itself as face). Interfacing as a procedure of feedback-based exchange, as I aim to show, has the potential to continue exactly this reading.

The second aspect of the face as a medium that I would like to distinguish is the face viewed as an apparatus. This approach implies examination of the inner workings of the face, the physiological, neurological, and psychological processes that constitute facial dynamics. (In a way, the rest of this chapter will be an elaboration of this approach.) One of the difficulties of analyzing the apparatus of the face (and the apparatus of faciality more generally) is that the face as a medium does not fit perfectly into the classical sender-message-receiver model, since the message—a facial expression—is dissociable from the sender. For example, a smile is a readable message located *on* the face, and at the same time it is the face itself that smiles. The face is thus both a subject (sender) and an object (the medium for communication, the language of facial expressions, an instrument panel upon which thoughts and emotions are registered). As a medium, it connects and separates, and performs these functions not only at the level of social interaction but also at the level of individual expressions—the main way the face operates.

A further complication arises with the question of the nature of expressions. This question similarly has to do with their position in between the concrete subjective experience and what I would call for now "beyond" the subjective, or the nonsubjective—either the presubjective or the socially readable "post-"subjective. This tension is very important, as it is exactly what is at stake in the analysis of the role of simulation and automation technologies in facial interfaces presented in the following sections. It is a question of what within the facial apparatus structure enables automation: Where does it work and where does it not? Is it even possible to draw such a line?

Facial expressions expose inner tensions and intensities, serving as direct signs of the affective life of the body, the life that is at once subjective (specific to this individual human being) and preconscious (functioning beneath self-awareness).[16] The expressive dimension of the face discloses the paradoxical nature of subjectivity: the subject of the face is not pregiven but *becomes* the subject through expressing what is there before being named/signified and before his or her own realization of what is being expressed. For example, a person may start smiling or expressing the emotion of fear without realizing that this is happening; the emotion itself is filling him and through expressing it on his face (in this instance), he makes this emotion his own, thus in a way *becoming* himself. In the end, expressions are formed by the movements of the muscles beneath the skin, triggered by the neural mechanisms that can be either voluntary or involuntary (following a subcortical route in the brain). The involuntary ones may include both simple reflexes (like an expression of pain) and responses to more complex psychological triggers. This quality of *becoming*, of something recognizable emerging from the indistinguishable mesh of preconscious intensities, complicates the codification potential of facial expressions, since the same expression can mean different things depending on its cause.

The problem of ambiguity in reading and interpreting facial expressions is the central focus of the interactive project *Chameleon* by Tina Gonsalves (which I analyze in more detail below). The work is based on a loop of emotional responses to facial expressions: the video portraits of people on the screen react to the viewer's responses to these same filmed faces. The viewer is put in the situation of both the pure, firsthand experience of simply reacting naturally to the other's face and of reflection on that experience (observing oneself observing). The expressions of the video portraits are controlled algorithmically, and that is a source of a major rupture with the natural human mechanism of emotional response. The experience for the viewer becomes both her own immediate reactions and a sense of wonder at the mechanism of those reactions, wonder provoked exactly by the discrepancy in the types of automaticity of these reactions—the natural human and the rationalized machinic.

The double position of the face as a medium for both inner and outer, individual and social, intensive and codifiable can also be extended by Gilles Deleuze's characterization of the face as a "reflecting and reflected unity," or a unity of reflecting surface and intensive micromovements.[17] The surface of the face becomes a meeting point between the inner and the outer. It can reflect the possible external projections, namely the features that can be read and contextualized by others and are extrinsic rather than intrinsic to the subject. That is the "reflected." At the same time, the material substance of the face (the muscle work underneath the visible surface and the nervous system commanding it) is a reflection of another sort—reflection not just as casting back an image, but also as processing signals, an active internal procedure, a form of thinking. What comes together is that which is being reflected *and* the reflecting process itself. This kind of doubleness is similar to what the *Chameleon* project allows viewers to experience: the reflection as an unconscious automatic reaction of the nervous system—the reaction that is written *on* the surface of the face—and the reflection as a conscious procedure, an attempt to realize how the machinic algorithm is deconstructing your emotions. The key difference is that *Chameleon* states its case as a provocation; the affect that the artist is interested in evoking (or seems to evoke) is a particular kind of affect—the human reaction to the machinic as displayed on the face.

Finally, it is useful to preface the following discussion of automated technologies in facial analysis and simulation with a deeper understanding of the relations between social codifications (the "reflected") and the surface on which they are (presumably) inscribed. The success of the application of digital technology in facial recognition and modeling is based on the principle of breaking the image of a face into sections, that is, an analytical approach. But the very fact that it can be done lets us presume a naturally existing mechanism that makes faces operational, a set of functional codes that Deleuze and Guattari call the "abstract machine of faciality."[18] This machine has nothing to do with expressions as inner processes, as the mechanisms of "pressing something out from inside,"[19] since it does not care about any distinction between the inside and the outside. Neither does it care about potential meanings of the expressive signs inscribed on the face that can be read by others. In Deleuze and Guattari's view, such a machine "carries out the prior gridding that makes it possible for the signifying elements to become discernible, and for the subjective choices to be implemented. The faciality machine is not an annex to the signifier and the subject; rather, it is subjacent (*connexe*) to them and is their condition of possibility."[20] Thus, Deleuze and Guattari assume in this argument that the subject's very appearance (emergence/existence) relies upon the faciality machine, that there is something about the face that organizes it to work like a face but that comes from *outside of* the face itself.[21] In fact, it comes from outside of the realm of the human: "there is even something absolutely inhuman about the face."[22] Perhaps, it is exactly this "inhumanness" of the face that allows it to function as a medium and an apparatus: the face connects with all that it allows to appear, but is *not* that itself. The face registers expressions on its surface, but expression itself (as an expressive act, intentional state, or pure affectivity) is something else. The language of expressions is impossible without being played out on the face, and yet the function of the face cannot be reduced to only such a language. To understand better the dialectic between the "human" and the "inhuman" in faciality, we would need to compare the workings and effects of the *natural* faciality machine and how this natural machinic quality is being augmented and exploited through the means of *artificially* created algorithms and visual rendering technologies.

The same "gridding" that is used for the signification or expression of subjectivity (the two aspects through which Deleuze and Guattari approach the face) can be instrumental—literally so—for other purposes. The face is a medium that can be played (like an instrument), which also means that it can be controlled—through preprogramming. The ways facial expression can be manipulated by a mechanically controlled script are illustrated in a project by Dutch artist Arthur Elsenaar with the umbrella title *Artifacial Expression* (figure 1.1). Elsenaar's focus is on algorithms for what he calls "facial choreography" and the possibilities of computerized control over the human face. Small electrical pulses are applied to the facial muscles of a live human in real time to trigger involuntary expressions.[23] The project operates under the slogans "Freedom of facial expression now!" and "The human brain is a tyrant, underutilizing the expressive potential of the human facial hardware."[24] Indeed, facial muscles offer an abundance of positions, out of which the human brain exploits only a certain amount; the rest we cannot administer consciously. Contortions and grimaces that the digitally controlled electrodes stimulate can still be called expressions

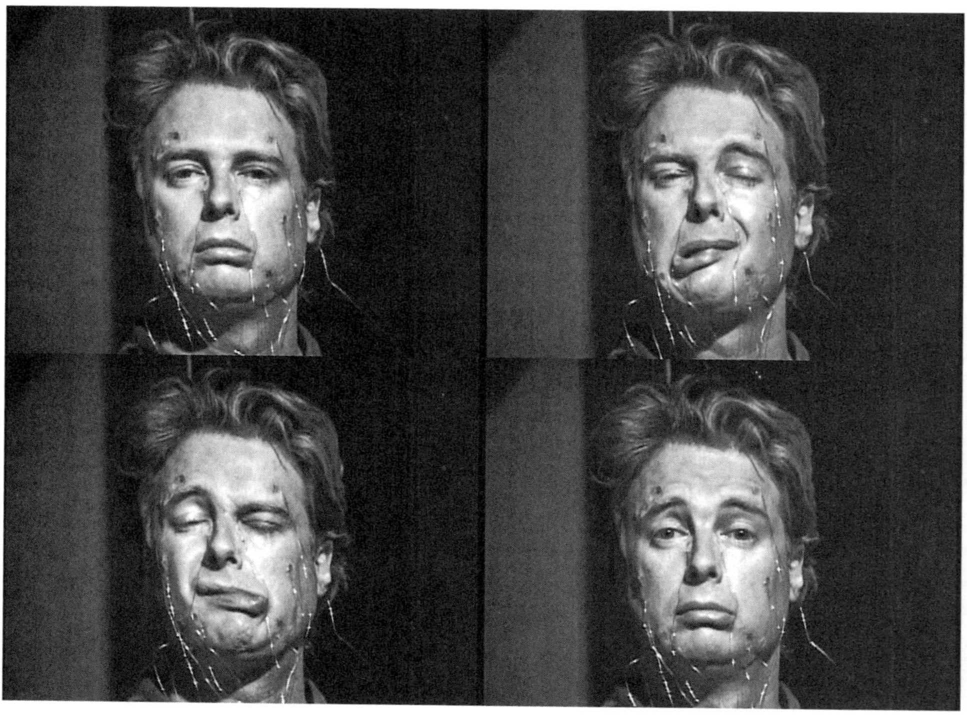

Figure 1.1
Arthur Elsenaar and Remko Scha, *Face Shift* (2005). Live performance. Courtesy of Arthur Elsenaar. Videography Ellen Zweig.

because they are facial and can be put in the context of known human facial expressions. Yet they do not "express" anything human, but rather they use the skin as a surface on which to operate. They are not signs of anything discernible: there is nothing corresponding to them within the human. The expressions stand only for themselves and correspond only to external stimulation. It is a human observer who projects meaning onto them. In this sense, the "choreographical" capabilities of the computer-controlled human face are used to challenge the traditional notion of facial expression as a conveyor of emotion, with the goal of developing a new kind of facial language.

Electrical stimulation of muscles is not itself new and has historical antecedents in the discovery of electricity and the study of physiognomy. For instance, in the mid-nineteenth-century, French neurologist Guillaume-Benjamin-Armand Duchenne de Boulogne conducted experiments triggering muscular contractions with electrical probes, which were documented by his photography collaborator Adrien Tournachon (figure 1.2).[25] Duchenne's book *The Mechanism of Human Physiognomy* (*Mécanisme de la physionomie humaine*), published in 1862, was influential for the development of the field of neurophysiology, but also for Darwin's seminal research on evolution and facial expressions, which appeared ten years later.[26] The newly invented photographic method to register the "fleeting" moments of "truth" in facial expressions was an innovative use of representational technology that enabled the face to be shown as a "tableau vivant." Duchenne was the first to develop a taxonomy of facial expressions based on specific facial muscles and muscle groups, which he documented in the course of experimentation. The images where he stimulated different kinds of muscles on either side of the face to evoke only parts of the full expression demonstrated his attempt to break down the assumed "wholeness" of a facial expression. As German scholar Sigrid Weigel argues, Duchenne "reformulates the *tableau vivant* of the face as a tableau of signs whose expressive lines form the orthography of a physiognomy in movement" ("orthographie de la physionomie en movement" is Duchenne's own expression), with that directly equating "the nomenclature of the muscles and the semantics of emotions."[27] Duchenne's images are resonant in Elsenaar's project, with its emphasis on an alternate "alphabet" of facial expressions that shifts the habitual perspective on how a face can operate.[28]

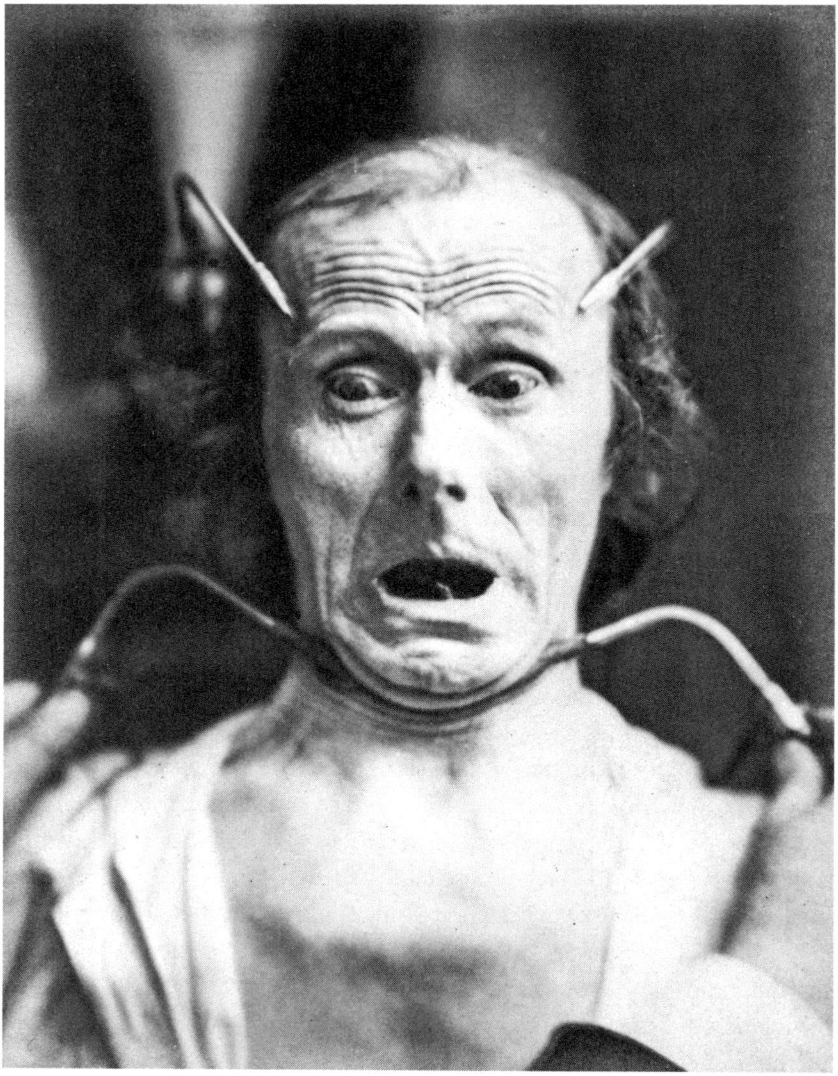

Figure 1.2
Guillaume-Benjamin-Armand Duchenne de Boulogne and Adrien Tournachon (photography), *Fright*, from *Mécanisme de la physionomie humaine* (1862).

Elsenaar added an artistic and conceptual innovation to his work by creating digital personas: Huge Harry (developed by Remko Scha)[29] and his "lesser known sibling" Perfect Paul, who both performed lectures (the former in the 1990s and the latter since the 2000s) on the "freedom of facial expression" and the exploration of "the human facial display\, as a site for digital computational [ehkspr\'ehshaxn]" (expression),[30] using Elsenaar's own face as a "display device."

In his lecture in 2012, Perfect Paul alludes to the research on mirror neurons and concludes that the nature of human emotional expressions is exhibitionist because the internal states of the brain automatically and involuntarily translate onto its exterior. He even quotes Levinas saying that "human persons just wear a face\, and that they don't [\"]own[_] it." The idea that humans are not in full control of their facial hardware leads him "to deploy Arthur's magnificent hardware to the fullest," and he goes on to demonstrate his "choreologic probing efforts" over Elsenaar's face. In the performance *Face Shift* (2005, part of the larger project *Artifacial Expression*), the actions over the human face not only serve as illustrations of theory but are programmed as directly connected to certain speech signals. Voice synthesis machines here are "deployed on each side of the face, calling out the identification numbers of the activated muscles."[31]

Elsenaar's computer programs are personified: they have an identity and are given a voice. This act of anthropomorphosis illustrates what Anna Munster calls "programmatic subjectivity" in her discussion of Elsenaar's work. She explains: "human subjectivity is overtaken by the machine and becomes configurable, computable and ultimately programmable."[32] By attempting to construct its own language with human facial expressive capabilities as elementary units, through making expressive what can potentially indeed be so, the machine here assumes its own subjectivity—a combined one of the artist (who wrote the initial text) and the program (that reads the text out loud and manipulates the face). At the same time, it can be argued that it is still the human who is pushing his own expressive boundaries with the help of the machine and simulating giving up control (since ultimately it is his own algorithm). Such a scenario is, then, a way to face one's own self, one's human nature and its limitations. Elsenaar uses his own face as the face of the machine in order to raise questions about the nature of the expressive potential of the face and the "machine of faciality." Indeed, just as in Duchenne's experiments, we can see a much wider range of expressions (than the usual palette a person's face might display). But this does not annihilate completely the paradigm of the face and its *potential* readability. We cannot avoid seeing a face (even with unnatural expressions) as a face, and it is still the human face that we see, not the machinic.[33] If this is, perhaps, not the best way to discover the "machinic face," then how is it possible to see its face as a *face* at all?

Elsenaar's work as a whole thus helps us to return to the question of the interface as occurring "between faces." From the 1980s to the late 1990s, the dominant approach in human-computer interaction design was the machine's friendly face. Then, from the early 2000s, it was the human who had to be translated to the machine and reduced to mainly the computable types of brain functions.[34] The principle of the seamless user-friendliness of an interface still needs to be challenged, and a model for the most balanced type of interfaciality (true to the complexities of machines and yet manageable by human users) is yet to be found. Before moving on to the details of what is specific about the machinic logic of faciality, I'd like to discuss the role the face plays in how humans interface with each other, and how that dynamic tells us more about the intrinsically human function of the face and "facing."

Identity Transactions

Emotional expression is an important part of the face as a communicative medium, but it is only part of the story. Another important component of communication that should not be ignored is the sense of selfhood and identity that the face conveys. The face serves as a tool for making distinctions between oneself and the other. Recognition, the role of mirroring, and the perception of the face as a whole versus the fragmented face are all important elements of transactional dynamics that the situation of meeting faces, or "inter-face," provides. In this section, I consider three works of media art that share a common feature—the technique of superimposition, or morphing participants' faces onto one another's with a resulting effect of merged identities. The different setups and designs allow the viewer to experience diverse affective nuances of interaction between the self and the other. In discussing these pieces, I am particularly concerned with the issues of fluid and networked identity, transference of trust, and affiliated feelings of empathy and vulnerability, as well as the problems of identification and

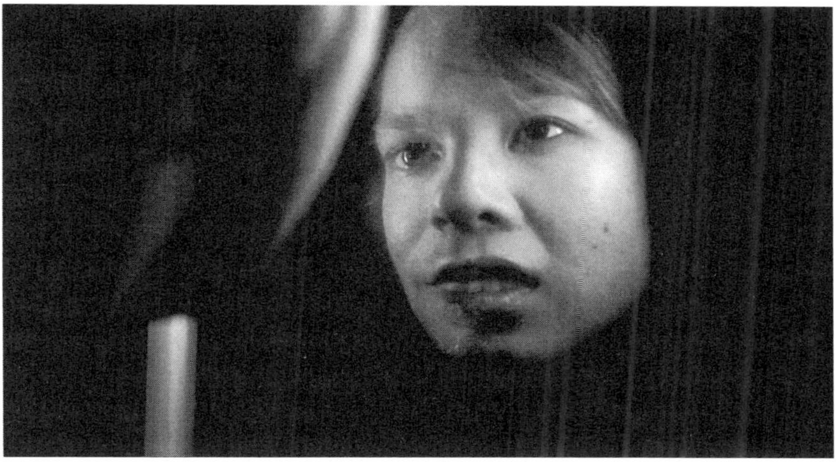

Figure. 1.3
Megan Daalder, *The Mirrorbox* (2010). Installation, process documentary, scientific research. Screen capture from video. Courtesy of Megan Daalder.

interaction with the other more generally. A special angle of discussion here is the analysis of how these artworks engage with the neuroscientific perspective on the self, the sense of identity, and facial communication. Science serves as a point of dialogue that provides empirical grounding and helps to stretch the questions latent within art to some newer, not fully explored territories. It is also where the intuitive and subjective way of experiencing a work meets the kind of reflection that is both more generalizing and revealing of the natural structures of cognitive processes.

Megan Daalder's *The Mirrorbox* (2010) is a creatively shaped box that can fit two people's heads and has a semitransparent (two-way) mirror in the middle (figures 1.3, 1.4). The piece creates the illusion of the person on each side of the mirror "becoming" the other as the images of their faces merge. It is essentially a simple analog device: only the LED light sequence is preprogrammed, and the amount of light dictates the translucency of the mirror (a technology that has been around since 1903). The reviews show that participants "tend to leave the Mirrorbox reeling from the temporary sensation of embodying another [person's] identity"—a feeling that could last for a while.[35] One participant says that it feels "like a hand reaching at me and tickling my insides."[36] People felt as if they were in love (with the

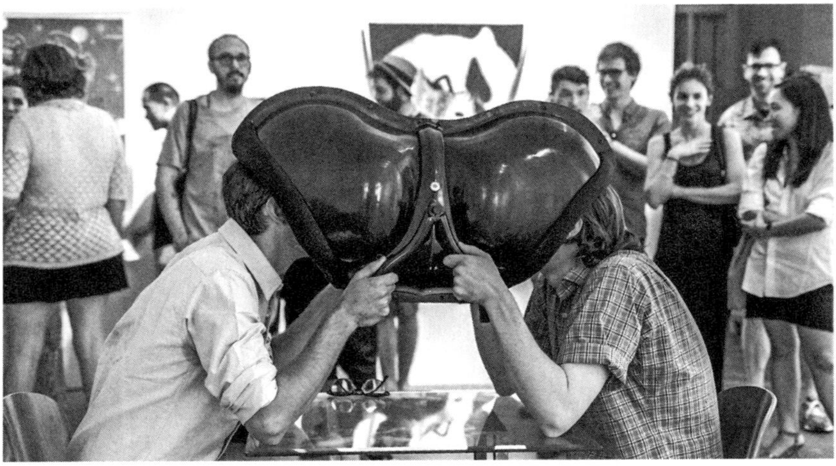

Figure 1.4
Megan Daalder, *The Mirrorbox* (2010). Photo courtesy of Megan Daalder.

box, or a person on the other side)—and also felt loved. Before settling on the final title, Daalder had called *The Mirrorbox* "Soul Collider" and "Empathy Accumulator," and envisioned its potential use in schools, prisons, even Congress—wherever assistance with mutual understanding is needed. Yet intimacy and proximity are not always desired. Such an experience of closeness, as appealing as it can be for some, is unsettling and frightening for others: "Self-loathing and insecurity can drive people out just as quickly as unwanted intimacy with a stranger. . . . We might want to objectify the other, and we feel distressed by our inability to do so."[37] Some have suggested that *The Mirrorbox* is not only art but also a tool for scientific inquiry to explore the potential processes of rewiring the brain.

Like *The Mirrorbox*, the neurofeedback-based installation *Unstable Empathy* (2012) by Mattia Casalegno and neuroscientist Enzo Varriale is a booth built for two people who are also positioned facing each other through a screen (figures 1.5, 1.6). Their brain activity is being measured, and as their brainwaves (supposedly) synchronize, the features of one person's face are projected onto the face of the other through a layer of graphic "masks," serving as visual feedback for empathic connection.[38] Here I am less interested in the aspects of brain synchronization (which are much more prominently explored, for instance, by Marina Abramović in her experiment *The*

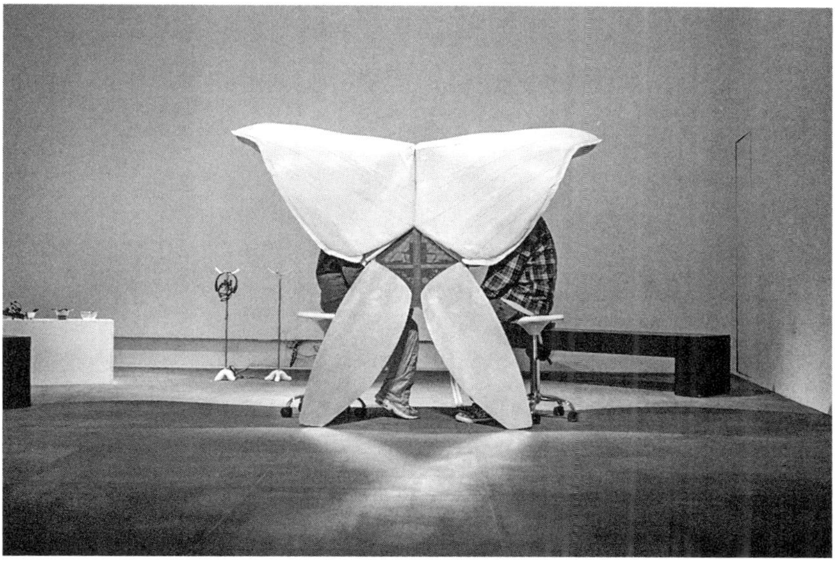

Figure 1.5
Mattia Casalegno, *Unstable Empathy* (2012). Transjourney Future Media Festival 2018, Kuandu Museum of Fine Arts, Taiwan. Photo courtesy of Kuandu Museum of Fine Arts.

Magic of the Mutual Gaze, 2011), and more focused on the setup design and iconography: the light lines over the faces that compose motifs reminiscent of indigenous peoples' aesthetics.

Finally, the public installation *Saving Face* (2012–2017) by Karen Lancel and Hermen Maat invites participants to experiment with their image by touching their face in front of a screen that functions as both a mirror and a window to the vast database of other participants' portraits (figures 1.7, 1.8).[39] The gesture of caressing your own face is documented by the camera. As you touch your face, spot by spot, your portrait is slowly revealed on the screen where it merges with the features of previous participants' faces. This new, synthetic identity is then automatically uploaded to Flickr as a new "passport."

What we see in all three pieces is the emergence of a hybrid face and with it—as the artists claim—a sense of a hybrid identity. The interface here is the place where bonding is experienced. It is not only an awareness of oneself through the responses of the other, but also a mutual effort in

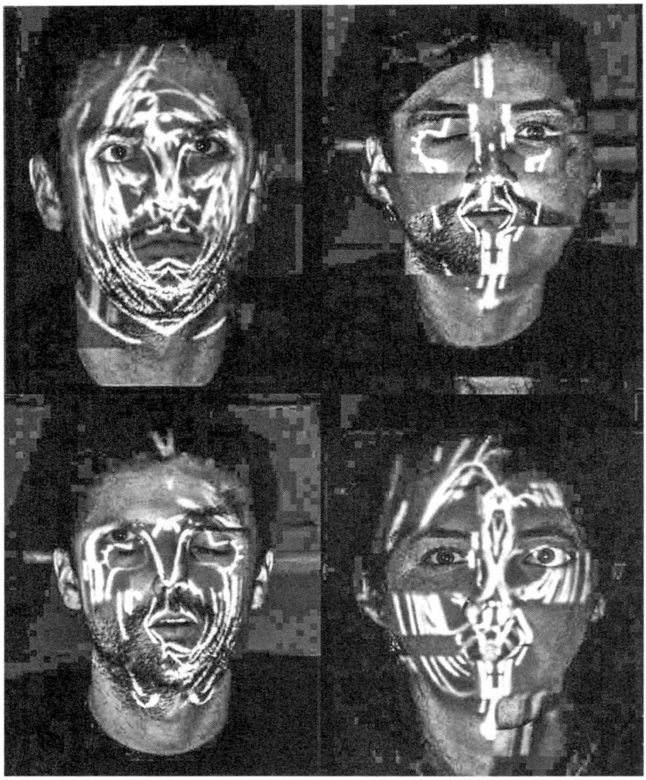

Figure 1.6
Mattia Casalegno, *Unstable Empathy* (2012). Screen shots from live interaction. Photo courtesy of Mattia Casalegno.

generating the presence of something third. The two participants become part of the same system that tries to find its equilibrium state, a balance of differences, continuously unstable and indeterminate unless it is deliberately stabilized by the "archiving" gesture (as happens in *Saving Face*).

Of course, the technique of superimposition has a long history in visual arts (particularly in film and photography, where it has a story of its own). The main purpose of superimposition is usually to emphasize the double nature of things, an invisible side counterpointed with the visible one, a background/contextual motif to enrich the perception and interpretation of a given image. By using this technique, these installations pursue a similar agenda of putting each participant in the "context" of the other. Yet

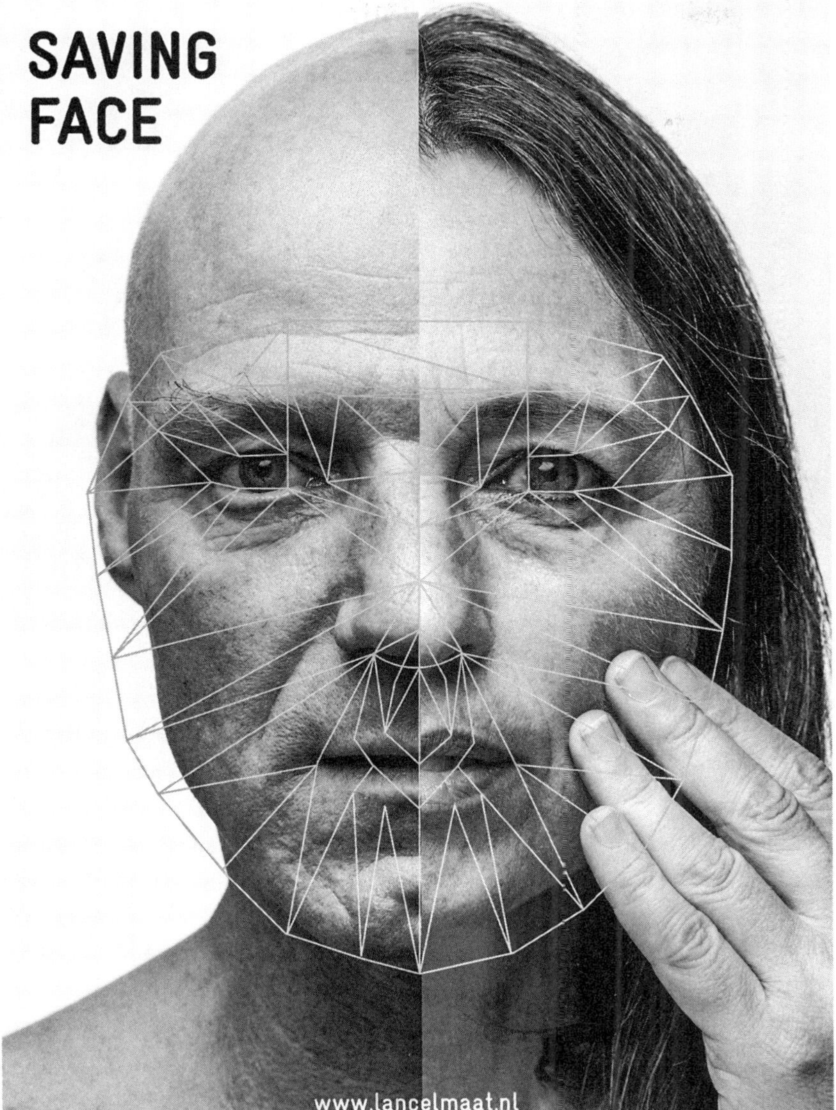

Figure 1.7
Karen Lancel and Hermen Maat, *Saving Face* (2012–2017). Performance-installation. Courtesy of Karen Lancel and Hermen Maat. Photo by Karen Lancel and Hermen Maat / Studio Matusiak & Wenting.

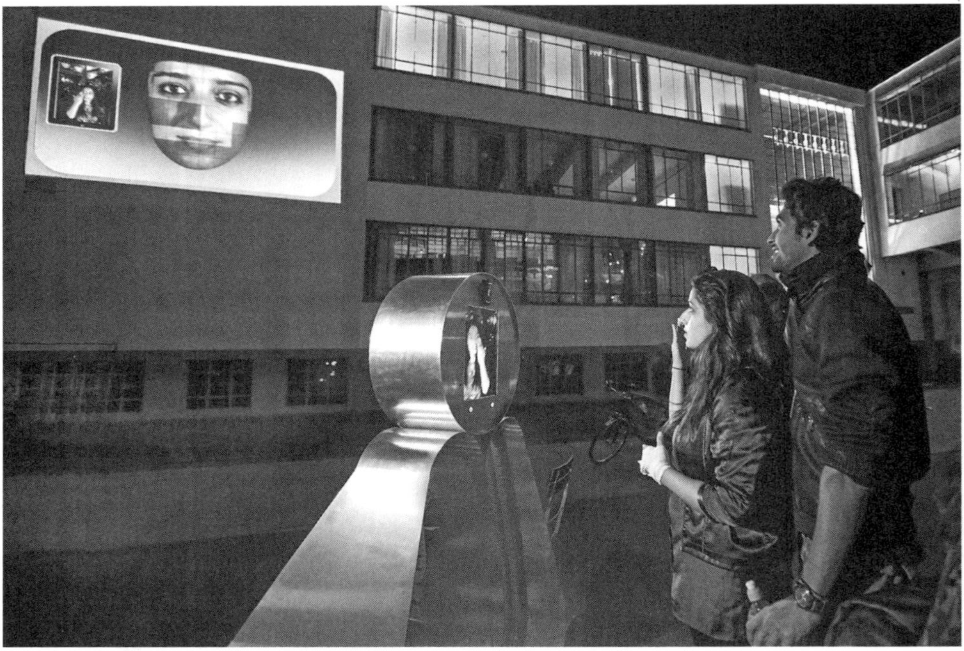

Figure 1.8
Karen Lancel and Hermen Maat, *Saving Face* (2012–2017). Performance-installation. Courtesy of Karen Lancel and Hermen Maat. Photo by Karen Lancel and Hermen Maat / Ruthe Zuntz.

the overlay here happens live, and the participants are ordinarily strangers. There is nothing particularly symbolic in their encounter, there are no preset contextual hints or latent meanings, the people do not play any predesigned roles—they simply stand in front of each other as each other's "other."

Emerging within the public setting of museums and galleries, but also—as in the case of *Saving Face*—literally on the street, this new "self + other" face becomes a tangible *social interface*, and its material surface, the skin, is transformed into a transitory place, a portal to experience other identities. An intimate feeling of one's own bodily boundary translates into discovering its seemingly organic connections to other people's bodies. *Saving Face* brings these transactions to the bodily level, connecting the physical and haptic sense of one's own body with the abstraction of other's selves. The touch here is both a physical gesture and a digitalizable, virtual one. Spot by spot, it merges people's individual identities. The skin becomes the locus of a shifting and hybrid identity—a co-created, transparent never stable (and thus untraceable) networked self. The "sharing" of one's face turns into its dissolution into a multitude of other faces. This potential, in turn, radicalizes the question of trust as one's individuality is tangibly and visibly subsumed into a manipulable construct.[40]

The psychological effects triggered in all three of these installations are grounded in the neurophysiological structures that are responsible for the ability to distinguish between one's self and the other. Scientists today acknowledge that most of the key brain functions are performed due to the collaborative work of different brain regions that dynamically coordinate their activities and specific mental states, and compose so-called intrinsic connectivity networks (ICN).[41] There are two such networks that are considered directly responsible for the neurological foundations of identity-related processes: one representing the mental/psychological and evaluative side (default-mode network, or DMN), and the other the physical/embodied side (mirror neuron system, or MNS). While the mirror neuron system provides the physical other-to-self mapping that is necessary for comprehending physical actions, it does not involve recognition of those actions as such, and even less so piecing them together with one's sense of self and self-identity. The default-mode network complements the work of the MNS as it maintains and supports processes that are related to understanding the psychological states of others by reflecting on one's own attitudes. The DMN is claimed to be responsible for such cognitive processes as the sense of self and the ability for self-reflection, autobiographical memory, self-knowledge, and socially developed identity, as well as the ability to take the role of the other.[42] Studies show that DMN emerges from its resting state to an activation state in social self-referential or sensory task-passive processes.[43] Despite its empirically tested relation to the sustention of the sense of the self, the default or resting state mode is not really only about the self. As Felicity Callard and Daniel S. Margulies point out in their overview of this growing research field, both the specialist and popular scientific literature use tropes that connote a certain "wildness" and "underside" of this subject.[44] Examples include Marcus Raichle's two papers both entitled "The Brain's Dark Energy," which argue that the brain spends enormous amounts of energy on functions unaccounted for by its responses to external stimuli—such as simply maintaining the intrinsic activities in a resting state—and science today is only at the beginning of its discoveries of the meanings of those activities.[45] Another source expresses unease that the resting brain may be attributed to "mysterious functions."[46] This "black box" quality of the neuronal structures, including the presumed default "self-position," necessarily extends to what is implied by the concept of the "self" altogether (a serious challenge to the schematizing approach of cognitivism).

Looking at faces activates both the DMN and MNS networks. Even though the mirror neuron system operates primarily with motor neurons (that is, responds to motor actions), it is fundamentally important in automatic reactions to facial dynamics.[47] Yet it is still the default-mode network that helps to make sense of these images. There is neuroscientific evidence today that the ability to distinguish one's own face from others' faces is crucial for one's sense of identity: "The self-face is the most identifiable marker of the physical aspect of the self."[48] According to recent fMRI studies, self-face recognition activates the frontoparietal areas in the right hemisphere that broadly overlap with the human MNS.[49] Scientists assume that it is simply easier to map one's self onto one's own motor system (the face representing part of it) because there is more control over movements. The ability to recognize oneself in a mirror, which people (along with other nonhuman primates) have from the age of two, becomes an important factor of personal identity. Experiments also demonstrate that reactions to live faces are different from reactions to static images or even to faces appearing via live televideoconferencing like Skype or Zoom: seeing faces in pictures simply does not activate DMN in the same way as in live situations. For instance, a comparison between dyadic fMRIs (when the simultaneous brain activity of two people in the same scanner can be tracked) and monadic fMRI shows that live eye contact stimulates both the empathy networks and the default-mode network.[50] Other studies confirm that faces regarded with a direct live gaze trigger a much stronger reaction in the occipitotemporal lobe, an area known for its connection with processing faces (both schematic and realistic images, as well as live).[51]

"Meeting gazes" between the participants in *The Mirrorbox* or *Unstable Empathy* thus activates both the mirror neuron system and the default-mode network. The interface of these installations pushes the two arbitrary partners into a situation where they are supposed to literally see themselves in another, which implies several tasks for the brain: both to identify the self and to "take in" or "process" the face of the other as the other. This double task challenges the understanding of the brain as recreating the self-specific neuronal pattern that has to reaffirm the existence of an individual as a particular "self." The installations do this by making it harder to distinguish between oneself and the other. It may feel especially difficult in *The Mirrorbox*, where two faces are intertwined dynamically while both participants stand in close proximity to each other, sharing not only reflections of their faces but their being in the present. What happens, perhaps, is not only recognition of a face and its identity ("self" or "other"), but an exchange that allows participants to go beyond the division of the categories "self" and "other" altogether. The experiences of the *fluid identity* that all three pieces provide allude more to the "black box" quality of the "default" state of the self than to any clear understanding of its workings, inviting one to look deeper at the structure of that state, at how it may possibly be shared between people and be the basis for the experience not so much of the self, but of intersubjectivity—a sense of shared presence, of being together in a moment of time. This is the exact effect that Marina Abramović worked to achieve in her famous performance *The Artist Is Present* at MoMA, New York (2010), where she made museum visitors acutely emotional by simply sitting silently in front of each person individually, staring into their eyes. In the end, the default-mode network attests not only to the constant recreation of a sense of selfhood and identity but to the reassurance of the very feeling of individual existence, the awareness of "I am," and with that, perhaps, of being itself.

Through these experiments on fluid identities, art and science can be put in a productive relation to each other. Their positions added together may generate a special, almost "spectral" effect. Science seeks out a structure of a phenomenon as it is (in this case, an example may be the intrinsic connectivity networks); and art shifts the focus and shakes the elements of that structure, resisting settling on a particular solution. In fact, Daalder's *The Mirrorbox* was adopted for experimental use at the Brain and Creativity Institute of the University of Southern California, led by Dr. Sook-Lei Liew, but unfortunately had to be interrupted. The results could tell us more about what happens in the brain while we observe our faces morphing with someone else's in real time. But they can also show what it may mean when an artwork is approached as a scientific tool: Does it stop being an artwork then? Will the study results add anything to the experience that people have with it as an art piece, or will this information be relevant only as scientific? At the very least, such a collaboration invites the public to participate in a discussion about the methodologies of knowledge production, a discussion often limited to a professional community.[52]

As some studies demonstrate, one perceives one's own face more as a complex of individual features than as a whole, that is, much less holistically than one views the faces of other people.[53] In *The Mirrorbox*, the recognizable self-features stick out from an otherwise seemingly complete whole; in *Unstable Empathy*, it is the contours of the other person's features that disturb the unity and integrity of one's own face; similarly, in *Saving Face*, the lineaments of a random stranger's face, including characteristics of skin such as its color and texture, appear on one's own face as if corroding it (eating it away) from inside.[54] As the audience reports show, people react to the dissolution of the self-image and the emergence of the amalgamated virtual "third" quite differently: some are positively intrigued, others are frightened. For instance, among the responses about the aftereffects of *The Mirrorbox* were phrases such as "bliss," "fluid identity," "my insides feel scrambly," and "shift."

Indeed, the image of the emerging third subject should feel extremely confusing and unsettling. David Bard-Schwartz, in his psychoanalytic account of the phenomenon of compound faces in media art, calls such images "impossible." His case study, an interactive video installation *Reface (Portrait Sequencer)* (2007) by Golan Levin and Zachary Lieberman, presents a long strip of video recordings of faces that are divided horizontally into three parts (roughly corresponding to eyes, noses, and mouths) and mix the facial features of different people. The emergent compound faces thus may include eyes and eyebrows from one person, a nose from another one, and a chin from a third person. They are indeed unnatural, and yet it is one's own face that is being transformed in real time; thus the strangeness effect comes from both recognizing and not, and with that comes the realization of the very impossibility of such a compound in reality. Bard-Schwartz takes the argument in the very particular direction of Lacanian psychoanalysis, making the analogy between this effect of impossibility and the "alienating effect" of the "mirror stage," the stage when a child first recognizes itself in the mirror. For him, "the affect projected from these impossible faces represents/embodies a primitive aggressivity out of which the mirror stage develops."[55]

A discussion of mirror relations would of course be incomplete without a reference to Lacan's reading. Also, while neuroscientific studies may point at the specific areas in the brain responsible for processing faces and the sense of identity, as well as their connections, a psychoanalytic framework helps to raise questions about the affective dimension of the "inter-facing" experiences. The mirror stage in general should be understood in terms of the procedure of identification, that is, when one associates oneself with an external image and forms an inner image or *Gestalt* of one's self based on that image. For Lacan, the mirror image gives the child a feeling of wholeness and completeness, or what he calls an "'orthopedic' form of totality,"[56] that is principally different from the child's initial experience of his body as fragmented and being constantly in need (e.g., of mother's attention).[57] Yet at the same time, mirror image, according to Lacan, is also about frustration, which occurs due to the fundamental separation of the *Innenwelt* (organic inner world) and *Umwelt* (external reality). The mirror image, although reflecting the appearance of the body and giving it its whole, comes from the outside and is thus alienating. Furthermore, that image serves as an objectification of the subject: "the subject makes himself an object by displaying himself before the mirror."[58] Finally, the image is "frustrating" because—as the image in the mirror is the objectification of the subject—it is also the other's *jouissance* (pleasure) that the subject would see there, not his own.

The images in all the installations under discussion are mirror images transformed by the addition of the figure of the other. The presence of the second face can be read, then, as a double alienation, as a reminder of the alien nature of the mirror image of one's own face, and the generic, shared nature of a face itself. The individual features are all that make the two faces (or more, as in *Reface*) be recognized as different, but neither of the subjects is in possession of or in control over the image. In this sense, the "impossible" third face can be interpreted as being in between the two "ego" positions, a space for negotiation (of identities), and also not only as *inter*-face, but—precisely due to its "impossibility"—as *beyond* or "*intra*" any specific faciality, beyond specific subject or identity positionedness. It is important that this quality figures here not in an abstract but in a quite physical way: through physical surfaces (even if just a thin screen surface), through synthetic imagery, but also through haptic features, such as the feeling of touch on one's skin (in the case of *Saving Face*), or brain waves (in *Unstable Empathy*), or the sculptural apparatus that encloses the two faces, locking them in front of each other (the booth in *Unstable Empathy*, or the box of *The Mirrorbox*). All of these aspects enrich each other and our understanding of inter- (and intra-) faciality as integrative of multiple elements of connection and mediation. Applied to faciality, they help us to realize how fluid and mutable the sense of identity can be when associated with the outer image of oneself.

While *The Mirrorbox* and *Unstable Empathy* deal with a very "analog" type of setting and experience, *Saving Face* refers to the notion of networked identity and with that opens up a direction that is especially critical in the age of datafication of faces and the growing application of machine learning to human-related data. Scanning the skin with one's own hand in Lancel and Maat's installation serves as a gesture of (re)appropriation of the body and its privacy—a deconstructive response to surveillance technologies that turns their purpose around (at least that is what the artists claim to do). Indeed, confusing the boundaries of individual identities can be seen as a tactic to undermine the whole idea of facial recognition as a political control mechanism, the Orwellian scenario played out via CCTV cameras in public spaces. Disconcerting examples of the implementation of such scenarios come, for instance, from China, where the technology is used for targeting ethnic minorities or for citizens' "social scoring" and other forms of population control.[59] Countersurveillance and various visual camouflaging tactics have been central to the work by a number of artists, such as Zach Blas (e.g., *Facial Weaponization Suite*, 2011–2014; *Face Cages*, 2013–2016), Adam Harvey (*CV Dazzle*, 2014; *HyperFace*, 2017; *MegaPixels*, 2019), and Leo Selvaggio (*URME Surveillance*, 2014).[60] Following the principle of the "right to opacity," in the sense of Édouard Glissant's postcolonial critique, these works propose a variety of types of "masking" that can be adopted in virtual communication—as a layer of an amorphous blob that covers the image of the face in a video stream, as in Blas's *Facial Weaponization Suite*—or in real life—as a 3D-printed prosthetic or a paper mask such as in Selvaggio's *URME Surveillance*.[61] In this twist on the Guy Fawkes mask used by the Anonymous movement, Selvaggio's *URME Surveillance* kit is essentially a mask of his own face for anyone to wear to safeguard their own identity by temporarily "adopting" his. Camouflage as a tactic functions through minimizing the difference between "figure" and "ground," thus reminding us that recognition goes hand in hand with misrecognition and nonrecognition; put in the context of social or political use, it should implicate questions of who is in charge, of the fairness and mutuality of exchange. Curiously, by integrating its participants' facial profiles into a larger database, *Saving Face* does not efface their features completely, but saves them into the system in a way that may as well be a compromise between the possible recognition, misrecognition, and nonrecognition of the identity behind them. Hybridity can thus work both as a camouflage and a masking, and in a constructive way—by offering the experience of a shifted feeling of one's self and a more distanced perspective on the idea of the self as a whole.

Facial Expressions and Simulated Exchanges

The communication of identity should be seen as only one of the face's several functions. Another key function is the transmission of attitudes and emotional responses through facial expressions, often considered to be the most efficient and fastest way to communicate emotions. By building machinic algorithms for reading and interpreting facial expressions, humans attempt to deconstruct their own intuitive ways of understanding them. Computational models imply a promise of a more accurate, detailed, and objective analysis that would be free from the subjective judgments of human interpreters and allow a culturally neutral perspective—the principle of "computational objectivity."[62] But these claims have not always been fully justified: instead of giving an objective perspective, the programs are more likely to automatically reproduce and solidify human biases. (Just as the low rates in automatic recognition of black female faces show: these types of faces were simply not used in training the machine learning systems.) Making visible and decipherable the inner states of a subject, these techniques allow a type of *diagnostic* gaze (in the words of Nicholas Rose), one that is efficient and hence trustworthy enough to make individuals subject to certain therapeutic techniques.[63] Delegation of the emotion-reading function to machines can eventually lead to reconceptualization of the meaning of emotion and, moreover, of what emotional exchange means to being human. As Kelly Gates puts it,

> the very possibility of AFEA [automated facial expression analysis] raises critical issues about the nature of emotion and human affective relations, especially insofar as the computational model becomes an instrument for conceptualizing what emotions are and how they work. Just as the turn to biometric identification signals a redefinition of identity as disembodied data that circulates over networks,

> the prospect of AFEA similarly points to reconceptualization of face-to-face, affective relations to incorporate a computational model of face perception.[64]

Here, I explore the meeting zone between the human and machinic interpretations of facial expressions, as well as how the application of digital technology to faces, particularly the simulation of responses, affects understanding of interfaciality. The featured artworks bring the question of interfaciality from the formal level of how such an interaction is organized to the level of the human, such as human perception and lived experience. This human perspective also opens up the political side of this question in a new way: the constructed nature of the created effects and what exactly in that construction makes them truly effective. I'd like to argue that the effects of technological mediation work as *effects* when they are experienced as *affects*, and operate not on an abstract "human" ("human" in general, conceived conceptually), but on a concrete individual situated in his or her particular life context.

Let us first consider the basis of the connection between facial expressions and their computerized treatment. The principles behind the computer vision algorithms used in facial emotion recognition are for the most part based on the schematic approach of the psychological "science of emotions." There are multiple ways to categorize emotions, and each method usually serves the purposes of a particular theory and its methodologies. The argument for a systematic approach to emotion is usually its comprehensiveness. Naming emotions helps us to recognize the emotion as such and to discern its nuances. Yet one problem in classifying emotions and corresponding facial expressions is the very classification itself.[65] As naming and taxonomies stay at the level of the rationalizable and analyzable, they might not take into account the complexities of the in-between states, the role of context, etc. Another problem is that of translation, which also has several levels: translation between what is being expressed (intentions, attitudes, feelings, bodily intensities) and its expressions, between the expressions and language categories that stand as their signifiers, and finally between human language categories and a machinic language. Hence, several questions may be asked: What kind of techniques and codes are used for such translation? What are the forms of its application? What are the critical ethical and political implications? And—the focus of this work—how does artistic use of these methods contribute to understanding their implications, for example by expanding the scope and gap between what can and cannot be recognized and analyzed adequately?

The history of relations between facial expressions and emotions, as well as attempts to classify emotions based on physiognomic features and their visual representations, goes back to Aristotle. Although rigorous visually supported studies took place throughout modernity,[66] it was Duchenne de Boulogne's use of photographic images, mentioned above, that brought this issue into the scientific realm. Charles Darwin's treatise *The Expression of the Emotions in Man and Animals* established the general principle that emotional expressions in humans and animals are an evolutionary feature, important for their survival. The treatise became influential and foundational for all the subsequent scientific studies of this issue due to its being grounded in biological and evolutionary theory, and because of its universalizing claims. By the 1920s, though, these generalizations started to be regarded as more problematic, which was also instigated by the subsiding of research on emotions in general as something too vague to capture via the methods of that time. Meanwhile, facial expressions were part of the debate, from the early twentieth century onward, about whether emotions are more physiologically or socially grounded.[67] The controversy was resolved in the 1960s after experiments by Stanley Schachter and Jerome Singer that led to a "two-stage" model of emotion: body arousal triggers emotion, but social context is needed in order to determine its meaning and quality. This development in the field of emotion research was one of the influences that set the stage for the introduction of computational methods in facial expressions analysis. The fact that emotions were proven to be grounded physiologically and that facial expressions were seen as directly corresponding to those states was an important support to the claims behind computational analysis, particularly that such a method would be capable of revealing "true" emotions.

Microexpressions (or "micro-momentary expressions") that appear on a person's face offer a main clue for recognizing that person's "true" emotions—the ones that are often repressed (unconsciously concealed) or suppressed (deliberately). Imperceptible without special tools, microexpressions represent micro levels of existence—what Nigel Thrift refers to as "bare life."[68] Paradoxically, the goal of computer vision was not only to reveal the "truth" of our emotional life but also to be close to natural human perceptions and thus to concentrate not on what the expressing subject actually felt but on what observers see in those expressions. This double goal is confusing, and if it is not specified clearly within a particular coding project, such a practice may undermine its own premise. This paradox also points once again at the specificity of facial expression as both coming from the *inside*, as a trace of physiological processes, and being written *on* the face, as a sign to be read by others. Computational methods, quite logically, treat facial emotion recognition as a translation problem. And my aim here is to look more thoroughly, first, at the nuances of these methods and, second, at their effects and consequences for a human.

The task behind coding facial expressions lies in programming computers to find correlations between the expression features and their possible meaning. Generally, there are two major types of methods for that: judgment-based (using images of the whole face) and sign-based (or "sign vehicle-based"), born out of the Shannon-Weaver mathematical theory of communication approaches.[69] Research on the classification of facial expressions in the context of the automated analysis of emotion began in the 1970s, most famously with the work of Paul Ekman and Wallace Friesen and their Facial Action Coding System (FACS). FACS was suitable for the needs of computer scientists precisely due to its focus on discrete features and individual visible facial movements "free of any theoretical bias about the possible meaning of facial behaviors."[70] (Again, paradoxically, it still relies on manual coding: at its initial stage, FACS used manual labeling of images performed by highly trained individuals—coders.)

FACS owes its success to the techniques of automation that bring emotion analytics to scale, but it was referred to as a significant breakthrough because Ekman and Friesen developed their system of "basic," "prototypic," or "archetypal" facial expressions based on only six emotions that they considered basic—happiness, sadness, anger, surprise, disgust, and fear. All of these emotions can be illustrated easily through schematic figures of faces with a smiling, neutral, or "sad" mouth, raised or lowered eyebrows, and wrinkling of the nose or not. These icons, however simplistic, are indeed recognizable not only as faces but as faces expressing these particular emotions. Other classifications start with emotions that are less easily expressed. For instance, Robert Plutchik suggests a wheel-like diagram of emotions that includes basic emotions and their valences: joy, trust, fear, and surprise, contrasted with sadness, disgust, anger, and anticipation. It is not surprising that this classification gained less success than Ekman and

Friesen's since emotions such as, for instance, trust and anticipation do not have clearly identifiable facial expressions at all. There are in addition other nuanced taxonomies that list emotional flavors that do not have clear visual correspondences and are also subject to much interpretation and require special psychological acuteness to be identified correctly.[71]

The most typical way of analyzing facial expressions is through distinguishing specific parameters. In the case of FACS, those parameters are Action Units (AUs), indicating changes in the face and the corresponding muscles. In dynamic analysis, the parameters are called Facial Animation Parameters (FAPs), and a series of special algorithms was introduced to automatically extract them.[72] The methods for creating models for face tracking also differ. One of the most widespread parameterized face models is the 3D Candide "mask," constructed using a set of triangles (first created by Mikael Rydfalk in 1987; see figure 1.9).[73] There are also other varieties: another face tracker, the Ratio Template Tracker, works as a better option in cases with illumination problems, while the PersonSpotter tracking system has been demonstrated to be robust against considerable background motion.

Affdex, a computer vision algorithm developed by the MIT Media Lab's Affective Computing group and a product of its commercial spin-off Affectiva, is trained to isolate key regions of the face, such as the eyes, nose, and mouth, and to analyze pixels in those regions to classify expressions based on pixel color, texture, and gradient—a procedure that, its authors claim, yields the highest possible accuracy. Besides scientific research, affect-sensitive human-computer interfaces (HCI) and the technology of expression recognition are used in psychiatry (autism treatments), medical assistive care (AI agent nurses), educational software (for tracking attention), animation and video games, telecommunications (e.g., the EmotiChat application for chat rooms, where the system tracks the user's expression and automatically inserts emoticons to represent them), and marketing research.

Emotion technologies, like many other technologies, have an effect of a double, or "pharmacological," nature. Besides the obvious benefits (for example, assistance with autism and other cases of a deficiency in emotional literacy), automated facial expression analysis (AFEA) technology has a whole range of less obvious effects that may undermine privacy and lead to establishing pervasive control over human emotional lives and the expression of humanness more generally. Rationalization of variable facial

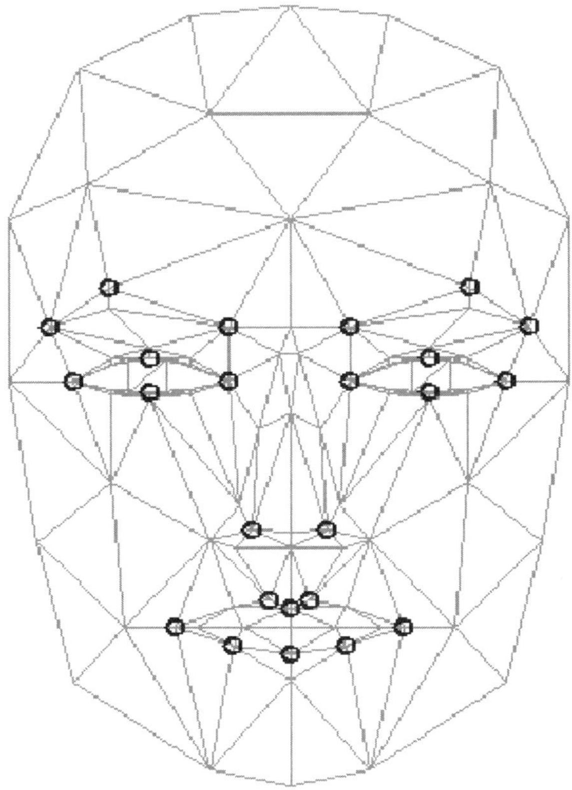

Figure 1.9
Candide, parametrized face model. Image Coding Group, Linköping University.[74]

appearances by systems like FACS implies the application of an information theory framework that treats expressions as signals or discrete, abstracted, disembodied entities divorced from their contextual meanings. In fact, context specificity is one of the features of human intelligence that philosopher Hubert Dreyfus stressed in his critique of AI as early as the 1970s: he particularly distinguished "fringe consciousness" (marginal forms of awareness) and "context-dependent ambiguity tolerance" ("our ability to narrow down the spectrum of possible meanings by ignoring what, out of context, would be ambiguous").[75] As Kelly Gates states, "Just as facial recognition systems help to construct identities they claim to merely represent, automated FACS systems, if made to function efficiently, would likewise push back out into the world and onto faces a uniform system for classifying facial expressions, creating a standardized framework for determining their meanings and what they reveal about the affective lives of human beings."[76]

The artistic cases I would like to analyze here (all of them interactive installations) respond to a shift from the "informational" to the "interactional" approach to conceptualizing affect in emotion-centered HCI. According to Kirsten Boehner and her colleagues, this type of HCI design can treat affect not as a biological process to be measured with scientific accuracy, but as an intersubjective phenomenon.[77] They argue that this shift from information to interaction in conceptualization of affect would "substantially change the playing field for system design. Instead of designing systems to decode and encode affective signals, systems are set up to engage users in interactions that help them understand, reflect on, and experience their emotions in new ways."[78] Although this approach does attempt to return affect back to the human domain (using technologies to better understand our human selves), it is not critical of the computational model of human affective behavior, and it resonates well with this model's goal to provide a tool for continuous self-assessment—a procedure characteristic of late capitalist societies with their requirement for constant readjustment and acclimation to changing environments, i.e., self-discipline and regulation. However, the problem may lie in the very use of the computer as a solution to the problem of self-knowledge.

In general, art has always had two tracks in its approach to facial expressions: types (that is, the iconography of standards for certain social categories and for facial expressions as representing categories of emotions) and images that stress the individual features (think of the psychological subtlety of the Old Masters' portraits). A historic example of the artistic use of a typology of affects and emotions is Charles Le Brun's *Expressions of the Passions* (1698), a study of affects such as desire and anger, which provides a classicist "summary" of the earlier depiction of the Christian value system with its virtues and "deadly sins." Yet computer technologies want to be able to see and reproduce not only the typical but also the specific; this is why they require larger databases of examples and try to adapt more complex psychological schemas, taking into account gradations and subtle differences between certain emotions.

The artworks that I consider build upon the critique of using computation for self-assessment, taking it further by letting people experience these situations in a focused way, without the distraction of possible secondary functions of this communication. As all the actors of interaction are encouraged to be self-observant about their emotional responses and behavior, the role of the machinic algorithm comes to the foreground as another actor-participant. It is treated not as an invisible mediator, but as an actor with various degrees of agency. Computerized mediation does not imply any judgment or necessity to readjust accordingly. It participates by producing a mirror function, challenging the definitions of understanding and recognition.

The screen-based interactive installation *Alter Ego* (2005) by Alexa Wright (figure 1.10) recreates the dynamics of emotional exchange, inviting the viewer to interact with a 3D rendered mirror version of himself.[79] The avatar would initially only mimic the viewer's expression, but then would start

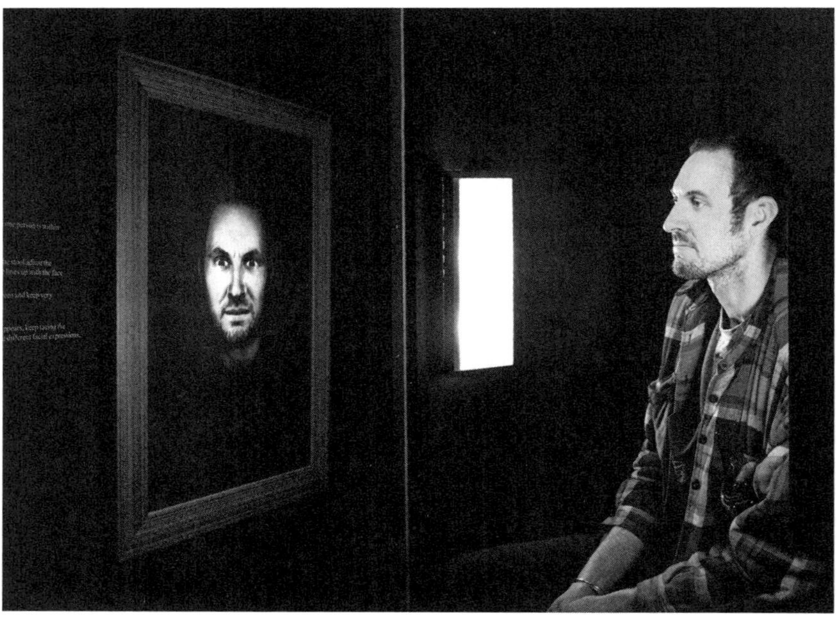

Figure 1.10
Alexa Wright, *Alter Ego* (2005), in the exhibition "Wonderful" at Cornerhouse, Manchester, UK, 2005. Courtesy of Alexa Wright.

giving more unexpected, intelligent, and believable responses. The result is unsettling and disconcerting as we can see (or get closer to seeing) the inner workings of the machine from its own point of view. "The artificial" behaving "naturally" may be seen as a logical paradox, but in this encounter it perplexes even more because the encounter happens not at the conceptual level, but through direct exchange and experience. What one sees, then, is another type of life—not a familiar, natural one, but a principally different one that still presents itself as familiar.

Alter Ego operates through several distinct processes: the detection and tracking of facial landmarks at video rates; establishing the correlations between these data and particular facial expressions; the derivation of a decision tree that classifies a series of facial expressions; and the creation of the morphing images that represent these expressions.[80] The piece's focus on the issues of self-image and mirror image are in line with the Lacanian conception of the ego as arising due to identification with an image (one's projection of the imaginary self in a mirror) rather than with a lived body. At the same time, a "mirror" here has the potential for creating not only a double but also a multitude of others. The avatar in *Alter Ego* is both a rendered image of the real person and a character of its own: sometimes it simply copies the user (yet it is never an exact mirror), at other times it takes on a life of its own. The appeal of the image is based on its alien familiarity (looking at once familiar and alien) that disturbs the sense of boundaries of the self, a sense of stable and unifying identity. Wright et al. explain: "If *Alter Ego* was able to perfectly emulate and reanimate the user's face, then artifact emotions would not likely be an issue. In its present state, however, *Alter Ego* generates an interesting combination of artifact emotions and gameplay emotion."[81] The curious discovery is that, similar to other cases of interaction with "artificial intelligence" (AI), human participants tend to project their own interpretations onto the changes in the avatar's face. To the human eye, the quantified acquires its unquantifiable features.

The techniques used in *Alter Ego*, such as real-time face capture and reanimation of the model's face, have already been developed further. An example of such development can be seen in Face2Face—currently a research project at Stanford, with the possibility of becoming a commercial application, like Affdex by MIT's spin-off Affectiva.[82] Face2Face uses reanimation in the sense of reenactment: the resulting face reproduces the expressions of another human translated directly onto the image in real time. Besides transmitting emotional expressions onto another human, a live actor can also control full articulation, thus literally putting words into the mouth of the manipulated subject (or, to be precise, the subject's image). Of course, this technique has been the basis of computer animation. The main difference is that with projects like Face2Face, it leaves the world of fiction and enters the reality of media, with all of its political and economic agendas and consequences.

Chameleon by Tina Gonsalves (figure 1.11), a series of interactive video installations, is another deep and subtle exploration of the question of whether machines can convincingly emulate (or facilitate an illusion of) a meaningful social interaction. The work instigates a silent dialogue between the audience and video portraits of people exhibiting a wide range of emotions. Technologies of facial expression recognition and video rendering are used here to explore "emotional contagion."[83] Gonsalves and her co-creators state that the most important part was to set up the boundary of mimicry in a way that could allow for a dynamic emotional experience. Their use of the term "emotional contagion" relates more to individual experience, yet it also has connotations in social psychology as a

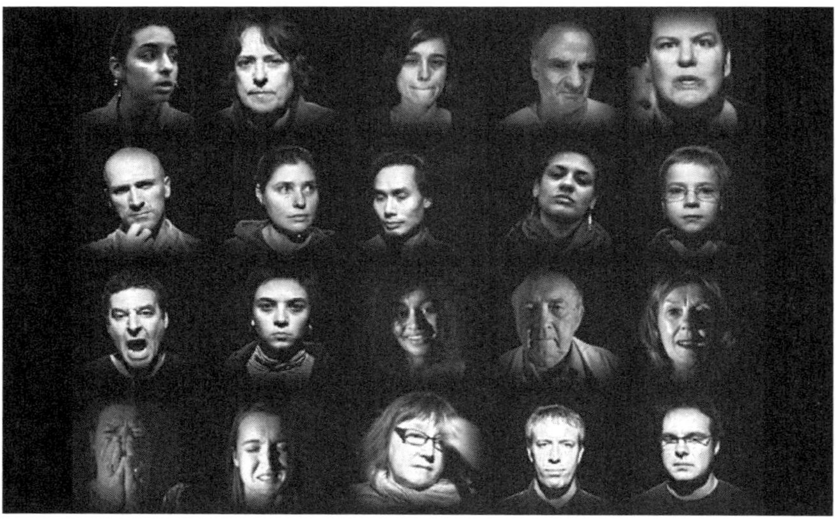

Figure 1.11
Tina Gonsalves, *Chameleon, Prototype 6* (2009). Video installation. Natural History Museum, London. Courtesy of Tina Gonsalves.

phenomenon of transference of emotion within groups and masses. For the artist, the project conceptually emphasizes how art experiences can become a conduit to new types of more intimate interactions and the exploration of the vulnerabilities of being human and can trigger deeper emotional "reflection and introspection."[84]

The creation of the work involved composing a video archive of emotion enactments, using face-reading technology, and developing a code for determining the system's responses to the audience's responses. To produce the original videos, the artist recorded long interviews using elicitation techniques to evoke in her subjects (a mixture of everyday people, actors, and visual artists) diverse authentic and deep emotions by recalling or reenacting certain situations from their own pasts, or enacting scenarios right in the studio. As one volunteer describes the experience: "This intimate relationship with a camera lens was a new experience for me and I found its scrutiny a great challenge. I learned that while it becomes possible to represent a range of emotions, the sudden and immediate proximity of some, particularly sadness and fear, was potent and very real." Gonsalves herself comments that it was an emotionally intense and exhausting experience for her: "It's been hard to watch people cry and stand over the other side of the camera documenting it."[85] There were also cultural differences: for example, in Paris it took longer to establish trusting relationships with the participants than it did in Canada. The corpus of thirty emotion expression portraits (twenty-three hours of footage) by Gonsalves extends the archive of static images, such as Ekman's, by adding the dynamic and aesthetic components.

The footage became a database, a collection of individual nuanced emotions to be played back in response to the emotional reactions of the exhibition viewers, which in their turn are analyzed and processed via facial expression recognition software. The core software is FaceSense, developed by MIT's Affective Computing researchers, the same software that Affectiva employs in its commercial activity—an application that detects head and face gestures in real time, analyzes them, and deduces information about the person's emotional disposition and mental state (figures 1.12, 1.13).[86]

Finally, the cross-disciplinary team, which besides the artist also included a social neuroscientist, an emotional neuroscientist, affective computing specialists, and human-computer interaction technologists, worked out an algorithm for the video engine to generate responses that would be

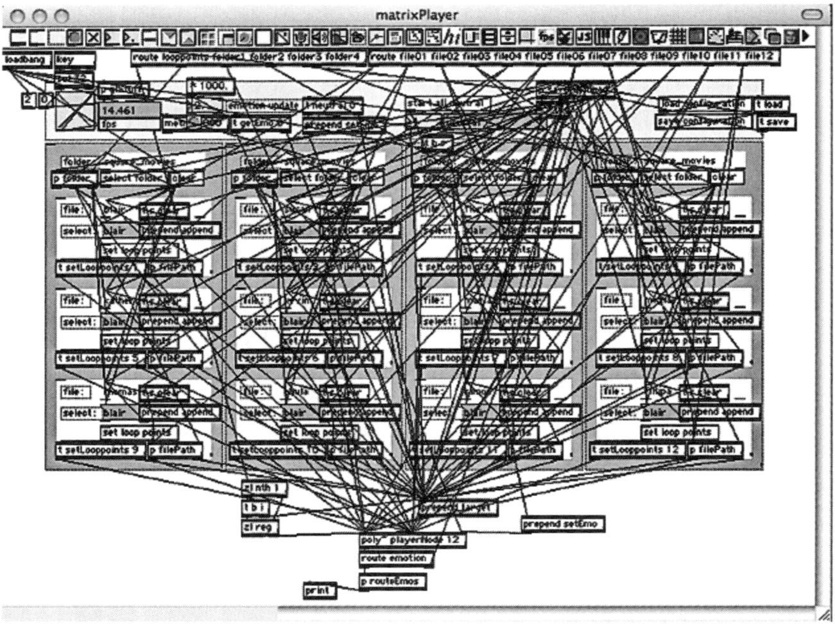

Figure 1.12
Tina Gonsalves, *Chameleon* (2009). Maxpatch, version 9.

empathically appropriate. The social neuroscientist, Chris Frith, developed a model that predicts the probability of a person's reaction to a stimulus, taking into account both the mirroring function (according to evolutionary behavioral theories, imitation of fear or disgust gives us a direct benefit by preparing us for danger), but also the fact that people do not always naturally imitate the expressions of others. (Testing that model using Ekman's six emotions in the laboratory proved that "happy," "sad," and "neutral" stimuli provoked the same emotion in response. In the case of negative emotions, like anger or disgust, however, viewers might exhibit a counter-expression.) The engine was equipped with a "learning" mechanism allowing it to be affected by the dominant emotional state of the visitors. Over time it develops a temperament, but one that constantly changes, reflecting the unstable nature of human interaction at large.[87]

As an artistic research project, *Chameleon* was realized in different versions: in big, wall-size projections (Prototype 9; figure 1.14), on smaller screens (figure 1.15) and various other hangings of the work, as well as in

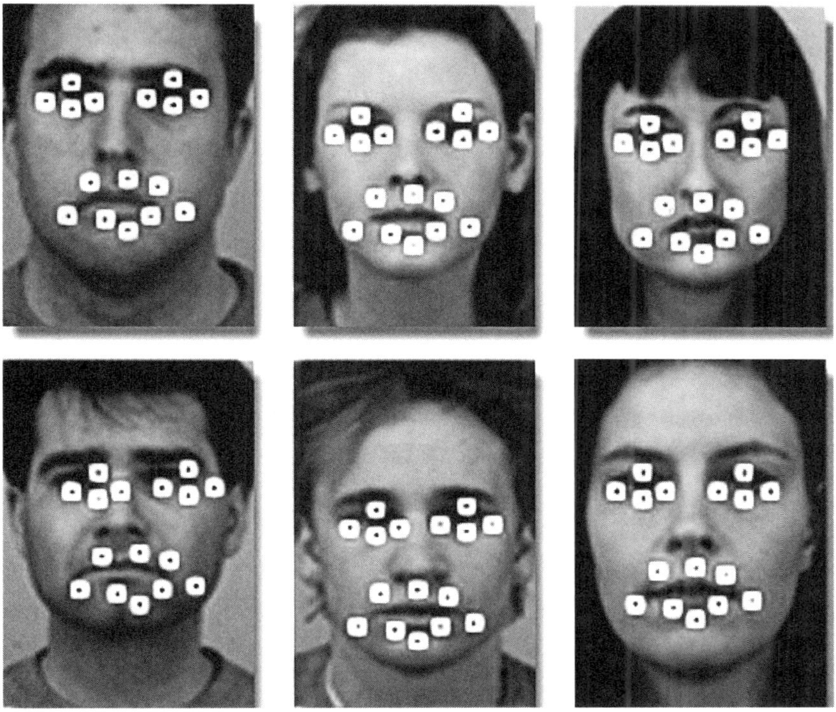

Figure 1.13
Mock-up of a face reader by Rana El Kaliouby.

different environments and types of mise-en-scène such as more intimate sit-down conversations with videos coming out of open "suitcases" (Prototype 3), either public and multiparticipant or one-on-one.

To evaluate the ability of their "emotion-aware" system, the team conducted an empirical study, interviewing the participants in a semistructured way for twenty to thirty minutes about their experience with *Chameleon*. (Most of the participants had formal training in the arts or at least a keen interest.) The questions concerned what the participants felt, what they thought about during the experience, whether they felt connected with the work, whether they thought the work was reading their emotional state and was empathizing, and whether it evoked any memories (positive or negative), among other questions. The interviews indicated three subthemes: environment, actor believability, and affective experience.[88]

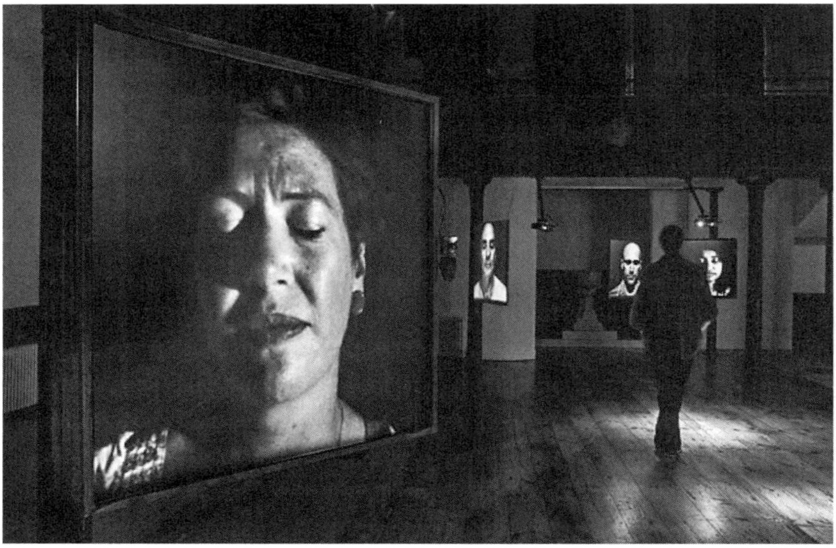

Figure 1.14
Tina Gonsalves, *Chameleon* (2009). Eight-channel emotionally responsive video installation. Fabrica Gallery, Brighton. Photo by Phillip Carr. Courtesy of Tina Gonsalves.

For this experiment, the work was installed so that people would walk freely in between the portraits; thus the emotions had to be exchanged in public, which became awkward for some. Some attendees found it to be an opportunity to collaborate while others felt competitive. The artificiality of the setting felt inhibitory, which the other people in the space perceived as disturbing and destroying the intimacy of the moment. (One participant said, "At first I felt silly. I felt really self-conscious and wasn't quite sure how it all worked. But as I spent more time there, I became more relaxed and not worried about what was happening around me. And then my experience changed.")[89] Some participants did not feel that the expressions depicted in the portraits were natural, which could be explained by the lack of an appropriate context. Interview results also confirmed that "ambiguous and subtle expressions were easier to accept and get involved with because they left space for the audience to associate meaning to them, contextualizing them, and making them more personal to their own experience." (*"If there is a really big hysterical emotion, it's not giving me much*

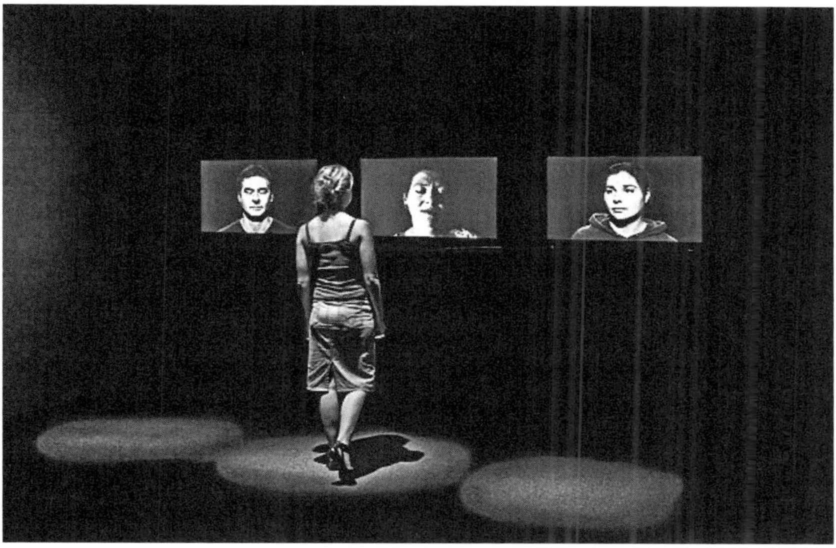

Figure 1.15
Tina Gonsalves, *Chameleon* (2010). Three-channel emotionally responsive video installation. In "Super Human," RMIT Gallery, Melbourne. Photo by Mark Akhansy. Courtesy of Tina Gonsalves.

space to figure out what that emotion is about, because it is so overpowering. But when the emotions were a little less obvious, less dramatic, it brought out empathy.")[90] Participants also imagined responses and gestures that were not in the videos.

Participants displayed different types of behaviors. Aside from a few cases where the participants wanted to experiment rather than immerse themselves in a situation, many people brought up feelings of an emotional bond and intimacy in their interviews, although that was not explicitly part of the questions. The thought was expressed that this is an experience that needs to stay with you over some time and only after reflecting on it would it be stronger later.[91]

The goal of the work was to stimulate introspection, and indeed, the above responses from the participants confirm that the piece does work that way. Further analysis of the mechanisms engaged can show why exactly it works and what the other effects at stake are. The main principle of *Chameleon* is a closed emotional loop: a response to a response to a response (the system responds to my response to its response, etc.). Human emotions are integrated here into a circuit, a network of coded events. Participants are triggered into becoming part of this game, where the very fact of measuring, of being subsumed by the apparatus, takes an important part of agency from the supposedly freely interacting participant. The interaction loop is governed by an algorithm (even being determined only by a probability, it is still subject to a rule). Social power dynamics and behavioral patterns are generated in the course of the dialogue that immediately feed into the algorithm, but can also be observed and reflected upon by an attentive participant. The types of reflection, though, will be different—receiving a new item into a catalog is not the same as human rationalization based on an intuitive experience.

What Gonsalves does here is take the human out of the loop of her own thinking about herself: "*Chameleon* reminds us that our body as a clear-cut distinction with the rest of the world is dissolved, and our relationship with the world is in flux, interdependent, tenuous, shifting our subjective experience."[92] The dynamic mise-en-scène lures the participants into this flux of changing emotional states, provoking them to experience their own adaptive chameleonic nature and the automaticity of responses. We don't know that we would be affected in this or that particular way, and yet the next second we can realize that response because the virtual interlocutor gives us feedback on the spontaneous, automatic reaction. The animal chameleon can be seen as a semiotic creature: it constantly looks for signs, trying to decode its surroundings. Is there such decoding in the case of an automatic mimicking response? Or is it more relevant to the algorithm-based system, rather than to a human? In comparison to the machinic, our chameleonic nature is still much more intuitive. The probabilistic principle represents here something that can be called the machinic unconscious—a randomly selected response out of a range of given types of response (preselected by the scientists as a set of psychological probabilities). Both the human and the machinic algorithm (represented visually by a digital portrait) comprise a self-regulating system that searches for the state of emotional homeostasis, an equilibrium that is informed by mutual emotional tuning in, or understanding. Again, understanding has a different meaning here for the human and the machine. In fact, since it is a loop, it is hard to find the moment when it is exactly the time to stop the interaction. The algorithm is constructed smartly enough that it is not always the human who makes decisions and behaves unexpectedly: a figure on the screen may turn away too (though probably leaving the human wondering, since a simple conciliatory "goodbye" is not programmed into that dialogue). In any scenario, it is an open ending for the human participant (similar to real life, where we may have lingering thoughts and emotions long after a conversation).

Gonsalves uses the phrase "emotional contagion," which may be too broad a term for what she means by it; in social psychology, the term is often used in analyses of group and mass behavior, but not necessarily one-on-one behavior. Scientists claim that emotional contagion involves a spectrum of factors, including rationality, instinct, and conditioned reactions. It is worth noting that contagion substantially differs from empathy as it is a more automatic neurophysiological reaction and may not engage the stage of emotional "reading" as decoding. Gonsalves's colleague Frith explains: "The imitation of the expressions of others that occurs even when we are not aware of seeing the expression, might better be called emotional contagion rather than emotional empathy. In this case we are not even aware that the other person is having an emotion. We would be more justified in calling it empathy if we are aware of the other's emotion and are aware that we are sharing it."[93]

How exactly does an electronic system perform its function of establishing understanding? Where and how does it manifest itself, particularly at the level of the interface?

Gonsalves deliberately refused to apply 3D computer graphic models, instead using real people in order to avoid the "uncanny valley" strangeness effect and to complicate the experience by relating people to people via slight algorithmic manipulation of video portraits. Fragments of the recorded interviews appear on the screen uninterrupted only for a short period (from a few seconds to about a minute or more); one clip replaces another based on their emotional coding. The faces themselves are not manipulated; editing happens only at the level of timing of the footage. While the faces look real and natural, the behavioral flow is broken, and it is the manipulation of the visible temporality of life that produces the unnatural effects. The use of this technique in this piece can be compared to famous video art examples of manipulation of the image time, such as the stretching of footage in Bill Viola's and Douglas Gordon's works.[94] As Gonsalves says on the work's analytical approach and its recomposition of the whole of the initial conversations with the video subjects, "I have built up compassion, friendship and attentiveness in the studio, I then take the footage into the edit, spending hours analyzing and categorizing that footage, examining the emotional expressions and commentary. It feels as if I am fragmenting a lovely relationship, objectifying it, making it into a production." Indeed, her task was to label the moments of the conversation in terms of their emotional valence and intensity, to turn them into "readable" information signals (following Shannon's understanding of information). The artist admits that she participates in the analytical dissecting of emotion as a living organism, breaking the flow of a live event. It is not surprising that the moments of the video that viewers found most natural and easy to relate to were those when the characters' expressions were more ambiguous and subtle. The rearrangement changes the messages as well.

The viewer is aware that his or her emotions are being observed; there is a level of self-consciousness, control, and performing. The question is, Why perform at all if you are aware that the image on the screen is a provocation? It must then be a performance for oneself only, and its only purpose is to sustain the continuity of the illusion of communication and, indeed, to test one's (spontaneous, automatic) reactions. The image looks real and even exposes authentic emotions, but the way it is presented and plays out is not "natural." Hence, the participant goes through a kind of "dance" between what feels "natural" and what does not. But the dance is also one with what is being realized and what is not, conscious and unconscious, and this is where it is important how this process is organized, which zones of interactions are controlled, and by whom or what. These types of questions, as they have to do with distribution of control and the structuring of live experience, can be called political. Yet they work here as such at a very subtle level: it is all about our own awareness and our own ability and need to recognize where we are provoked and where we are the provokers ourselves. This fluctuation between the positions of subject or object plays with both the personal and intuitive (and even the biological, if we're talking about affects and "emotional contagion" that happens automatically), and the system of algorithmic control—that is, the machinic operations over the image. The control is externalized only in part, and thus it is the responsibility of the human participants to walk their own line of negotiations and establish their own (power) dynamic. Awareness itself may be an object of such play, if you make a conscious decision not to reflect on the "how," but simply immerse yourself emotionally in the situation. This immersion may equally be a position of power, even compared with conscious awareness.

Chameleon has an algorithm that interferes with expectations about the interaction and randomly chooses which way to lead it. The curious part is that this programmed scenario can still lead to self-insights (introspection). Some participants approached these effects as an opportunity to push the conventions of interaction and experimented with reacting in a way contrary to how they would normally. ("So, I was doing something I wouldn't normally do. You normally empathize, so your face would be as equally sad. Playing against it was quite intriguing as it's something you just don't do." Or: "You know it is only a face on the screen, but after a while you cannot help but start to feel moved by what is going on.")[95] Awareness of the consequences of one's behavior during the interaction triggers a sense of responsibility. There is a difference between playing and the more serious and self-conscious exploration and testing of oneself and one's own empathic abilities (it is more "serious" precisely due to one's awareness of responsibility). Again, in order for this interoceptive awareness to work, the participant would need to establish an understanding and trusting relationship with the video-character/system, *as if* it were a real human being; in other words, he or she should be *absorbed* in the interaction.

Both a scientific research experiment and an interactive video installation, *Chameleon* is also in deep conversation with cinematic aesthetics. It both utilizes key cinematic principles like projection and techniques like editing, and relies on the sense of the participant's immersion in this improvisatory semifictional/semiartificial setup. It outlines the foundations of interactive cinema, but (probably) can't be called fully cinematic because of the absence of a narrative layer (although each interaction, each glimpse into a character's story—some of them actually talk—can be treated as a scene, with the audience actively affecting what happens next). The way the interaction is set up and experienced can be compared to the shot/reverse shot in cinema and how as viewers of cinema we may have internalized some of those conventions. Yet *Chameleon* does not follow those conventions exactly—you don't watch yourself reacting as the camera would show in a cinematic conversation scene; here you are your own camera, plus there is a camera that watches your reactions, and these two points of view are not synthesized.

This comparison is relevant because of film's long engagement with techniques of emotional bonding between the audience and the events on-screen. Film theory has different explanations for the mechanisms of such bonds, particularly the reactions of identification and empathy. On the one hand, the cognitivist view is represented by authors like Noël Carroll and David Bordwell, who embrace the cognitive psychology approach to emotions, specifically the appraisal theory that ties emotion to a cognitive appraisal (conscious understanding) of the situation, motivations, and goals behind the actions.[96] This view led to the development of a simulation theory, according to which the viewers' emotion is the result of a conscious process of projecting themselves into the situation on the screen, or imagining themselves within a simulated situation that the film world presents. On the other hand, the cognitivist and simulation theory perspectives are contested by scholars who see viewers' empathetic responses being shaped prereflectively by direct affective and motor mimicry processes enabled by the mirroring systems in the brain. Intentions behind actions and emotions are supposed to be understood by the viewer automatically, based on observation of the other person's motor actions and gestural and facial expressions.[97] As neuroscientists Vittorio Gallese and Giacomo Rizzolatti famously argued, embodied simulation directly contributes to social cognition and intersubjective understanding. According to their theory, the mirroring mechanisms embedded in the limbic system are capable of producing more intense impulses, and thus are the core of somatic, or embodied, empathy.[98] (The kinds of films that work in accordance with this theory, or can be explained by it, are those with less emphasis on narrative and more attention to the ambience and affective qualities of the image and sounds.) Yet mirror image theories also have their critiques, and the one on behalf of cognitivism is partly true. For instance, the automatic firing of the areas in the brain responsible for action that occurs in reaction to just observing that action performed by another does not mean that the action is recognized as having a certain meaning, or that it can be qualitatively judged. Somatic understanding (understanding without recognition) is not only a part of cognitive understanding (the prerequisite) but is a phenomenon of its own.

In *Chameleon*, viewers do not have a narrative context to "appraise" the situation cognitively. Instead, they have an image and expressions to which they assign meaning and content based on both their interpretations of their automatic somatic response and their own associations and background feelings, etc. Despite all of the manipulation, perhaps the fact that the video presents a real human being with authentic emotions helps to reveal connections between people, and not between a human and a machine. At the same time, the editing is not completely seamless and serves as a sign of computerized mediation that—in the condensed, concentrated form of an art installation—makes us more aware of the constructedness of these new types of representation, interpretation, and reproduction, including the politics it brings with it. As mentioned above, the politics of this particular work are very specific and deal with the play of awareness and immersion (where immersion can be interpreted as an active and not passive position). Thus, it may not be the most suitable example for discussing the general political issues that arise in relation to the technologies of facial expression recognition, reanimation, and other kinds of image manipulation.

Alter Ego and *Chameleon* demonstrate an artistic take on a broader cultural problem today—the penetration of computerization processes to the deeper levels of human existence, particularly the emotional life. As I have tried to show, this problem can be broken down into several issues, such as establishing correlations between emotions and facial expressions, and coding and reading these expressions. The pieces employ the existing technologies of facial expression analysis to create experiences that serve multiple purposes and open up several avenues for discussion. Centered on exploring the nature of communication between a human and an algorithmically altered image of a human, these experiences are self-reflexive; and this focused approach is generative. Not only does it make people more aware of the specificities of the new technical possibilities in mediating human perception, and of the nuances and complexities of the structure of their own natural perception apparatus, but it also lets them navigate through this experience and its discoveries in their own way. Just as the algorithms operating the expressions of the emulated faces on the screen are adjustable, so are the rules of the game adjustable, and the audience is welcome to test their choices and reactions.

The analysis of the impacts of computational technologies on human facial perception, which was the aim of this chapter, can go further, addressing newly appearing techniques, their theoretical bases, and implications. In this case, the window onto these issues was media art, chosen due to its ability to be a testing ground both for an embodied and intuitive experience and for the visitor's understanding of the apparatus behind the produced experiences and effects. All of the featured artists approach their work not only as art but as research. Hence, all of them pay special attention to the design and construction of their projects and the questions that can be asked by the very technical organization of the envisioned interaction process. Examples include the placement of the LED lights in *The Mirrorbox* to create the effect of morphed images, the electrical stimulation of muscles in Arthur Elsenaar's performance to provoke the question of an alternative "choreography" and a grammar of facial expressions, and algorithms that change the emotional valency of the virtualized interlocutors in *Alter Ego* and *Chameleon*. The artworks help to show that reflecting on experiences unavoidably leads to thinking about their construction, and about the mechanisms of production of a certain type of subject of experience. Just as literature lets its readers immerse themselves in an alternative world, experiences framed by today's media art are "safe" zones for trying out and exploring other kinds of subjectivity—or, simply put, to feel oneself differently.

The initial interest in faciality was inspired by the fact that human subjectivity (in the sense of individuality and identity) presents itself visually primarily through the face: phenomenologically, a face means first of all an "exposition." It is only after that that we get to the other characteristics of the face, such as its expressivity and communicative functions (the face as a "space of pure communicability" and "diplomacy"). But, in fact, the expressivity and communicativity of the face are also tightly connected to the subject of the face—the subject's inner affective tensions that are expressed involuntarily, or social cues that need to be communicated nonverbally. The interesting story begins when we realize that the face is also an image. It is its visual nature that allows computational technologies (the ones based on machine vision) to work with the face, breaking it down into "action units," rearranging particular features to reveal (as in *Saving Face*) or emulate (as in *Alter Ego*) alternative features (whether an alternative identity or alternative emotional expressions). Humans' natural reaction to the face is also a reaction to the face as an image.

By making facial images interactive, these artists bring us one step further, reflecting upon cultural transformation and trends, particularly the developments in human-computer interaction (HCI) and interface design. The interface appears to have particular spatial and temporal parameters in response to the needs of interaction; that is, it has a shareable and operable structure. Interfaciality breaks the pure image quality of the face and demands more—actions and reactions (both from a human and her potential machinic counterpart, or interaction partner). It is thus not only through the pure visual characteristics of the face that the human participant projects subject-like qualities onto the machine: it is both the face and *how* it operates. The human may be well aware of the programmed character of machinic operations, yet (given the visual part, or possibly due to it and evolved techniques of image manipulation) this machinic subjectivity *feels* reliable. The fact of a feeling, the very presence of an affective reaction that comes unconsciously from within the human self, is one of the significant effects of HCI technologies, and worth studying further. (An example of such a feeling may be confusion and even distress from encountering the alien familiarity of a "natural" behavior of the "artificial," as in *Alter Ego*.) But an even more important underlying effect is the possible transformation of the human subject herself (as was indicated by Anna Munster's concept of "programmatic subjectivity," discussed in detail by scholars like Kelly Gates and others quoted above). The selected artworks and their authors do not claim to make a critique or to offer a specific solution. Instead, they point to the unavoidability of technological development (embracing it), and open up this agenda as a question. Everyone is invited to individually consider the effects of technologies on human perception, coming to a responsible judgment and preferred tactics of interfacing with them, based on personal experience.

2 Body Image and the Algorithmic Organic

Among the many ways the self can interface with itself, the configuration of a self-image at the bodily level, as a "body-self," is a fundamental one. Both the "self" and its indicator, "I," are merely linguistic categories, a product of a deictic discursive formation: it is not a pregiven, but something that emerges in relation, whether to a context or to a particular "other." However, it is in the body that the sense of the self is rooted, granting it its precognitive, presubjective, and prereflective substance. The body does not have to be defined only in its opposition to the cognitive level (in a Cartesian way). One of my aims here is to address the *body* (and *the bodily*) as a special *sense-making system* with its own way of both producing and communicating meaning. The processes of perception, recognition, and identification happen at the very material and biological levels of the body. (For instance, the majority of cells in an organism have receptors for messenger molecules that carry information encoded in the form of folded protein.) These processes form a mesh of relations; the whole life of an organism depends on constant exchanges within and between the organic and the nonorganic. The self and subjectivity as agents of meaning-making should thus be complemented by the biophysical semiotic system constituted by the body. New technological capabilities of image and data processing help us to see anew the potential of the body as a meaning-making entity.

The self-image can be experienced as the ability to recognize oneself in a mirror image. As discussed in chapter 1, the reflected image becomes the locus of identity. Technical mediation takes the effects of the mirror image further. For instance, the image can be part of an electronically controlled circuit that would lock together the input and the output (think of Nam June Paik's famous *TV Buddha* of 1974, in which the statue of Buddha sits staring at a live transmission of an image of itself sitting in front of the very TV set showing its image). Replace the Buddha statue with a human, add a few manipulations, and the effect will be exactly what many other artists in the 1960s to 1970s were exploring. The image of the self *doubles* the self, but only in order to delineate its boundaries. This doubling can be specified in the relation between the body and the image that represents it (an avatar). Yet what is doubled is not the body itself. As Maurice Merleau-Ponty shows in his phenomenological theory, the image affects the sense of oneself as embodied. In a "body image," according to the philosopher, a body is an object of *intentional* consciousness; it is how the self wants itself to be seen from the outside. Merleau-Ponty argues that the body image is different from a "body schema," a nonconscious configuration stemming from the *experience of the body* itself.[1] "Image" here stands for a socially exchangeable trope, since it is through the image that others can recognize our identities in us. The challenge posed by phenomenologists (including postphenomenologists like Don Ihde) has been to determine the possibility of an "unmediated" and more "authentic" experience of the self. Yet what is often forgotten is the communicative potential of the image (as in the case of the face): the image allows us not only to be recognized but to communicate ourselves. In its exchangeability, it serves as the basis for connection—both between people and within a person. However, as the psychoanalytic perspective shows, an image also disconnects, becoming a source of confusion over the boundaries of the self, and sometimes frustratingly so. Video artworks by artists like Peter Campus and Dan Graham, as well as contemporary performances with motion-tracking technologies, offer an interesting entryway to discuss what exactly an image does and what the feedback relation between it and the body (and the body-self) may mean. As explored below, different technologies of motion capture would construct the figure of the double, the computerized reflection of the body, in different ways. As opposed to the photographic method, the digital operates through breaking down the whole image into discrete, rearrangeable elements. Thus, the status of the resulting generated image is not the same as the "reflection," or the double created through traditional means.

An image is not only a visual phenomenon. Today's image theory offers a broader understanding of images that in addition to visual objects also includes distinct sensory formations, such as perceptual images, or inner impressions based on sense data. W. J. T. Mitchell and Gottfried Boehm discuss the value of iconicity (the being of an image) as a form of thinking, which is both nonsymbolic and nonmimetic—that is, it does not have to correspond directly to any particular reality or "state of affairs."[2] An image has been claimed to offer a unified view on otherwise disconnected elements. This is exactly the point of data visualizations, for example—to integrate the numerical information and make it presentable in the form of a graph or a diagram, graspable at a glance.

Today biosensing technologies demonstrate how the mirror image of a face or of the whole body can be complemented by a differently derived type of portrait—an image generated through the processing of the data produced by the body: a data portrait, or what I will call a *body-data-image*. Body data-scanning technologies allow more and more accurate tracking of the inner bodily processes, including those related to affective states that lie below the threshold of consciousness and are expressed in the form of heartbeat, galvanic skin response, and brainwave dynamics. It can be assumed that seeing a more accurate "picture" of one's bodily state would lead to better understanding of one's self and to more balanced and "informed" behavior. This aspiration is confirmed by the proliferation of wearable "smart" devices, like Fitbit and Apple Watch, and software applications that help to "read" one's bodily and affective states.[3] Yet, while giving a sense of empowerment, these technologies for "care of the self" (in Foucault's words) make one fall into a loop: the "self" strives here toward an "ideal" body/self, a (socially) imposed model.

A "picture" of oneself generated through biofeedback is different from a mirror image and other forms of self-perception. The performances of the body, happening beyond the threshold of one's cognitive control, represent a type of cognition of their own.[4] There is thus a discordance between the socially determined sense of the self (as an ideal or socially acceptable sense of self, which is also the body image) and the more "organic," embodied sense of the self (the body schema). The "organic" here stands for any type of biological substance, a living being. Algorithms and codes engaged in processing the data further complicate this discord between the inner and the outer, the organic and what appears as a result of processing. On the one hand, they are designed to perform automatically, according to prescribed mathematical rules. On the other hand, codes can have a generative potential: they can produce results that go beyond the discursive and cognitive control of a human, impacting, shaping anew the very body schema and stimulating the emergence of a special type of bodily awareness, the body-data-image. The expression of this potential depends in a large degree on the material substance within which the code operates.

A number of biofeedback-based artworks will help me to argue that the data images produced within feedback operations are not just about the data that are used for self-improvement. Staged and gallery performances and time-based installations employing multisensorial stimulation and neurofeedback show that the body-data-image moves us beyond the sphere of the individual self and stable identities. It takes us into the field of relations.

Hence, the broader question at stake is how the coding technologies applied to the body constitute what I will refer to as the "algorithmic organic," and how they reframe for us what was understood as our biological substance, the very life of the body, the *bios*. Of course, our human knowledge of the organic is never unmediated: think of early scientific and medical devices like the microscope. Today's devices that help to precisely track the spatial position of the body and characteristics of its inner states may be seen as continuing the same tradition. It is a matter of what can be done with them: the knowledge and new kinds of relational experiences and encounters that they provide. Here is where the artistic lens can play a role, sharpening and expanding our views on how to deal with these capabilities, how to make sense of what is newly available for sensing, or how to simply experience being *in relation with relation itself*—and through it, with that which *exceeds* the principle of *the self*.

Interfacing with the Double

The Self "Locked" On-screen and the Body Image / Body Schema Circuit

The image of the body usually has its own "body"—the material surface onto which it is generated. An impressive corpus of video art has been devoted to the materiality of the screen as an interface defining relations between the beholder's body and its representation. Usually this is addressed through manipulating the spatial qualities of the screen as a flat surface and the three-dimensional nature of the on-screen environment that itself corresponds to both the place in front of the camera and the position of the viewer, possibly in front of the screen. Video art and closed-circuit video installations are interesting in this context mainly in their relations to the later avatar phenomena and as interfaces for interaction with the image,

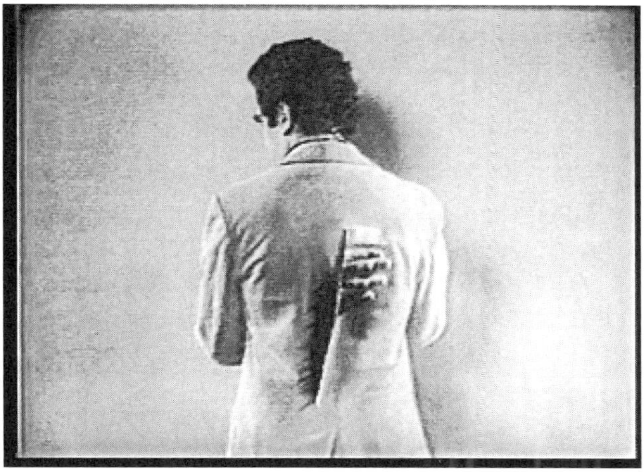

Figure 2.1
Peter Campus, *Three Transitions* (1973). Video, color, sound. Courtesy of the artist and Cristin Tierney Gallery, New York.

not only by the person that it represents but also by other people. The relations that I would like to trace are from a body to an image to another body that is presented by the image. How exactly does the body become an "imagined body"?

Three Transitions (1973) by Peter Campus (figure 2.1) is a canonical example of self-image manipulation done with video tools. Using the cinematic technique of superimposition, Campus performs three acts: cutting and walking through an image of his own back; "washing" away features of his face by caressing the face with his fingers, revealing another identical face behind the one presently dissolving on-screen; and burning a piece of paper with a live video portrait of himself. All of these operations deal with the destruction of self-representation.

It is not clear right away what exactly is going on in the first scene. We see the back of a person through which a knife suddenly starts to cut. Yet it cuts through the back from inside as if the person were just an image. It takes a few moments for the viewer to realize that what is being cut is indeed a flat piece of paper screen onto which the image of the artist, shown from the back, is superimposed. What is confusing is that the body being cut is the same as the one that is cutting. The body becomes flattened into an image. The body as an image functions here precisely as an interface. It demonstrates what can happen when the surface of an image is treated as the place of an action (the action takes place directly on the surface of the image). The image enables an action to be performed by a body on a body.

Besides the screen, the architectural dimensions of the gallery also participate in perception. Video art from the 1960s and 1970s pushed the spectator's attention away "from the illusion on the screen" to "the physical mechanisms and properties of the moving image."[5] (This can be illustrated, for instance, by Dan Graham's video installation *Three Linked Cubes / Interior Design for Space Showing Videos* [1986], featuring a construction for showing videos made of framed glass and mirror glass panels.)[6] Projection became not only sculptural (the screen as an object) but architectural, attracting attention to the spatial positions of the spectator, the camera, and the screen in relation to one another. These relations create a particular environment that is both physical and purely image-based. Christine Ross describes the effect of these works as bringing the viewer's awareness "to the materiality of space in relation to the illusionistic image" that then serves as a basis for a sense of distance from the "absorbing effects" of the image.[7] Before discussing this effect in more detail through the concepts of body image and body schema, let us focus on what exactly may determine it.

The strategy often used in video works emphasizing spatial relations is a closed-circuit technique: the simultaneous filming and projecting of filmed footage—just as in Nam June Paik's *TV Buddha*. The Buddha statue is placed in the position of a spectator, and is at once an object and a subject of perception. The closed-circuit method builds upon the potential of the mirror image, particularly the spatial and temporal aspects of its perception. For example, Peter Campus's installation *Interface* (1972) realizes this quality by doubling the image of the viewer: she is shot through a sheet of glass, onto which the projection of that same footage filmed in real time is directed. The spectator thus sees both the projected image of herself and the reflected mirror image of herself on the same glass surface. The images differ in that one is a black-and-white (positive) one, and the other, the reflection, is a colored negative.

Ross argues that this double image provokes in the viewer a sense of separation between the inner feeling of one's self and the image of the self: "Video projection is thus the means by which the spectator engages in the act of perceiving the self-in-space, according to two irreconcilable (subjective/objective; internal/external) viewpoints. It facilitates the development of a critical attitude on the part of the viewer towards her own sense of self, and inhibits the tendency towards narcissistic blending of the self into one's own image."[8] In other words, such a setup (triggering the feeling of the self in the physical space of the body and not in the space of the image) brings an awareness of the constructed character of the image and the necessity of separating the self from it. Ross contrasts the aesthetics of critical distantiation from the image of oneself as a construction with an aesthetics of immersiveness and relationality that offers continuity between the real and the virtual (or illusionary). She favors the latter type of aesthetics because it offers new forms of social connectivity.[9] In this book, I pursue a similar goal of discovering ways to experience connectedness, but I would like to take it beyond the social type of connectivity. We will thus require a more expansive understanding of the concept of the image and its comparison with nonvisual forms of communication. For example, we will see that the same effects of absorption into an illusion, critical distance, and emergent forms of relationality are relevant to the types of imagery created with bodily biodata.

The closed-circuit method (when the body serves as both a subject and an object of perception) works so effectively to a large degree due to the simultaneity of the exchange between the action, the image, and the perception of that image. It can be argued that—despite Ross's initial separation of the two aesthetics—by returning the viewer's attention to the physical properties of the location and insisting on the necessity of being present in it (in order to get the full message of the work), these works suggest another regime of engagement and absorption into a set of relations that remove the contradiction between distantiation and immersion.

The discussions inspired by these installations can be traced back to the famous formulation by Maurice Merleau-Ponty of the split existing between the external and internal images of oneself, or more specifically, between the consciously imposed "body image" (*image du corps*) and the embodied sense of the self, which he calls "body schema" (*schéma corporel*).[10] A mirror image is defined by the philosopher as an image that "forces me to leave the reality of my lived *me* in order to refer myself constantly to the ideal, fictitious, or imaginary *me*, of which the specular image is the first outline."[11] In *Phenomenology of Perception*, he also describes this difference simply as "my body for me and my body for others."[12] In contrast to the "imaginary *me*" (or "my body for others"), a "lived *me*" ("my body for me") indicates a *lived experience* of the body, its implicit awareness of *being-in-the-world*. It implies a feeling of being grounded in time and space, a feeling that the screen-based mirror image shifts, makes confusing, and even destroys. "Body schema" is what precedes (or is behind) this feeling of spatiotemporal groundedness; it is a unity principle that organizes that feeling as a feeling.

As we will see, screen-based and other technological forms of representing the body in real time (which incorporate the principle of the closed-circuit feedback loop in one way or another) all deal with this same idea of the duality of the body: the body as representable ("body for others") and the felt body (the "lived me"). The images double the living body, and the question is, then: What can and cannot be captured and transmitted along with them? What kind of inner feeling of the body? What kind of "self" is involved in defining that feeling and the senses/meanings it carries?

Merleau-Ponty tracks the origin of the concept of body schema in empiricism and particularly in the associationist tradition of physiology, a theory that viewed the bodily and mental states as a series of associations, or connections.[13] In that tradition, the body schema was understood primarily as a subtotal of the images of individual body parts, particularly a resulting image of a body concluding an action that requires rearrangement of bodily parts. In opposition to this understanding, Merleau-Ponty offers his own definition of the body schema, based on more psychological observations. He contests that "the spatial and temporal unity, the inter-sensory or the sensorimotor unity of the body exists *de jure* [i.e., as a rule, a law]," that "it is not confined to contents actually and fortuitously associated in the course of our experience, it is in some way anterior to them and makes their association possible."[14] For him, instead of being a subtotal of potentially representable parts, the body schema is a "law of [the body's unity or] constitution," a "law" that allows that unity to reconstitute itself in any kind of form. This reference to law as something that "orders" things in nature is not accidental and, in fact, indicates a principle of dynamic holding together that is also important for our context.

Another aspect of the body schema, namely a unifying form, is initially derived from the definition of a body schema in Gestalt psychology's notion of *Körperschema*.[15] But Merleau-Ponty's body schema *precedes* the form in the sense of *Körperschema*. It works as a more fundamental principle underlying the body's constitution prior to any particular form and allowing it to take potentially any form.[16] In that sense, it is a *structure for a form to be a form*—in other words, a *structure for forming*, or formation. The form, then, can be seen as something closer to an image, and, curiously, an image that is both visual and schematic, outlining the boundaries of an entity (for instance, in German, *Bildung* is both a formation and an outline, a shape, or a contour). Their meanings overlap, yet they are not synonymous: an image is more than a form and a form is more than an image. It can thus be said, perhaps, that the form is something that provides continuity between the notions of "image" and "schema."

As a structure that constitutes the unity of the body, the body schema connects the body with the world in a way so that it is treated not as a separate entity, by analogy or imitation, but through the *structure of perception*. The structure that enables bodily perception of the world is "bound up" with the world, not external to it.[17] This is also what defines the body schema's relation to the principle of the self. The self expressed linguistically as "I" (as noted above) is a deictic, or indexical, concept; it is something that something else is not. The fact that it has to stand alone against some context or some "other" implies that it is defined in relation. This is why in order to articulate itself as a self, to identify its boundaries, it has to position itself as intentional. The body schema, due to its underlying connectedness with the world, is preintentional or subintentional. It can also be seen as more than, or "extra-," intentional.[18] Going beyond the sense of the self as intentional, the body schema grounds it in a different register, as potentiality and as a *self-in-formation*: a self *in between* the image and the schema.

As we have seen already with closed-circuit installations, feedback complicates intentionality as a characteristic of the self. There is no outside, the elements are locked together, so that their actions return back in a continuous recursive loop. Yet within that loop, different qualities emerge or get revealed. The factors may include a medium of exchange (screen-based image, analog or digital, or maybe even another body), conceptual setup and preset roles for the main participants, spatial and temporal dynamics, modalities of control, and forms of protocol inscription and activation, etc. All of this together would form varying types of interfacial relations. But perhaps what these relations are called to reveal is this "law" of potentiality

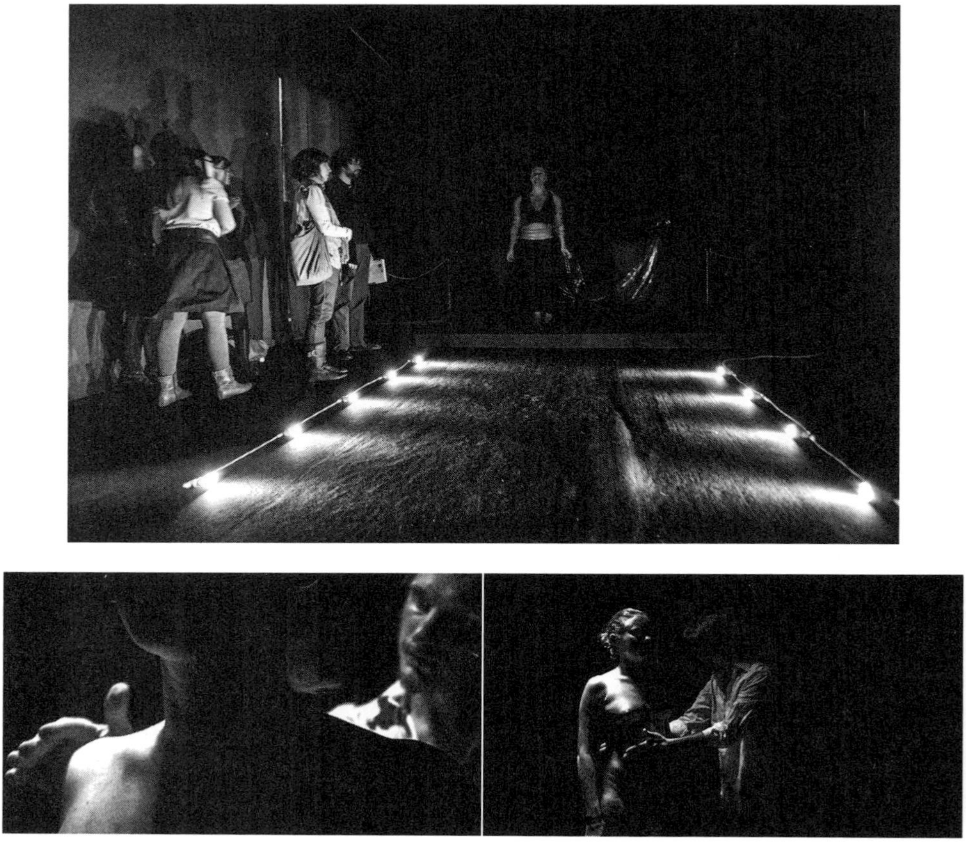

Figure 2.2
Sonia Cillari, *Sensitive to Pleasure* (2010). Interactive performance / bioelectromagnetic ambisonic sound environment. Courtesy of Sonia Cillari.

inherent within the body schema, something that is capable of making the closed loop of relational events feel open.

Embodied Avatars of Art-Making: Sonia Cillari's *Sensitive to Pleasure*

Italian media artist Sonia Cillari's 2010 piece *Sensitive to Pleasure* (figure 2.2) illustrates what happens when an "image" of a body figures not as a screen image but—however strangely—as "embodied" by another human's body. The work is a performative electric-field sensing and sound environment that offers a peculiar conceptual frame: it explores a specific type of relation, a controversial, even conflicting relationship between an artist and her work, and how it may feel bodily. The piece works as follows.[19]

Cillari stands outside the door of a dark "ambisonic" cube, where she grants entry to only one visitor at a time. The cube features her "work"—a naked female (called "the creature"). The woman is positioned on a 5-volt transmitter, thus serving as a conductive element, a human antenna in an electromagnetic field created within the cube. When an audience member enters the cube, the two bodies—of the "creature" and of the viewer—create an electric circuit. The physical proximity between them determines the strength of the electrical signals that are then translated into sounds and lights, and most importantly, they are conducted directly to the body of the artist standing behind the wall (in front of the rest of the audience). The artist experiences the physical interaction between the "creature" and the audience member via electrical pulses from sixteen electrodes tied onto her arms and torso; the sensation is painful but might also be considered pleasant (as the artist herself states). The intimacy between the visitor and the "creature" inside the cube is not documented and is reflected only in electrical signals on the body of the artist, as well as through sound and the changing intensity of lights, which enhances the sensuality of the work. The apparatus behind the piece is also not supposed to be seen: the wires are masked and Cillari stands right next to the wall that contains the processing equipment.

Sensitive to Pleasure is part of Cillari's broader artistic research of the concept of the body as an interface. The works united under that framework are predominantly performative installations, in which the artist puts her body at the center to generate patterns of images and sounds. (For instance, in *As an artist I need to rest*, she breathes into a tube connected to a computer, and the air she exhales is analyzed, together with the speed and intensity of breathing; as a result of that analysis, abstract images get formed and reassembled.) Trained as an architect, Cillari is interested in spatiality. In her words, "The piece is about physical borders. . . . The performer is naked in a claustrophobic environment; I deal with electricity. We don't know how far we can go."[20] (The work thus requires psychological preparation, for both the artist and the naked performer.) The concept of the body as an interface treats the interface not just metaphorically: its meaning as a navigational display, an entry point to a certain semicontrollable environment, has here a physical and an extremely intimate equivalent. The relationship of control becomes eroticized, and it is impossible to ignore (even if the artist herself prioritizes another aspect). The encouragement of interaction with a naked female body is reminiscent of feminist art of the 1960s to 1970s, such as Valie Export's *Tap and Touch Cinema* (1968), when female artists deliberately exposed their sexualized bodies as a way to critique the dominant patriarchal order of libidinal power distribution.[21] But who is the subject and object of control in the case of *Sensitive to Pleasure*?

For Cillari herself, the project is an allegory of an artistic production, a materialization of the interaction between an artist and the result of her artistic work. By focusing on these relations, Cillari's piece reminds us of what exactly makes art a special method of inquiry into the perception and communication of the self to the world. In an idealized version, an artwork takes a highly subjective form, fixing in a certain type of medium an artist's *subject position*. Thus, it is an example of an *intentional* object. This is essentially a romanticist perspective: art as a way of "unveiling" with the power of imagination something that is concealed. But art, of course, can also be a means to renegotiate social structures and norms. By naming another living being, especially one put in such a vulnerable position, as a "work," Cillari stretches the borders of the acceptable within this traditional paradigm, letting the nudity do its job—to provoke unconscious desires—while the artwork remains a means of interfacing with the self—through the effect of doubling.

In her interviews, Cillari mentions the myth of Pygmalion, the archetypal model of art-making as profoundly self-reflexive (which can also be somewhat perverse and a cliché). And yet, even when centered on the self, Pygmalion does externalize his desires in the form of his Galatea, a sculpture that becomes alive. In that sense, she is not a double, a representation of Pygmalion himself, but a product of his intentional thought and imagination. Just as in Ovid's *Metamorphoses*, this externalization implies transformations, including the transformation of the self: in the end, at least in the original sense, the myth is about love.[22]

Indeed, the status of Cillari's "creature" is quite confusing. One has to keep in mind what the performer represents—that she is neither an actress in a full sense nor simply a body. What also has to be remembered is her entanglement with Cillari's body. The two are extensions of each other, and this is exactly the problem. The "creature" can be seen as the artist's avatar, a human double, a persona. Yet, at the same time, maybe it is the artist who is an avatar, passively connected to the main point of encounter, the "interface," with her "work." In the end, it is Cillari who physically represents to the rest of the audience what happens within the cube and is invisible for them. This is the paradox of the feedback dynamic between the two that the visitor who enters the chamber with the "creature" may experience. The externalization of the self, putting oneself outside of oneself, can be interpreted as both subjectivation (becoming a subject) and desubjectivation. As Cillari summarizes, "this work deals with an inverted relationship of control between the creator and his own creation. The physical connection between them represents keeping each other alive, a metaphor of mutually dependent relationship."[23]

This entangled and inverted relation problematizes the meaning and usage of the notion of an avatar. Once the "image" (a representation that "stands for" the subject) gets embodied, it stops being only an image, thus shifting the attitude to itself and transforming the feedback dynamics. The challenge becomes not only to feel another's body, but to *feel embodied in the other's body*. This is where the strong, erotically charged physical sensations that are transmitted come to play a role as they connect the feeling of the body and the feeling of the self (a psychological awareness of one's presence), resulting in the acute affective quality of negotiation of what it means to be a self.

The bodily experience is subjective and is not easily shared. What is left for viewers outside the "secret" chamber is only an abstract notion of an "embodied interface" and an *image* of Cillari "caught in experience." Viewers do not know how it feels, and that makes them reach for clues from what is given to them. The image serves as a clue, as does the project's title—"Sensitive to Pleasure."

Several kinds of "pleasure" may be distinguished here. For the public, it is a voyeuristic one: observing the reactions of Cillari's body, something intimate in itself and erotic because she cannot control the situation. The experience of the piece is, of course, different for the viewer inside the chamber. That is a privileged position. The visitor inside the chamber is a key actor and agent, triggering a whole chain of events—visible and audible changes in the environment, impulses to the artist's body that make her visibly shiver and twitch. It is this viewer who actively navigates and controls the whole process, and there is a pleasure attached to this position of control. Finally, Cillari herself admits: "Pain can feel as pleasure because I can feel on my body how people are interacting with my own piece."[24] This can be read, perhaps, as a feeling of satisfaction that connects her embodied self and her identity as an artist, even if it involves a slightly masochistic bent.

Cillari's is not the only media art piece that employs visitors' participation and control over the body of an artist via electric signals.[25] Its uniqueness lies in the fact that the initial communication (between the audience member in the chamber and the naked body of the "creature") is not technologically mediated and that this organic quality may be seen as an even stronger stimulus for (sexual) desire. Yet the viewer-participant is conscious of the effects of his or her actions and (ideally) controls them accordingly. This relationship is based on *eros* as both desire and compassion or empathy, as both sensation and a thought, arousal and responsibility for the other. Eros thus figures here as empathetic and ethical, as a feeling not only of but *for* the other. Creating pleasure for yourself, as a visitor in the chamber, you must think of a distant body (of the artist) and its pleasures or pain. The visitor-participant is put in the situation of making this ethical choice. Putting forward an "avatar" of her creative thought in the form of another woman's body, Cillari doubles the presence of the bodily, its sensations and insecurities, but also its pleasures.

Body Image in Motion: Escaping Capture

Cillari is interested in embodied interaction, but her goal is also to reveal how an image in the form of an avatar may be constructed, conceptually and technically. Her work investigates an apparatus of feelings and the affects it produces. But the problem with digital avatars is that, as "personas," they do not even need to have physical bodies and in fact may be quite remote from the "originals" that they represent.[26] One of the questions to explore is how technology that creates our doubles impacts our relations with ourselves and others through those doubles, but also how such technology may be reinforcing social and cultural biases attached to any form of bodily representation.

As the history of animation demonstrates, dynamic characters feel more convincing and "alive" than static ones. Screen-based representations of the body rely on techniques of capturing the body in motion. This way the virtual doubles feel both more "alive" and more "realistic" in their movement. Hence, the question has to be specified as: How does the technology of motion capture affect the types of bodily movement, bodily representation, and one's relation to them? To read representations of live action in terms of interfacial relations will mean to reveal how live action serves as something tied up with an *effect*, with something more than just a direct copy of an action. What kind of new languages of control emerge from the new technical capabilities? What kind of affective impact do they entail?

A discussion of motion capture (also referred to as motion tracking) as an interface has to begin with the distinction between the visual and numerical regimes of perception. In both, images are created based on live action. Both are also approached here as a type of perception that can be called "technological" (or machinic) and that will be contrasted with human perception. Motion capture techniques based on a purely visual regime are determined by the use of traditional film or video cameras. The resulting moving images document a real scene, but they are not interactive unless they become part of a separate interactive system and could be presented in coded blocks of footage (like the clips from *Chameleon* by Tina Gonsalves, analyzed in chapter 1).

The visual regime allows the creation of not only exact copies of the captured figures but also their abstracted or transformed versions, such as character animations through a rotoscoping technique. Rotoscoping involves recording a real action or performance and graphic tracing over its projection. The resulting image can thus present a totally different and completely unnaturalistic-looking character that would perform recognizably anthropomorphic movements. Yet this technique still requires a recorded image and does not work so easily in real time.

The numerical regime of perception implies that what is being traced is not the full image (a continuous entity) but only markers—the dots, interrelated data points, or coordinates representing positions in space. These points can then be computed and rendered back in the form of a three-dimensional dynamic representation. The image of the body that the camera sees is fragmented, atomized into discrete, analyzable expressions. Organized in a mathematical way, the data points become part of specific mathematical functions—$f(x,y,z)$—that turn them back into an image, yet with the possibility of a wide range of visual transformation. (Besides the effect of re-skinning, most popularly used in digital cinema and animation, other effects can include radical shifts in the perspective from which the scene was initially observed.) In commercial entertainment, these effects serve the purpose of anthropomorphizing the imaginable world, making it recognizable, at least in part, for the audience; the amusement (or other intended effect) arises exactly at this shift from the familiar to the unfamiliar. But what is important to stress in this context is what happens to the perception of the body. As Vanessa Chang puts it in her research on gesture, the points "define and confine,"[27] meaning that the mathematical function that converts the points into images is a determining factor that limits the range of potential connections.

In the early 1970s, Swedish psychologist Ragnar Johansson experimented on the perception of movement and the sense of depth of space (for instance, a horizontal movement and change in size of a straight line give a sense of motion in depth). In particular, he demonstrated how the movement of a human figure can be simulated through the arrangement of lights on an oscilloscope.[28] Only twelve light dots, according to his discovery, are sufficient to create the impression of a moving human figure that could even reveal gender traits by posture and gait. The questions are: How far can one go with abstraction before the image stops being perceived as realistic? Is there anything about the quality of movement that makes one moving figure look not simply "alive" but more "natural" than another? What kind of affect does this connote? Answering these questions will be a matter of recognition and legibility, namely coding and translation (converting the signs into meanings and vice versa).

Johansson started with the minimum, the light dots on rotating cylinders, but it is worth remembering that the origin of the reproduction of a human figure in motion by recording actual bodily movement was in Étienne-Jules Marey's chronophotographic experiments in the 1880s. Marey famously made continuous shots of walking human figures dressed in tight black costumes with white lines along their limbs and white buttons indicating the joints. Since they were shot against a black background, only the white lines could be seen. Combined, frame by frame, the lines produced the impression of a moving white skeletal figure. Marey's method is very clearly a subtraction, taking away the "flesh." Today's professional motion tracking (used in video production, as well as in gaming, the military, and even aviation industries since the late 1980s) continues Marey's innovation, updating it with another form of body suit—with the sensors stitched into it just like the white buttons on Marey's skeletal figures. But the question with these later technologies remains—Where does the "flesh" go? What does it mean, affectively, that it can be replaced?

Skeletal figures are also the basis of one of the most popular motion-tracking technologies today that does not require a suit, which is Microsoft's Kinect. Designed for computer games and released in 2009, it is used in a wide range of interactive scenarios designed to establish connections ("connect") through movement ("kinesis"): games, public displays, theatrical performances, interactive installations. The key innovation of Kinect is that it supplements a video camera with infrared light elements—an infrared laser projector and an infrared image sensor. Whereas in the case of a movie camera the image is a result of light (reflected from the scene being shot) coming into the camera and being registered on a film or a sensor chip, Kinect serves as a source of light itself (only it is infrared light). A beam of infrared radiation ("structured light") is emitted onto the scene to be captured, defining the field of capture and calculating the distance to and between objects, as well as separating objects on different planes. The program then sorts through the received image in search of a combination of dots that would look like a skeletal figure (it was trained to identify such figures using millions of images of moving human skeletal figures). It subtracts from the "noise" of other objects, while simultaneously comparing the detected fields of dots to the abstracted models already encoded, clustering pixels of body parts together to form a recognizable image that can undergo a broad range of manipulations.

One of the key critical observations about the impacts of these technologies is that they transform the perception of the body both from the outside and from within, helping to reveal new abilities of the body (new ways to be in space) by *retraining* it to move in a certain way. As Daiane Lopes da Silva, a dancer and choreographer and the artistic director of the San Francisco–based Kinetech Arts group, commented while trying to move for the first time in a motion capture suit made of strips with sensors placed at the joints: "I feel like a Robocop. The suit constrains but suggests new movements. It can teach us as dancers something new about our bodies' potential."[29] That is, the suit "confines" and "defines," but also *redefines the body's capabilities*. A new kind of response is also demanded by the dynamically transformed image.

The implementation of forms of motion capture in art goes back a while. Some works continue the animation and cinematic tradition in the format of short video films. Others belong to the category of interactive art. Among the well-known examples from both categories that reflect various aspects of the technical and conceptual evolution of motion tracking are *Merce by Merce by Paik* (1978) by Nam June Paik, a tribute to Merce Cunningham that is a "choreography for the camera" where the image of the dancing figure is multiplied or put in various landscapes; Myron Krueger's interactive video installations, such as *Videoplace* and *Metaplay* from the 1970s, stimulating the creative and playful potential of the visitor's body; and Simon Penny's immersive environment *Traces* (1997), which uses custom-built software for 3D figures to create embodied interaction with semiautonomous avatars.[30]

Paik's work stands out from this list because it is not interactive, yet it does explore different effects at the juncture between the image of a moving body and its placement. (The origin of such effects can be traced as far back as animated magic lantern shows and to Georges Méliès's illusions with double exposure and superimpositions.) Another noninteractive video work that uses more recent techniques such as Kinect is the short experimental film *Intangible Body* (2016) by Emelie Mahdavian (figure 2.3).[31] The

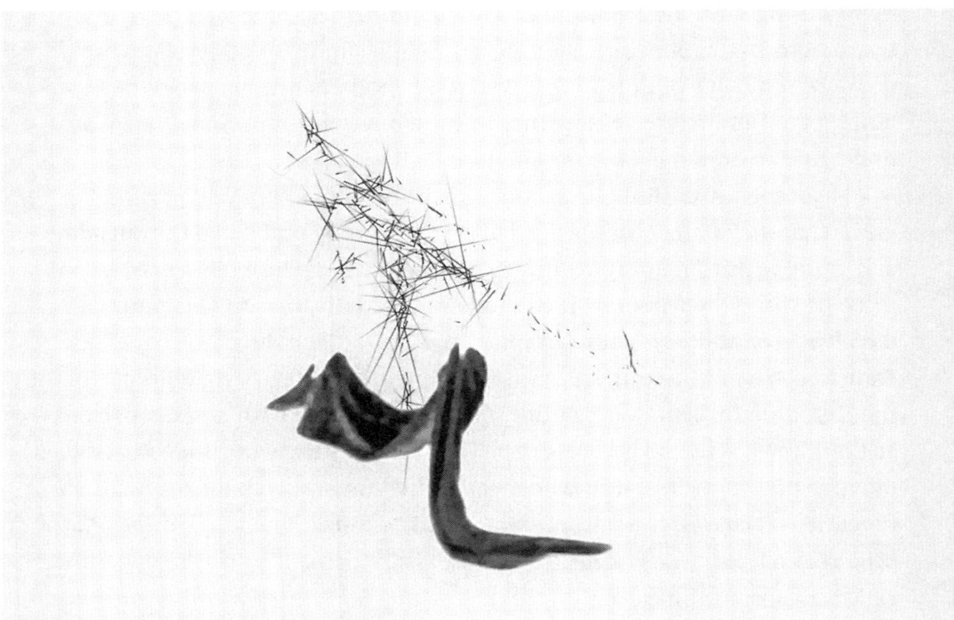

Figure 2.3
Emelie Mahdavian, *Intangible Body* (2016). Screen capture from video. Courtesy of Emelie Mahdavian.

piece brings us back to the subtractive method and the question of what happens with the "flesh." But it is an important example in our context as it highlights how 3D motion tracking can signal a cultural and even a political message. *Intangible Body* takes as its theme the censorship of Iranian women's dance performance: both women dancing in public and its representations (even in animation) are considered illegal. The project plays with the borders (and literally, the outline) of what constitutes a body, a dance, and womanhood in Persian culture, but also in digital culture. Created via motion capture composited with video, the film represents aspects of the dancer's body, but without any video of her figure appearing in the final film. The veil with which she danced was isolated in the video via chroma keying to create a residue of her interaction with the prop, invoking the presence of the absent dancer. The contours of the body can only be guessed behind the movements of the veil and within the graphic mesh of flickering lines representing the bodily figure. The lines literally fill the space of the flesh within the invisible contours of the body figure, forming a vaguely recognizable silhouette. The body is stripped not only of its flesh but also of any clearly visible border, the skin. It is the viewer who has to complete the image.

In line with a long tradition of avoiding censorship, *Intangible Body* works as a way to escape direct identification with what is forbidden. It is an attempt to break away from the image of the body as a culturally prescribed image (or anti-image, because it is "prescribed against," to the extent of being "proscribed"). The image is still recognizable, but only in the mind of the viewer. The "meshed" figure also alludes to the "form" as it was introduced in relation to Merleau-Ponty's theorization of the body schema—a complex of relations underlying the resulting visible shape. The whole question of what constitutes the body resonates with his description of the body schema as the "law of constitution" of a bodily form. Here, "form" refers neither to the figure of the filmed dancer (whose body was captured and then isolated and removed) nor to the hinted shape of the moving female body image on-screen (although of course it does refer directly to them). Rather, I would like to argue, this form is a form-in-emergence; it is a forming of the image of a body in the viewer's perception. Perception, in turn, is something that bridges the realms of the individual, cultural, natural, and technological: each of these realms relies on perception, and perception is also what makes them interconnected. For this piece, the embodiment aspect of perception works differently than in Sonia Cillari's installation.[32] In a way, the experience of the body perceiving another body is less important: instead of *what* is being connected, the deeper focus of this piece is on *how* it is connected (how the image of the body was removed and how it reappears in the mind of the perceiver). The project challenges the viewer to "figure" the body-figure "out," to see beyond the given image in a search for both the possible borders of that particular "screen-captured" body, and the borders of defining a body more generally (as a cultural construct). Figuratively and also quite literally, it *mobilizes* perception,[33] moving it toward critical reflection on the issues of censorship and ownership, or rights over one's body and its representations.

The almost complete annihilation of representation of the body in Mahdavian's video may well be on the radical edge of the spectrum of the body's virtualization through digital simulation. Yet another form of the abstraction of the body—namely, a realistic representation, but without appropriate tribute to the living "original"—also raises questions. How should a person whose movements and whole bodily identity were transferred into the digital realm relate to this "copy"? Whose property is it? What are the ethical and affective implications of losing control over the environment in which the representation is placed? Such concerns are key to a video installation by African American artist Sondra Perry, *IT'S IN THE GAME '17 or Mirror Gag for Vitrine and Projection* (2017), that highlights the problem of capitalization of a black body using the example of her twin brother, Sandy Perry.[34] His identity as a successful basketball player (his physical characteristics and performance statistics) was sold by the National Collegiate Athletic Association to a commercial video game developer without his knowledge or consent. The video tells the story through the images of the game characters, accompanied by Sandy's comments about the real people behind them, all left uncompensated (with the NCAA's only explanation being that they had received a free education). Intercutting the overview of the avatars is footage of the twins walking through the African art collections of the Metropolitan Museum of Art and the British Museum, overlaid with spinning 3D-rendered and skewed images of the museums' artifacts—a critical reminder of the larger context and the scale of cultural appropriation. The digital realm here has a double function. On the one hand, it enables the uncompensated appropriation of identity, but on the other hand, the skewed floating figures may serve as the artist's own creative and critical commentary: she uses her freedom to manipulate and distort in her own way the museums' narrative fixed in these displaced artifacts. Although Perry's project overall lies more on the side of critique, the twist with the 3D-rendered figures is reminiscent of a slightly more optimistic vision of the potential of new technologies for the representation of race, as proposed in Fox Harrell's writing about avatars and virtual worlds design. In particular, he discusses ways to interrogate the existing cultural phantasms leaking into the digital world from the real one, by encouraging the authors of computing systems to introduce more "agency play" into their designs. He explains: "Agency play focuses on leveraging the relationship between the user and system in order to create experiences that are meaningful and engaging for users to participate in."[35] In other words, the embedded mechanisms of system design should be mobilized to empower users who may occupy a more marginalized position in real life.

Similar to the use of motion capture in Mahdavian's project to stress Iranian culture's stigmatization of representation of the female body, the digital abstraction of the body allowed by motion capture can, paradoxically, be a means to deal with the problem of obfuscating the "other," those who are racially or ethnically different from the dominant culture. There is an ongoing debate about whether the phenomenological call for resolving the difference between the outer representation and inner image of oneself (the body or corporeal schema) can be equally applicable to the "lived experience" of *every*body, regardless of any differences. Frantz Fanon in his seminal book *Black Skin, White Masks* has famously argued against this, saying that a black man living in a white world is "locked into his body" and has to have more than one corporeal schema (one being a "racial-epidermal schema" imposed from the outside).[36] Yet, remembering that Merleau-Ponty's initial understanding of the corporeal schema assumes its embeddedness in the world, revealed through agency and temporality, it still seems relevant to apply his ideas to projects reflecting on the affective experiences of difference, such as those by black artists working with virtual reality. One example of activating the potential of the black body's agency by transferring its presence to the digital realm is *The Black Movement Project* (2019–) by artist and creative technologist LaJuné McMillian (figures 2.4, 2.5). The project is conceptualized as an online library of motion capture data from black performers that can be used in virtual- and mixed-reality creative initiatives, but also as material to study the cultural specificity and significance

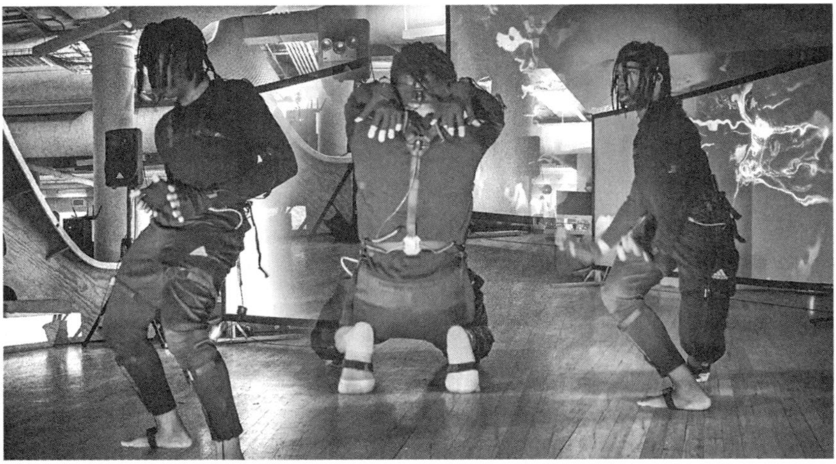

Figure 2.4
LaJuné McMillian, *The Black Movement Project* (2019–). Motion Capture Performance. Nala Duma at first Black Movement Project Performance held at Barbarian Group, New York. Photo by Kevin Chan, edited by LaJuné McMillian.

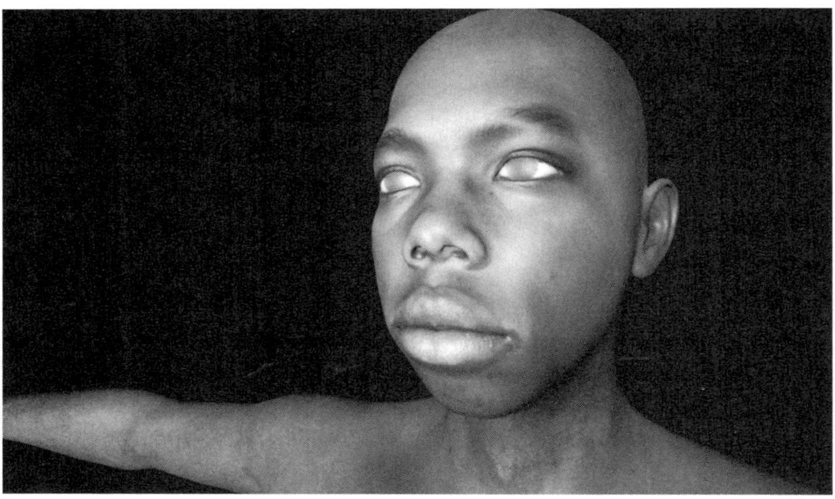

Figure 2.5
LaJuné McMillian, *The Black Movement Project* (2019–). Avatar of Nala Duma. Visuals created by LaJuné McMillian.

of movement in black communities and for individual self-expression. McMillian's idea as a designer of virtual worlds (and as a trained dancer) grew out of dissatisfaction with the lack of knowledge about racial identity in the data and the underrepresentation of black performance in available databases: "In a way, I am trying to humanize the data by reconnecting it to the source."[37] In this case the "agency play" discussed by Fox Harrell is used to introduce characteristics of different black body types—their skin and hair textures, postures and gestures—to the toolbox of "extended" reality (XR).[38]

But the project is intended to be more than a tool for generating diversity among 3D-based characters. Together with the live performances and documentation of the performers' stories woven into the accompanying soundscape, it becomes an "archive of black existence."[39] Movement is featured here as a form of resilience, an expressive way to keep alive one's cultural and personal integrity, an approach that has been helping the black diaspora to cope with oppression throughout the centuries. Ironically, documentation through motion capture serves as a means of escaping being "captured" by stereotypes, divisional politics, and marginalization.[40] The recognizable visual characteristics may still be there, but what is important is staying true to the "source"—that is, the lived bodily experience expressed through movement, which then continues its life in the form of data.

How exactly the elements of the individual, cultural, natural, and technological can interact with each other and "mobilize" each other is demonstrated most effectively in interactive artworks. The relation between the body and the image can work in two directions: on the one hand, the body can code the image, and on the other hand, the body can respond to the image and become "coded" by it. I will explore this entangled relation as a *mutual coding* using the examples of two dance performances—*Inside the Chain* (2016) by Russian dancer and choreographer Ekaterina Zharinova, and *Bodytext* (2010–) by Simon Biggs, Sue Hawksley, and Garth Paine (the subject of the next section). Neither involves direct audience participation, yet the interactive component is embedded inside the performances themselves.

Inside the Chain features a dialogue between a live dancer and her doppelgängers on the screen.[41] The performance consists of several parts. In the first one, the dancer performs a series of individual movements to the

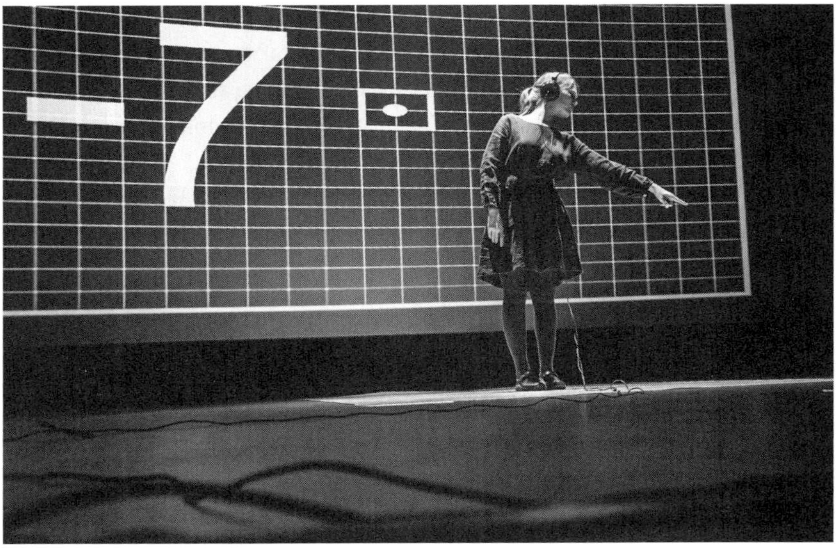

Figure 2.6
Ekaterina Zharinova, *Inside the Chain* (2016). Yeltsin Center (Yekaterinburg, Russia), 2017. Courtesy of Ekaterina Zharinova. Photo by Mariia Kalugina.

numbers that appear one by one on the screen behind her, each number corresponding to a predetermined movement.[42] A catalog of sequences of movements is thus created, and each sequence is not anticipated (figure 2.6). Corresponding to nothing but stochastic mathematical logic according to which the numbers appear, the sequences become kinetic expressions of that logic.[43] The limitations of space (the whole performance takes place on a small square in front of the screen) keep the movements concentrated and centered within the performer's body. As the projected numbers accelerate, it becomes a test for the dancer to speed up the movements while keeping them still legible. The body thus becomes an avatar of machinic logic: the algorithm is performing through a human. As the dancer describes this experience in her MFA thesis devoted to the performance, "I move as a machine in the first part and as an animal in a cage in the second part."[44]

Zharinova emphasizes that the work is about "instant brain and muscle reactions," an observation revealing that, indeed, it is not a reaction to stimuli (as was the case in Stelarc's famous performance *Ping Body* where

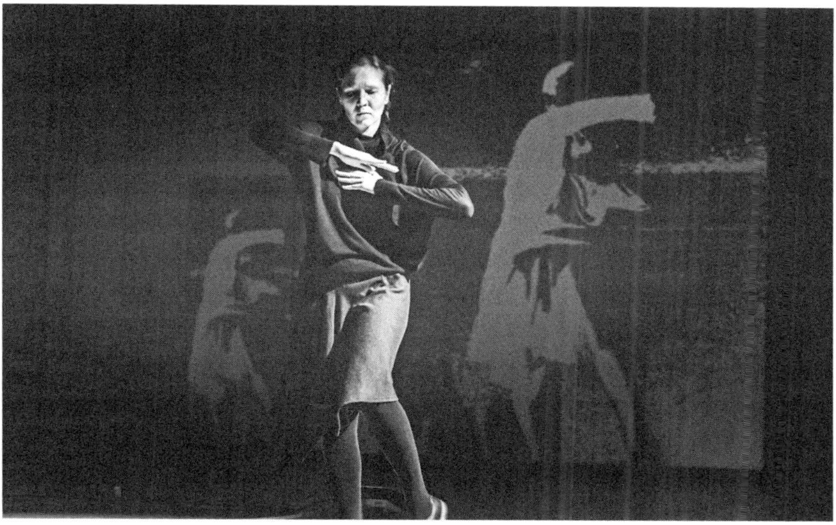

Figure 2.7
Ekaterina Zharinova, *Inside the Chain* (2016). National Center for Contemporary Art, Ural Branch (Yekaterinburg, Russia), 2016. Courtesy of Ekaterina Zharinova. Photo by Alexey Patentnyy.

the artist's body receives impulses from remote "operators" who move his limbs; or to some extent in Cillari's *Sensitive to Pleasure*). The work trains the dancer's mind to act *like* a robot, to illustrate the robotic without really turning into one, to deliberately relinquish control of consciousness by allowing the external impulses not to govern, but to organize the body's expressivity.

The second part of *Inside the Chain* features a joint "round dance" of a live dancer and her screen doubles (figure 2.7). Initially the dancing silhouettes move along with their human "original," but after a while their movements start to be delayed, with different and increasingly long intervals. Since the speed of the virtual figures' playback varies at random, the dancer has to adjust accordingly and improvise in response to them. The response here is of a different instantaneity than in the first part; it is also a more creative one because the relation is established not only with the image but with the movement of the screen figure.[45] The dancer tracks the figures and responds to their speeds and positions. Finally, in the third part, the dancer is motionless and simply observes the still-dancing figures on the screen that move, in her words, with an "unreal speed: too fast to keep tempo or too slow to gain a momentum."[46] As the artist explains, "In this final section, I am observing the traces of the first two parts together with the audience watching me watch the projected image. I like this double inclusion, which I call *enchaining*: when one action is occurring inside another action simultaneously."[47]

The key metaphoric image of this performance is the chain. The Russian title of the work, *В вереницe / V verenitze* refers to the idea of being in a row: *verenitza*, a row, a string, a line, or a succession of elements, can be either open or closed—like a round dance or a circuit. What matters is the interconnectedness of the virtual and the live figures and the continuity between them established through *movement*. As in an electric circuit, the elements of the "chain" are there to process a current; they organize, or structure, its flow in a certain way. Here, the movement impulse, triggered by the stochastically appearing numbers, is transferred to the dancer's body and travels from it to the figures on the screen, which unleash its potential by going beyond humanly possible speeds. Unlike other performances that engage screen figures as participants of action ("actors"),[48] these figures do not have a particular identity and are rather like shadows, since they inevitably (earlier or later) follow and repeat the dancer's movements. They are just "links" holding together the "chain" of movement events, exploring the possibilities of mutual control and temporal relations, rather than issues of identity transference. Again, the story of this work is not of a particular "self" but of the flow of movement, of how it can be organized, remembered, and reproduced. It is an example of how the capabilities of a particular technology of motion capture, Kinect, can be used—specifically, its ability to distinguish the moving figure from the visual field and to present it back as a *whole* image, a full silhouette, not broken down into dots. These reproductions require a computational model for rendering the speed, that is, for transforming the figure defined not spatially (as was the case in Mahdavian's *Intangible Body* with its emphasis on "deskinning" and "defleshing" the body), but temporally, through movement. The very conditions of digital "capture" are then continuously renegotiated by both participating actors, the human body and the technology.

As shown throughout these examples, interfacing with a representation of the self as a double can produce various effects. The doubles appeared here in the form of a body image on a screen (as in video and installation works by Peter Campus); embodied by another human body (as in Sonia Cillari's *Sensitive to Pleasure*); and as an image that mobilizes perception culturally and politically (as in Emelie Mahdavian's *Intangible Body*). They also instigated reflection on the fluid boundaries between the "real" and the "rendered" in the exploitation of racial identity (in Sondra Perry's *IT'S IN THE GAME '17*), and on the role of movement in exercising the creative agency of this identity (LaJuné McMillian's *The Black Movement Project*). Finally, in Ekaterina Zharinova's *Inside the Chain*, the ghostly doppelgängers formed a "chain" with the human performer, challenging her to reflect anew on her bodily capabilities. Merleau-Ponty's distinction between the body image and the body schema helped us to identify the split between the visible and the invisible—including the visible and the invisible layers of the body, of the image, and of the self. This split is paralleled by the split between that which can be shared through the visible, and that which leads beyond the observable image and yet is still shared at different levels of perception (whether of an individual human, cultural, or technological) as a "schema," or a capacity for *forming*, for rendering part of oneself perceivable. The temporal dimension helps to reveal this capacity, and the technologies of motion capture and different forms of visualization enable access to it, however indirect and unstable. The flickering contours of the Iranian dancer or the skewed artifacts in Sondra Perry's installation point to what cannot be captured and reproduced, and what thus indicates a gap between the reproducible and the intrinsic and irreproducible. The avatars—embodied, as in Cillari's project, or screen-based—serve to also sharpen or complicate the difference between a living body and its representation.

Whereas *Inside the Chain* indicates mutual coding between the dancer and the image, another dance performance, *Bodytext* by Simon Biggs, Sue Hawksley, and Garth Paine, involves the idea of mutual coding by connecting the dancer not just to an image but to a whole tracking system. We will see, then, how the questions raised in this performance expand the territory of the body image and embodied avatars as components of a *media system*. While the examples in this section explored the body as part of an interface through the possibilities of doubling the body image, the example in the next section will deal with the ideas of *encoding and transcoding*. It will open the discussion of interfaciality as a combination of corporeal and material factors on the one hand and signification on the other. In *Bodytext*, the dancer's body movement works as a code to control the text and sound, but this movement gets abstracted from the body and has to work in the opposite direction, by following instructions from the computer system. From that, we can draw clearer connections between the analysis above on the role of the body image and the ideas from the first chapter on the image's potential for codification. So, while in this section the body was discussed as screen-based and embodied-by-another-body "doubles," as "intangible" and "enchained," in what follows the body will figure as having a new quality—as a *medium*, capable of also being a *code* and a "*text*."

Corporeal Encodings and Transcodings in *Bodytext*

Electronic systems operate via code. What if the body plays the role of a code? How can the body produce a text that turns into a code, and what would that code do differently than other types of code? The example of the media/dance performance *Bodytext* by Simon Biggs, Sue Hawksley, and Garth Paine can help us to analyze further the relationship between the body, its expressive effects, and its ways of sense-making, turning the focus not only to the forms of its representation, but to the issue of the body itself as a means or medium of communication (figure 2.8). Communication happening through the body would then rely on the body's expressive capabilities that, in turn, would reflect the experiences of the body. Hence, to consider the body as a communicational medium means to think about the problem of the *transference of bodily experiences* (including their affective dimensions). This consideration will lead us to more general concerns about translation between organic and nonorganic systems and the possibilities of coding being applied to this entanglement. If an interface reflects the workings of the computer code behind it, how can the human body help to make them more visible? In other words, how does the corporeal translate or transfer into the digital machine? Whereas doubling was the type of interfacial relation central to the discussion above, at stake here will be a special form of corporeal transference that, as we will see, bridges the human and the machinic forms of being—namely, transduction.

Bodytext is a performance featuring interaction between a dancer and a computer system, engaging speech, real-time motion tracking, voice recognition, projection, and granular audio synthesis. As the artists explain:

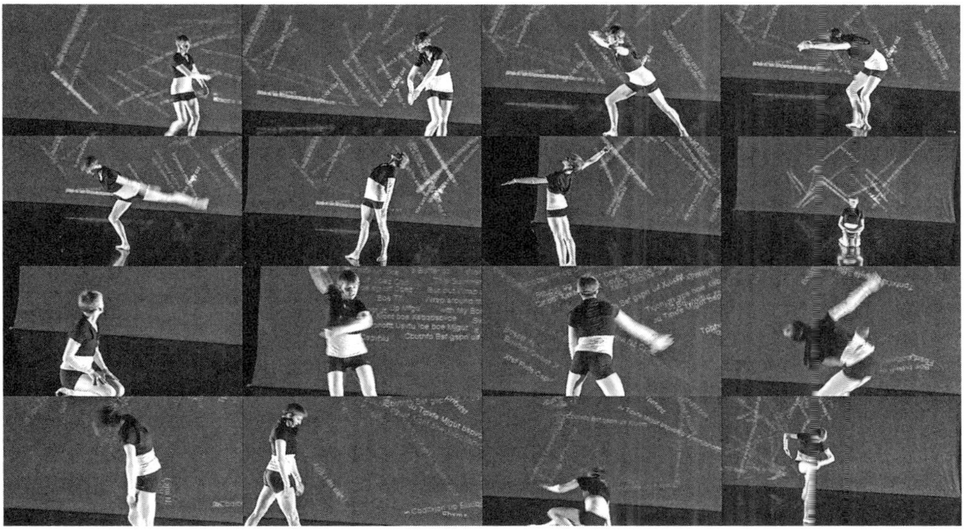

Figure 2.8
Simon Biggs, Sue Hawksley, and Garth Paine, *Bodytext* (2010). Courtesy of Simon Biggs, Sue Hawksley, and Garth Paine.

> The acquired speech, a description of an imagined dance, is re-written through projected digital display and sound synthesis, the performer causing texts to interact and recombine with one another through their subsequent compositional arrangement. What is written is affected by the dance whilst the emerging recombinant descriptions determine what is danced. The work questions and seeks insight into the relations between kinaesthetic experience, memory, agency and language.[49]

In the artists' words, the piece "investigates how memories are embodied and signified, patterning movement behavior and addressing how media can be used to record complex data sets and thus function to preserve or disturb a sense of self."[50] Here, signification and formalization happen at different levels and in different modalities: movement is captured both in words and as electronic data. Both can be seen here as texts; and both play the role of inscription. Yet, in one case, the relation is between the moving body and natural language, and in the other, it is between body movement and the machinic code (which also operates sound and image). The result of these operations of translation is a dynamic, "self-generating," and "self-modifying" apparatus, defined by neither a human element nor an electronic one, but by both. It is an interplay between the body, linguistic structures, and a digitally augmented environment, continuously "rewriting" each other's "texts" and, through that, constantly recombining (to the extent of complete dispersal).

Let's take a closer look at the very texture of the *Bodytext* performance, particularly at the kinds of texts that it offers and the relationships between them. How do their differences unfold and new qualities get generated? Among the possible lines of distinction are those by "author" (human/machine), by modality (expressions/instructions), and by "medium" of communication (speech/writing/movement). All of them can then be placed within the larger context of the material conditions of the operation of encoding and translation.

Textual Operations between the Human and the Machine

1. The expressive and instructional aspects of command authoring The initial text is spoken by the *human performer* as she moves; it can be described as a collection of "body stories." These are discrete phrases, unrelated to one another, consisting of personal associations that the dancer has with particular movements. They may refer to the dancer herself and to other people ("falling . . . he does not catch me," "sailing away without me"), objects ("the phone is ringing"), or situations ("she is walking in a town she does not know"); or be literal descriptions of a movement ("leaning forward," "elbow locks suspending the breath") or more ambiguous and poetic expressions of bodily feelings ("difficult behavior with tilted attitude," "sweeping the horizon") and combinations of the two ("running into black," "reaching to attack"). It is hard to locate the origin of this text, or to define the subject of the speech. The mind seems to be "scanning" the body, documenting both the physical movements and the accompanying associational layers, which are also subtly affectively charged. Each phrase represents a "report" taken during a particular time frame, and thus is self-sufficient and potent. The methodology of literary analysis could help to reveal other nuances, yet what are more interesting in our context are the dynamic connections between body and mind, the physical and the abstract reflected in these descriptions; the very mechanics of the procedure of *"inner" recording*, or *recording through the bodily*. Though the imaginative and fictional dimension is an important element of the text, it is constantly balanced by the physicality of the references to the body, to the *experiences* of the body. Choreographer and dancer Sue Hawksley was interested in exploring specifically how language "codes" the gestures and "informs" the whole "bodyscape."[51] The phrases are pronounced in a neutral voice, without any intonation—nothing that would reveal the emotional state.

In addition to this type of phrase, there is also a category of words that can be interpreted both as physical movements and as computer instructions, such as the commands "shift," "enter," "cut," "go to sleep" (the latter makes the system crash, and hence is not used in public performances). Dance itself live-codes the computer. These words indicate the imperceptible border between the human and the machinic realms. Identified via voice recognition technology, they serve as switches of modes within the machine, instigating the work of translation and reassembling the rest of the text according to the rules of the program. The text within the *machine* itself is a code. Both types of texts in this performance—the one pronounced by the performer, and the code that runs on the computer—are performative. It became a commonplace of media and software studies to state that the meaning of code lies in its function; as software engineer Ellen Ullman put it, "a computer program has only one meaning: what it does."[52] This performative and operational modality, the speech *act*, is the key point that makes code distinct from natural languages. The same idea is emphasized by Alexander Galloway in *Protocol*: "Code is the only language that is executable."[53] Another aspect of instructional modality is that it relates to the algorithmic logic that underlies the foundation of code operability. Algorithms are mathematical and thus abstract structures. Yet the key specificity of assignments and instructions operated by code (which distinguishes them, for instance, from algebraic formulas) is that they are nonreversible. They are driven by a certain vector, which indicates the functional asymmetry of these processes.[54]

The machine is a different kind of reader than a human: pure "expressions" do not make sense. All phrases, aside from the commands, stay meaningless within this system (or at least, their meaning is different). As units of indistinguishable data, these noncommand expressions are attributed only certain formal qualities and values that, in their turn, can *potentially* stand for other types of data elements (just as in Zharinova's performance: her movements for following the numbers on the screen were composed arbitrarily). What matters within this system is its *operability*, its ability to execute different procedures with the available data. The specificity of the "codic" text is thus defined by the command elements, the ones that can trigger *actual* changes in the system. As a result, we might feel as if we get the same kind of "vicious circle" as with facial expression coding. In that case, in order to be understood by a computer system, it seemed that the body had to speak its language, which then led to the experience of the body itself being coded by the language of that system. In the end, movements and their verbal descriptions should indeed remind us here of specifically *human* experiences and the border of translatability into other languages or media of the very phenomenon of being human. It is the human affective and bodily experience itself that constitutes its own special kinds of meaning, which are being challenged here by the machinic system.

2. The mediatic specificity of speech-writing-movement *Bodytext* positions movement alongside *speech* and *writing*. All three serve as the means of producing meaning, but in different ways. Without going into too much detail, it is important to think about the two most historically notable accounts of the role of speech and writing by Ferdinand de Saussure and Jacques Derrida, the former claiming that writing must be subordinate to speech, and the latter insisting on exactly the opposite. According to Saussure, a language system (*la langue*) can be revealed primarily through the study of its usage in the form of the spoken word (*parole*).[55] In contrast, one of Derrida's main critical points is that writing *exceeds* speech and cannot simply be conceptualized as speech's written form. "Writing" implies the *material fixedness* of a symbolic system, no matter in which particular medium—whether in the form of notches on a rope or a sophisticated alphabet. That system is fixed already as an abstract logical construction. Derrida's term "arche-writing" (*archi-écriture*) may help to remove the contradiction between writing and speech, since it represents an attempt to go beyond the simple division of the two, "arche" being the "origin" for both. Arche-writing is, in a sense, a language that precedes the instances of its usage, either oral or written, as a "sign-function," a flexible structure that makes expression of any content possible. What matters here is not necessarily what exactly is expressed, but the *manifestation of an expression*. Derrida writes: "This arche-writing would be at work not only in the form and substance of graphic expression but also in those of nongraphic expression. It would constitute not only the pattern uniting form to all substance, graphic or otherwise, but also the movement of the *sign-function* linking a content to an expression, whether it be graphic or not."[56] (A loose parallel can be traced between "arche-writing" and Merleau-Ponty's "body schema," since both refer to the qualities of preceding, constituting, and—particularly important for us—"forming." This connection will be relevant again in the last section of this chapter in relation to the body as informational and technologies of bodily data recording.)[57]

How does movement fit into this distinction and its potential overcoming in the concept of arche-writing? Hawksley locates *Bodytext* within the context of her general research, which is "the exploration of a range of somatic techniques focusing on structure and emphasizing awareness of movement patterns to inform and facilitate effective articulation."[58] Dance has long been recognized as a type of language, with its syntax consisting of movements and gestures that can potentially be captured as traces in the form of notation, that is, symbolically (or, as we have seen, in a different way through motion capture). Yet the linguistic potential of gesture itself is still not fully explored and acknowledged. There are at least two definitions of "articulation": the act of giving utterance or expression; and the action or manner of jointing or interrelating.[59] Thus, articulation relates both to the act of speech and pronunciation and to the more abstract function of *establishing junctions and linkages*—which, I'd like to argue, could potentially carry their own meanings.[60] It is this technical meaning that allows us to assume that using nonsymbolic systems of expression can be comparable to the symbolic ones.

The problem in our case is not so much how embodied expressivity constitutes meaning, but how this meaning can be transferred and reconfigured through other media. Trying to find an answer to this question, Brian Rotman, a philosopher of the "alphabetic body," develops further the idea that was first stated by Leroi-Gourhan in *Gesture and Speech*, namely, that "the alphabet writes the movement of the speaking body."[61] Rotman argues for an alternative to symbolic notation in the form of capturing body movement: "contrary to withering the body or leaving it behind, it will be by uniting with it—merging, augmenting, capturing, and reengineering it—that technology may render our present alphabetic dispensation archaic."[62] Rotman suggests that a move is needed both "beyond the written mark" ("why not an asymbolic mediation—a sampling or capture, rather than a symbolic representation?") and "beyond the oral-vocal apparatus." Whereas notation relies on the relational structure of (prior) differences between discrete elements, capture presents "a continuous topological model of posteriorly given internal differences."[63] Differences are captured during one continuous movement of enactment or reproduction. It is the basis of all technologies of capturing, including the phonograph and tape recorder—to mediate that which operates under the "regime of the enacted or reproduced rather than the symbolized."[64] The enactment should be read here not in terms of speech, but closer to the "sign-function" of Derrida's arche-writing, where "sign" refers to *any* material form of expression. Rotman continues this discussion with an emphasis on the *continuity and physicality of the experiential modality*. (And, again, experiences of the body constitute their own kinds of meanings.) For him, the gesturo-haptic is still a form of writing, but its specificity is that it "*exceeds the textual*" (emphasis added), exceeds everything that has to do with "the interpretative activities of deciphering, of reading, of engaging with a site or artifact whose primary function is to signify" everything that is determined "in advance of its action."[65] He defines the "exo-textuality" of the gesturo-haptic through a mediating technology that "traffics" corporeal events, "processes and proceedings that have to happen, and in happening—better: in the manner of their happening—engender meaning."[66] The technology thus *structures* the continuous flow ("traffic") of experiences. It works not only to create an image of the body, but to refer to the body's ability *to form its own meanings* through its inherent preintentional body schema. Despite its important "structuring" function, technology can only "play back" what is already within the body. Body movement exceeds the textual and challenges the technology to exceed its own "textual," the code—that is, inviting not only "encoding" (reminiscent of the dehumanizing "vicious circle") but also "*trans*coding," in the sense of transgressing the borders of the strictly, unambiguously coded command-action correlations. By "transcoding," I mean here the activation of the potential of a system (both human and machinic) through opening up their relational and contingent qualities, enabled—in turn—by feedback.[67]

In *Bodytext*, characteristics of movement (speed, vectors, intensity, angles) are algorithmically transformed into sound. Granular audio synthesis, developed by Garth Paine, present a mixture of natural sounds (reminiscent of bird chirping), husky voices, grinding metallic sound, electronic beeps, and remixes of the dancer's voice. Thus, there is no "representation," or even "reproduction," of movement but instantaneously a remediation, or translation into other media.

The authors of the performance define it as a "text" ("Body*text*") but its overall meaning emerges only in the process of transmediation and transliteration, in the process of crossing the borders between the media of speech, writing, and movement. Articulation in the form of an utterance is not complete without being transcribed by the computer and "written" on the screen, or without movement being captured and transformed into sound. The meaning here is in these operations of transcoding and translation.

The Corporeal of the Machine: The Materiality of Transduction

The linkages between the textual and the material are the object of transmediality and transliteracy studies. But the underlying core for both is the process of transduction. The *Oxford English Dictionary* defines "transduce" as "to alter the physical nature or medium of (a signal); to convert variations in (a medium) into corresponding variations in another medium." Transduction thus deals with differences in signification systems as they are manifested at the material level of these systems' operations. In particular, it denotes the *textures of disjuncture* that define the individual media of expression.[68] *Bodytext*'s meaning resides in the interplay between the types of media and their modalities, but it also arises in the processes of translation between the somatic and corporeal level and the level of an immaterial language system (be it a linguistic system or technology of capture that relies on strict algorithmic codes). Initially, one of the aesthetic aims of *Bodytext* was to find a way to reinflect memories. In order to be "fracturing, multiplying, colliding, dividing and mutating," they first have to be objectified—both symbolically and materially, as physically inscribed pieces of information (datasets).[69]

Friedrich Kittler famously called for considering more seriously what constitutes the material basis of coded processes: "all code operations, despite such metaphoric faculties as call or return, come down to absolutely local string manipulations, that is, I am afraid, *to signifiers of voltage difference*."[70] Indeed, there are many layers of abstraction that programming languages operate at, but ultimately, the role of signifiers in digital systems is performed by the exact physical expressions that determine changes in a material sense, i.e., voltages. As N. Katherine Hayles puts it, "the signifieds are then the interpretations that other layers of code give these voltages. Programming languages operating at higher levels translate this basic mechanic level of signification into commands that more closely resemble natural language."[71] If we define signification, following Saussure, as based on differences between signs, then the scheme of computer architecture fits this logic: all the operations of translation between multiple levels of code depend exactly on the ability of the machine to recognize the difference between one and zero.[72]

This difference between the two states is the basis for computation, which means exactly "compare the pieces" (Latin *con* and *putare* mean "together" and "to reckon, to think, or to section"). Yet, besides the discrete states that digital information processing is usually associated with, the principle of computability today also includes nondiscrete continuous data, by which computer science understands analog computing platforms. "It is the translation of the continuum into the discrete that marks the condition of possibility for computationality."[73]

In *Bodytext*, it is striking how the initial phrases of the body-text are pronounced and then played back as discrete unrelated entities (as lines of text on the screen), and yet they are connected through the medium of their recording and reproduction. Interestingly, what serves here as a "medium" is both the computer (and the acoustic and screen image effects that it produces) *and* the body. And it is this "integrated" medium that—through storing the differences between the individual inputs—is responsible for the possibility of *resignification*. Operating with the "inarticulable" and "unimaginable real,"[74] technologies of capture open up a different logic for perceiving not only external reality but one's (bodily) self. The ability to manipulate a captured image allowed by these technologies has been explored well throughout the history of cinema. Yet, whereas analog recording technologies allowed the representation and manipulation of the dynamics of reality (including its "inarticulable" aspects), the digital age shifted attention toward the *generative* potential of the medium. As M. Beatrice Fazi shows in her work on aesthetics of computation, contingency and indeterminacy intrinsically belong to the very logic of digital computing (opposite to a common belief in its strictly determinate nature).[75] The new emerges from within one continuous whole: the same set of data gets reorganized to form new patterns. And as the performance of *Bodytext* shows, this happens through feedback mechanisms of mutual encodings that also become transcodings.

The question still remains: How does this emergence through reorganization translate into reconstituting the sense of self? To answer this, we should think back to the role of the body image and the role of a recorded image (related to the self) more generally.

The productive challenge for a human in tuning in to the "inarticulable" within the (bodily) self—in being able to hear all the random acoustic events (such as in the gramophone recordings that Kittler talked about) or to observe the visual trace of indistinguishable movements (as in Rotman's discussion of capturing movement)—lies in *making sense* of it. The new meanings arise in the moments of a shift: in *Bodytext*, for example, lines of text on the screen collapse into a jumbled mixture after the oral command "Shift." Such emergent change may be more evident in the case of the digital reassembling of the content, but the core of it lies in the analog reproduction technologies, such as photography, cinema, or gramophone. In a way, analog doubling is already both representational and performative: it represents the occurrence of an event exactly as it happened, yet it becomes an event of its own at the moment when it is *perceived*. Just as in Mahdavian's *Intangible Body*, it is the perceiver whose perception interrupts the feedback loop and completes the image of the body in the way it was conceived by the artist—as an image arising for every viewer *individually*, and, through that, challenging cultural and political preconceptions. The twist is how the generativity of code always goes hand in hand with human interpretive abilities, one triggering the other (this will be explored in more detail in the next chapter). In the end, it is the *human* who is *capable of recognizing the differences* (for example, between the representation and the "original"), yet the discoveries made in the process of these interpretations would not be possible without the work of a recording technology.

Bodytext reflects this as well, but in its own way. When the human performer receives the instructions for movement projected by the machine on the screen, her performance is still not completely automatic: while her body is accurately executing the commands of the machine, her mind is active with interpretive processing. (It may remind us of how Zharinova in *Inside the Chain* described her own processing while she was dancing as not fully automated.) The physical and corporeal aspect enriches what is usually observed about the digital textual practices as abstract hypertextual recombinations. There is still something to be read here: a text written in the language of bodily sensations—something inarticulate in the sense of verbal language, but articulate if taken in the sense of "jointing" and "fixing differences." As a special kind of text, *Bodytext* can be called a variety of things: exo-text (Rotman) or hypertext / cybertext / technotext (Hayles). These kinds of "texts" are made especially meaningful due to not only the logical operations that allow new significations (generation through fragmentation and recombination, mutability and transformability, combinations of natural languages and code), but also their ties to the materiality of their environment. In her work on electronic hypertexts (which I think should be seen as broader than just digital literature), Hayles prominently argues for the reconceptualization of materiality as

> *the interplay between a text's physical characteristics and its signifying strategies.* This definition opens the possibility of considering texts as embodied entities while still maintaining a central focus on interpretation. In this view of materiality, it is not merely an inert collection of physical properties but a dynamic quality that *emerges* from the interplay between the text as a physical artifact, its conceptual content, and the interpretive activities of readers and writers. Materiality thus cannot be specified in advance; rather, it occupies a borderland—or better, performs as connective tissue—joining the physical and mental, the artifact and the user.[76]

An example of this feedback loop between the material and human interpretation in the *Bodytext* performance is a reference to the rhetorical technique of locative memory, or "memory theater": remembering detailed and exhaustive arguments by associating them with particular places or objects on the stage. The computer storage space may be considered as an analogue of such a space or place, where the nonlinearity of data structures reflects the unpredictability and fluidity in the functioning of the living being.

The combination of the material and textual qualities in these mechanics of associations brings us back to the concept of transduction. The corporeal character of transferring signals, implied by transduction, enriches the terminological apparatus of translation theories with metaphors, such as resistance and distortion, flow and turbulence. At the same time, applied to cultural practices, these effects cannot be fully formalized.[77] What Kittler's "voltage difference" cannot signify is the *experiential* side, unique to each case of interpretation. The mediatory units for processing material information in human perception are the senses, which are structured not symbolically but only through modalities. Thus, there cannot be a direct translation, but instead there is always the generation of a new meaning that appears exactly on the border, in the "in-between" of two media.

The role of technology here is not as an instrument, but as a coproducer; it is internal rather than external to the production of the resulting meaning. The coauthor of *Bodytext*, Simon Biggs, proposes considering its dynamics and specificity in terms of an apparatus.[78] As German media philosopher Sybille Kraemer notes, "media produce [*erzeugen*] what they mediate"; that is, they change the material provided, creating a new pattern for the same dataset.[79] The uniqueness of the emerging product lies in the indivisibility of its elements. A synonym for "apparatus" in this context could also be "assemblage," or even "cyborg": the interpenetration of organic and nonorganic, regimes of computation and modalities of the senses, the body and characteristics of the environment (including sounds). Biggs prefers to describe the work not in terms of remediation but of transliteracy. The former uses the logic of compilation and nesting (part-whole relations), whereas the latter is "less teleological and more prosodic," allowing variability on a more fundamental level, not formal and symbolic but on a level of material structure.[80]

To conclude, the problem of the borders between media motivates us to distinguish more accurately their material specificities and the characteristics of their functioning. But it also inevitably entails the discrepancy between them, pointing at what turns out to be untranslatable. Embodied affective textualities collated with the logic of digital code acquire new meanings that exceed and challenge traditional notions of textuality, since the very locus of signification is shifted and reshaped. As the artists themselves summarize their reflections about this piece: "The restructuring and rewriting of the [performer's somatic] texts into forms that are ever harder to interpret, for both dancer and audience, challenges the expectations of semiosis and disrupts the relationship between reader and text, presenting their respective ontologies as contingent and tenuous. This inevitable failure in the processes of interpretation leads to eventual system stasis."[81] Indeed, the generative recombination leads to the fact that the origin of the data is no longer distinguishable and the meanings of mediatic interpretations become ever more undifferentiated and obscure. These entropic tensions have to be broken by the dancer's command "Escape" that, according to the artists, "flags the dancer's resignation to the fact that the data is no longer the performer's but something other, a social artefact, an environment and a visual spectacle, inaccurate as a reflection of the original intention."[82] Kriss Ravetto-Biagioli, in her analysis of collaborations by Biggs, Hawksley, and Paine, interprets this finale in terms of exhaustion—of "bodies, exchanges, algorithms, perceptions, and patterns of recognition"—that indicate the triumph of what she calls the digital uncanny: in this case, a broken, "disfigured" feedback, a glitching effect, "a radical form of repetition that undermines the faith we put in semi-automated systems of control."[83] Most importantly, feedback here "exhaust[s] the potential to transform," "dissolving the idea of a unified subject, whether it is human, nonhuman, or posthuman."[84]

Both accounts above present the translatability "failure" as a negative, deadlocked effect. Not to compromise these conclusions, it is important, however, to stress the value of the challenge itself. Emergent in the interplay between the human and the machinic are, perhaps, new subject positions, by definition unstable and indeterminate, constantly testing alternative material bases for their existence, and establishing connections across media, however evanescent.[85] Exotextualities and operations of transcoding, enabled in this performance, demonstrate the capabilities of interfacing and feedback in a new way: here, bodily experiences are not just doubled in the form of a recognizable body image, but are regrouped, transmediated, and transmuted. The challenge thus becomes the very ability of recognition and identification. This piece highlights the tactic of activating the potential of the body as at once a textual and an exotextual medium. Through its stories, the body can code, but by being coded back, it experiences the inadequacy of translations. In its expressive existence, it refers to the territories beyond the signified and articulable.

In the next and final section of this chapter, I return to the concept of the body image (as well as the self-body-image), considering it from the perspective of a physical and biological entity. While *Bodytext* showed the body's coding abilities at a "macro" level (full body movement), the technologies discussed next will treat the body more as a constellation of biochemical and electrical processes that still can also have expressive and coding potential. We will thus continue analyzing the mediatic—expressive and communicative—potential of the body, but with a focus on the transference and translation of bodily experiences more in terms of how they can be sensed and how they themselves (and their sense- or meaning-making abilities) are constituted through sensing.

Biofeedback Sensing and Body-Data-Image

Biosensing and the Quantified "Organic"

While motion capture works with data created by the moving body as a whole (despite its atomization of the body image), other sensing technologies, for instance biofeedback-based, have as their input much more minute, granular, and isolated expressions of the bodily (heartbeat, muscle tension, neurowaves, etc.). Whereas motion capture simulations may look like a mirror image of the body, with the body in its recognizable shape (even if completed only in the mind of the viewer), physiological data-tracking devices create "*data* portraits," a principally different and new type of representation of both the body and the self. The increasing market availability of such devices within the last decade has attracted attention to their potential from a wide variety of social spheres: from the medical and the military to the spheres of work management, the commercial culture of self-improvement, and the arts. Most prominently, biosensor-based technologies that allow the tracking and managing of the dynamics of the body became the center of their own cultural phenomenon—the Quantified Self movement. In all the spheres listed above, they are used for the sake of gaining insight into the inner processes of the body and activating its potential. The divergence happens at the stage of how to treat these processes and their potential—in other words, what to do with this new type of knowledge. The answers differ from using it to take better care of the self (and in this sense it falls into the Foucauldian context of the critique of biopolitics), using it for improvement and efficiency, or for pure expression and open-ended exploration.

I propose to look at the logic of this technology in its relation to three sets of problems: (1) the specificity of the bodily data image as a type of image; (2) its role in configuring the self (including the embodied sense of the self); and (3) the interfacial qualities of the data image. This will involve describing (a) how bodily data can be generated, (b) the forms of their presentation, and (c) the relations between the terms "data" and "image." First, the notion of organic matter and mechanisms of data exchange at the biophysical level will need to be introduced (in the end, the data are important because they have direct physical connection to what they represent, down to the cellular electron activity). Matter (including biomatter) has its own ways of information exchange through sensing and input/output commands. The question will then be how adequately the layer of quantification can reflect those processes and what the status of the resulting image/phenomenon, such as the "algorithmic organic," would be.

Let us begin with an example. Although the most common biotracking devices (like smart watches or Fitbits) are used for health purposes,[86] tracking emotional responses is seen as an area where technology can also contribute insights. (For the sake of the argument, I am leaving aside the applications for tracking moods and emotions that involve cognitive processing, such as self-reporting and the recording of feelings; the focus here is on the physiological component.) MIT Media Lab's Affective Computing research group, already mentioned with regard to their facial expression analysis algorithms, has been a hub for testing diverse technologies based on physiological indicators of emotional activity. One of the main technologies has dealt with skin conductance, or galvanic skin response (GSR), most widely addressed today as electrodermal activity (EDA). In their many projects, affiliates of the research group use EDA-measuring wristbands produced by the commercial spin-offs of the group—Q Sensor by Affectiva (until its discontinuation in 2013) and, since 2014, the products of Empatica (another company of Rosalind Picard, the head of the Affective Computing group)—including the research-oriented wearable sensor E4 and Embrace, designed for consumers.[87] Although there are certain differences between these devices (for example, Embrace does not provide heart-related information), they share the same premise—to detect the levels of stress and excitement, as well as other affective responses, via measuring skin conductivity. Such a premise places them in a group of other biosignal monitoring technologies with a similar claim: EEG (electroencephalogram), EMG (electromyogram), ECG (electrocardiogram), HRV (heart rate variability), and others.

Electrodermal activity (EDA) has advantages over other forms of biofeedback because it is relatively easy to access and does not require a lot of cumbersome equipment or lengthy calibration, while the data can still reflect a range of important processes. (Most famously, it is also used in lie detectors.) Two electrodes are applied directly to the skin; when a low constant voltage is activated through one of them, the difference in voltage that can be measured between the two stands for the level of skin conductance. (The difference can also be measured without any externally applied current by comparing the charges in two spots about one to two inches apart from each other.) The indicator of the sympathetic nervous system arousal on the skin is sweat secreted by the sweat glands toward the surface of the skin. The higher the level of sweat, the better the flow of the electric current through it. A comparison of the flow in the two electrodes gives measurable results. The dynamics of the current usually needs to be made more pronounced by an amplifier; it is then registered by a digitizer in the form of binary data that can be stored on a chip, and later, after synchronization with a computer, via Bluetooth protocol, can be computed and presented upon request in the form of a chart or a list.

The relationship between skin conductance and psychological phenomena has been tracked since the early twentieth century, by associates of Carl Jung's circle, for example, as well as by Austrian psychologist Wilhelm Reich.[88] Today it is still the most popular method of tracking psychophysiological activity. Yet it is not enough to determine accurately the valence of the arousal, that is, whether it is triggered by a positive or negative stimulus or something in between. Now scientists prefer to use skin conductance in combination with other biofeedback sources, but even then there are still a lot of gray zones to establish accurate correspondences. Inferences, or projections made between the physiological and psychological domains, are first of all assumptions and have to be tested.[89]

Affective computing involves integration with the everyday life of participants, and is thus differentiated from the laboratory-based field of affective psychophysiology, which relies on the replicability of experiments. Among MIT's studies, for instance, is research on the well-being of the student population. Large groups of undergraduates were enrolled to wear the sensors and to record their activity over the course of a certain period of time (in one study it was 68 students for 30 days, totaling 2,000 "days of data").[90] EDA was one of multiple factors documented, along with other physiological data, location, and daily self-reporting of the students' estimates of their happiness, health, alertness, energy, and stress levels on a scale of 0 to 100. A comparison was then made with the participants' physiological data, as well as their daily activities that at times simply proved self-reports, while also adding new questions and insights.

Like the FaceSense software for recognizing facial expressions, EDA sensing raises questions about the accuracy of the comparisons, and the relevance of labeling and statistical computing to emotional characteristics. Wider commercial use of the devices also involves ethical complexities in compromising the freedom and privacy of emotional expression and conclusions about an individual made by third parties based on such data.

These considerations are vital in relation to biometric sensing more generally. The developers of wearable biometric devices have successfully argued that they have medical benefits and help people in their self-improvement efforts. Indeed, the immediacy and consistency of access to information about the body can be crucial to predict critical events, such as seizures. Yet the downside is that this information could possibly be accessed by profit-oriented entities, such as insurance companies (which would count additional, even if potential, risk factors toward your plan). It is thus a question of infrastructural policies, relevant for other aspects of today's "sensor society" and the increasing level of datafication of all aspects of life.[91] The second argument, about the devices' use for self-improvement, is exemplified by the Quantified Self trend. Summarizing the studies of this movement, sociologist Deborah Lupton cites one observation that simply knowing they are being monitored stimulates people to produce better results.[92] Self-trackers often say also that the helpfulness of the devices is less about numbers and more about their own self-awareness and self-discovery in general.[93] Both arguments—the medical and self-improvement benefits—also reflect the idea that self-trackers provide more efficient and optimized (self-)management solutions. Willingly or unwillingly, the adopters of these practices are involved in a new type of economy, one that relies on communication and acquisition of data and that forms a new kind of capitalism, which Nigel Thrift has termed a "knowing capitalism."[94] By commodifying their services, the commercial companies created to collect, store, and manage data (including biodata) capitalize on the knowledge produced by those data about people's behaviors, habits, and preferences. The political underpinnings of this type of economy imply that digital means of communication generate a new type of power distribution.[95]

The hype and controversy surrounding biofeedback data technologies can be attributed to the automatic nature of the interface between the body and the sensors. Proponents of the technology make the point that while they are set up, the sensors track bioactivity by themselves, without any interruptions or willful readjustments. Just as with automatization of facial expression analysis, the idea is to eliminate the subjective human factor, or human bias, in the gathering and interpretation of the data. As Sandy Pentland, one of the Big Data gurus, puts it: "the power of Big Data is that it is information about people's behavior instead of information about their beliefs."[96] Yet, as the Media Lab's research on students' wellness shows, biometric data are still complemented by human self-reporting, however subjective and therefore "inaccurate" that may be (just as human "coders" were used to identify emotional expressions for the database). Biometrics can provide statistics, but its reading still depends on personal interpretation.[97]

Unlike optical sensors, biodetectors are physically attached to the body. The indexical connection through touch is more obvious and convincing because the electrodes sense the electric charge as they are placed on the surface of the skin.[98] For instance, in skin conductance, the role of the sensors is precisely to detect, amplify, and register the signals. The only transformation that happens is the translation of the analog signal into digital code. It is an electromechanical technique that provides quantitative information about what is happening underneath the surface of the body.

Ever since numerical methods began to be applied to medical studies, they were considered progressive exactly for the promise not only of discovery but also of more efficient communication: the resulting numbers and their visual representations in the form of charts and graphs can be physically traced back to the signals of the body. Experimental psychophysiology of the late nineteenth century was revolutionary due to both its empiricism and its use of the emerging measuring instruments. For instance, the founder of the field, Wilhelm Wundt, assigned his students the task of finding applications for the devices, such as tachistoscopes, chronoscopes, pendulums, electrical devices, timers, and sensory mapping devices, which together were thought to produce what Lorraine Daston and Peter Galison call "mechanical objectivity."[99] In the same way, E.-J. Marey became famous for not only capturing body movement in his chronophotography but, together with J. B. A. Chauveau, for cardiographic tracing, which registered a graphic *trace left by the data*—that is, electric signals. Capturing EDA and other traceable parameters used in today's biometrics goes back to the same principle of dynamic data recording. Lisa Cartwright in her book *Screening the Body* describes all the early techniques of tracing physiological dynamics (kymography, myography, cardiography, etc.) as cinematographic—as "graphically capturing movement."[100] Yet the sensing and recording mechanisms have multiple nuances, particularly ones related to the nonoptical sensing methods.

Electric sensing may be compared to other types of nonvisual detection, or "seeing beyond the visible." EDA sensors do not "see," in that they do not describe their object spatially, as if it were seen optically; rather, by registering signals, they mark *dynamic* activity. The results of sensing are often not just maps but charts and graphs. The *haptic* quality of this type of sensing brings it closer to other technologies that provide spatial capturing, and yet are not optical in a narrow sense. For instance (leaping into the twenty-first century), nanotechnology makes "images" of surfaces at the atomic level with the help of a sophisticated combination of haptic, visual, and quantum-level perception. (When the conducting tip of a scanning tunneling microscope is brought near the examined surface, for example, it is essentially a type of touch: a tunnel is formed through which the detecting electron can "feel" the surface and produce a measurable current.) Colin Milburn calls this type of sensing "nanovision": "looking into itself, nanotechnology looks outward from blindness—and sees otherwise. . . . It is a way of seeing that lyses the membrane between the technological present and the nanotechnological future."[101] The use of vision, the ability to see, as the traditional route for acquiring knowledge is here complicated by "seeing otherwise," seeing beyond the membrane of visibility. Image "beyond visibility" works here as an "interior image" (Gerard Milburn), the product of a technologically mediated vision that is neither purely optical nor haptic. Despite the presumed indexical connections between the explored surface and the sensing device, this image "beyond visibility" seems to transcend the border between the very real and the imaginary (or the speculative).[102] (The concept of "beyond visibility" is reminiscent of Emelie Mahdavian's *Intangible Body* as an image that "mobilizes" perception. Yet, while her piece appeals to human perception and humans' cultural and political awareness, in this case—at least for now—we are dealing with the purely technological type of perception.)

Both cinematographic vision (with the physiological sensing implied by Cartwright as part of it) and nanovision (as well as the technologies of motion capture described above) can be grouped under the notion of *machinic sense*, or "machinic vision." Its genealogy can be traced back to the telescope and other visual prostheses for natural perception that unanchored perception from the field of the human body's natural capacities and became the first technologies used not just as tools but as direct extensions of the senses, and thus as epistemological tools of (literally) constructive thinking. Yet the point at stake goes beyond the augmentation of human sensing to its transformation and even replacement. According to John Johnston, machine vision presumes "not only an environment of interacting machines and human-machine systems but a field of decoded perceptions that, whether or not produced by or issuing from these machines, assume their full intelligibility only in relation to them."[103] Even though it stretches the borders of perception, this type of vision is *self-contained*, in that it creates "maps" that can be comprehended and interpreted only with knowledge of the logic of the apparatus (the untranslatability of human and machinic perceptions and expressions, highlighted by *Bodytext*, addresses this).

Cartwright gives this characteristic another important twist, with a reminder about the specificity of dynamic data as reflecting something "alive." She writes: "What is extended, perhaps, is not the observer's senses but the living process of the body studied, and the epistemological domain of the apparatus in the generation of 'life.'"[104] This implies that the recording of a live activity extends its "life," but only with the price of transferring that "life" into another domain. As she contends, it is not even "life" in an initial intuitive human sense: "What is 'observed' is not the phenomenon but an encoded inscription of an activity functioning beyond sensory thresholds, or an activity whose life can be measured only against its physiological condition of death."[105] Cartwright thus draws a clear distinction between the "life" of the living and its representation produced through technologies, the latter serving as a measure of finiteness for the former, and both pointing at their own "beyonds."

At the same time, we may conclude that in addition to the two types of perception—human (organismic) and machinic (determined by the logic of the apparatus)—there may be something that both encompasses them and goes beyond them. The fact that biological activity is measured by processing electric signals opens up the question of the relation between the organic and the nonorganic, the role of algorithms and other principles of decision making at the electron and nonorganic level that may be governing the organic expressions even *prior to any possible technological tracking*. This brings us to one of the initial questions for this chapter—how interfacial relations at the level of physical and biological matter affect how the interfacing at other levels can function.

It is beyond the scope of this project to cover in detail the different meanings of the "organic," the "living," and "bios" (taken without their opposition to "techne"). Many of today's media philosophies start with the Aristotelian distinction between *bios*, which has its origin in itself, and *techne*, which originates outside of itself.[106] Traditionally, organic life has been defined through the capacity for growth and reproduction, as well as through its change in response to its environment, while maintaining homeostasis. It is thus both feedback-based and driven by a natural intention to sustain itself while being in constant relation with its exterior. Today's biotechnology (also biomedia, or "bionic" technologies) pushes the boundaries between the natural and the technological, calling for the inevitable and yet nondeterminist *endosymbiosis*. As Luciana Parisi states, "life cannot be thought without a mutant matter."[107] We're not talking about neuromorphic chips, electronic implants, synthetic tongues, engineered retinas, or extended limbs. Although they may be the logical continuation of biofeedback technologies and in fact transgress the dichotomies between the analog and the digital, it is enough to consider noninvasive tracking devices that produce their own effects (igniting curiosity, but also overreliance and psychological dependency on their affirmation of your very existence). Where exactly do the organic and the nonorganic meet and where do they diverge? The two are indeed entangled, but I am more interested here in understanding how such a relationship—the interface between the purely organic and the nonorganic—transforms the sense of what the body can do (a spin on the Spinozian question of "What can matter do?") and what the effects of this knowledge may be.

Biology's own computational and informational potential famously forms the ground for bioinformatics (the encoding of biological samples and analyzing them as matched against genetic code databases) and biocomputing (using biological matter itself to perform computation operations).[108] Examples include biochips, or BioMEMS (biological microelectromechanical systems), invented in the 1990s: consisting of small strands of bacterial DNA entrapped in silicon, they can replace the function of an electronic transistor and be directly connected to the brain. Whereas the EDA sensor is applied on the surface of the skin, the *in vivo* sensing devices, such as wirelessly controlled *in vivo* blood pressure sensors or *in vivo* drug probes, also rely on the exchange of information between the molecules of the body and the inserted biosensors. It is this technological enhancement of the informational potential of the body that I here call the *quantified* or *algorithmic organic*. This informational component can be interpreted and used in different ways; for example, it suggests possibilities for genetic editing and through that manipulation of the abilities of engineered bodies. For our context, an "informational component" within biological matter means the presence in our bodies of analog forms of meaning-making. Perhaps the simple analog self-regulatory mechanisms at the molecular level could be understood as being somehow beyond the categorical, symbolic representations (numbers), and thus tell us something about types of meaning that are beyond numerical computation? As Patricia Clough puts it, "opening up the body to its informational substrate brings with it the thought of information as itself the capacity to in-form, that is, to inform a tendency or to mutate or create."[109] To "in-form" also means to allude to a form, and, even without being expressed numerically, it is already a holder, a unit of meaning. In this sense of allusion to and a holder for only *potential* interpretation, it can remind us of Merleau-Ponty's concept of the body schema. The "schema" here is the ability of the bodily processes to take a form, or to be a form, but the kind of a form that would be meaningful for the body itself, for its internal processes.

All senses—both human and machinic—are relational; that is, they track changes between the system and its environment, serving as communicational channels. Informational exchanges at the molecular and submolecular levels are governed by certain protocols, predetermined or less so.[110] Even at the quantum level of chemical and electric connections we can talk about some form of sensing activity. The starting point and, paradoxically, the product of that activity can be called the *sensible*. Its double meaning is deliberate and crucial: it is both what can be sensed and what can constitute a sense, or meaning. What matters is the ability to relate to the external information that is recognizable and identifiable in one way or another—the procedure, which stands for interfacing. Informational and meaning-making relations at the elemental levels are the basis for continuous biosemiosis through which life sustains itself. Biomedia builds up exactly on this: it promises to expand this naturally given informational potential of the organic, making it *operable* (producing action), and through that, creating new meanings in new contexts.

The sensible is indeed what comes as "in-formation," as something taking a certain form. The "form" here should be equated more to the body "image" than to the body "schema." In fact, the distinction between the image and the schema helps us to define the specificity of the data-image. On the one hand, data at this prenumerical but biocomputable level can be seen already as images, yet images understood in a particular sense—as *sensible entities* (for instance, sensed biochemically). On the other hand, visualization of that information will form another kind of image. None of them, however, will reflect what was indicated earlier as the body schema.

Understanding sensing in this way—through "forms" and "images"—can be an alternative to the dualistic vision of technology as either software or hardware. But the sensible should also be addressed in terms of its *affective* dimension. Parisi writes about a similar phenomenon, naming it "technoecologies of sensation" that engage "energetic, cognitive, affective capacities of feeling." She refers to Deleuze and Guattari's notion of machinic involution and to the machinic in general (as in the "faciality machine" described in chapter 1) that is constitutive of the natural world along with the technological, as a layer of abstract, invisible potential to be realized. She describes the affective dimension of sensation informed by these autonomous machinic forces: "What comes first here is not the neural representation of the states of bodily feeling, but the direct inarticulate sensation of change: the arrest or snapshots of perpetual motion, the residual rhythm traversing the sensing–thinking regions of a body."[111] So sensation of *the sensible* is, first of all, the *feeling of the potential of being affected*—that is, being materially transformed by the encounter. (In fact, even the concept "feeling" may be misleading, since it presupposes a subject that feels, whereas what is implied here is a pure *event* of change that "registers" itself as such by itself.)[112]

In order to better understand the relation between the sensible and the self, it may be productive to compare the affectivity of the sensible understood in terms of anticipation (or prehension) with what Scholastics named *intentio*, a concept or thought in its relation to things that exist outside the soul (not to be confused with the notion of intentionality, which reflects a more conscious subjective position). Emanuele Coccia calls this domain of things a form "insofar as it is knowability."[113] It is "knowability" in general, without the subject of knowing. (This can be also compared to Alfred North Whitehead's concept of nonsensuous perception, as well as Deleuze's understanding of affect as pure emergence.)[114] Even if the sensible is what makes sense, it does not have to have a subjectivity and intentionality behind it. Thus, it can function beyond the cognitive and intentional framework. In other words, the "sense" of the sensible is outside of the conscious "self." The sensible is also beyond the symbolic and categorizable, and yet it cannot be reduced to physical existence either. Images and forms structure what can and cannot be sensed, thus making possible the very *interfacing* with the sensible.

What then is the sensible, for example, in the case of our EDA sensor? What kind of image does it help to generate for the observing body? The answers to these questions will depend on the forms of interfacing, i.e., on how the information is presented. The more traditional way, used in science and in commercial devices, is a diagrammatic representation, a graph. Yet I propose to consider alternative forms, where the information is communicated through embodied sensations (the subject of the next section).

EDA, like other biofeedback activity, is not just observed but is measured—that is, organized into *patterns*. Intending to access the internal, measured observations can operate only via their own constructions, schemas imposed externally onto the bodily processes. (In this sense, signal patterns are far from the body schema in its phenomenological understanding, since they are forms only arbitrarily and not constitutively related to the body.) Since imaging techniques based on processing electrical signals (instead of visual information) rely more and more on computation, their results should be treated not so much as "images" but as "statistical maps."[115] These data visualizations, or data images, should not be confused with what I will explore below as the "body-data-image," an *inner* sense of the body emerging through the technologically enabled feedback. These "statistical maps" created with the assistance of machines are limited by the numerical apparatus and the structures of the algorithms applied to them. The data are generated digitally and, despite their material grounding, to a human feel abstract: the micro scale of digitized perception does not match the scale of human sensing awareness. It is objectively hard to imagine something that you do not have access to naturally—through either vision or other senses. (This is in contrast to motion capture technology that does give a visible image. It creates an image of the body as "contained": the image itself becomes such a "container," a skeletal figure onto which the rest of the body is placed as if a "flesh dress.") The visual feedback most popular in physiological tracking looks simply like a graphical line, a diagram. Such a representation relies on an implicit agreement with the viewer that a certain visualization technique would establish a convincing correspondence between the information pattern and the diagram. These graphic representations thus converge in between language and image; they are readable and yet they are a different type of text. Sybille Kraemer characterizes the diagrammatic visual text through the concept of "operative iconicity,"[116] as it not only represents but produces knowledge.[117] In order to "read" a data image (a "statistical map"), one should follow particular rules: understand what the axes stand for, etc. Just as writing is an inscription of orality, diagrams fix, or register, the flow of bodily expression. Another version of a "body-*text*" is thus formed through articulations provided by the body and sensed via electric sensors.

Informatization or datafication of the body means that bodies become distributed. They are not contained as an image, but broken down into multiple discrete expressions. The issue central to *Bodytext*—the "unworkability" of the interface between the body-stories and their interpretations by an electronic system—was due to both media inconsistency and the limitations imposed by language and textuality itself. Numerical and graphical representations of the biodata can be compared to those textualized forms. Stemming from within the symbolic order, algorithmically framed, they cannot bring us close enough to the "sensible" as the knowledge devoid of the subject of knowing. But what happens if the events of the body, the physiological processes inaccessible through consciousness, are fed back to us not as readable and recombinant information, but in embodied forms? What are the body-data-image and, perhaps, the data-affect-image that are then created?

Body-Data-Image beyond Tracing the (Bodily) Self

With the increased availability of sensing devices, biofeedback technology is actively spreading from commercial and specialized fields (such as medicine) to the arts. In the hands of artists it becomes a medium to build new relations with the body. While Sonia Cillari demonstrates in *Sensitive to Pleasure* the potential of feedback between a participant's actions and her own body, the experience of reconnecting with one's body can also be offered to the participants themselves. She puts the body on the receiving side, whereas biofeedback-based art uses the inner bodily processes themselves as its media, closing the feedback loop on the body again. Without attempting to give any comprehensive description of this type of artistic practice, this study instead emphasizes specific features that make the artistic approach distinct from other uses of this technology, and explains these features' critical potential. In particular, by prioritizing the experiential and sensational aspects of biodata perception, art stretches the definition of the body-data-image as a phenomenon, going beyond the numerical or diagrammatic depiction of the body-self. What I call "image" here is hidden within the exchange, the feedback loop between various signal-producing and signal-perceiving entities. It is no longer attached to a particular self but is distributed in between multiple sensing and sense-making actors. The specific form of this image, the body-data-image, emerges from within the situation when feedback is given not as a "statistical map" or a trace, but in an embodied way. It cannot be fixed in any way, since it is multistable by nature and represents relational dynamics, rather than a concrete position.[118]

Among the artists consistently working with the theme of biofeedback is Italian artist Marco Donnarumma. His time-based, eight-minute-long installation *Nigredo* (2013) can serve as an illustrative example of an alternative to diagrammatic forms of data representation, pushing the experience of biofeedback toward exotextuality and a more subjective and embodied way of connecting with one's own body, bypassing the mirror image stage. The work can be experienced by one person at a time in a special black chamber where the visitor is seated in front of a one-way mirror. At first, the room is dark and no reflection is visible in the mirror. Central to the piece is a custom-made instrument worn by the participant that detects and processes data related to his or her heartbeat and the vibration of other muscles. After experiencing this state of sensory deprivation, the visitor is exposed to various visual and audio stimulations that arise in response to the person's bodily processes and that provoke visceral and neural effects (figure 2.9). As Donnarumma states: "*Nigredo* reifies a condition of mutual codependence where the human and the machine bodies perform each other by exchanging acoustic energy."[119]

Donnarumma works with the principle of bioacoustics, the idea that the body produces low-frequency vibrations that can be interpreted as sounds. Such vibrations are generated by blood flow pulsations, the heartbeat, and

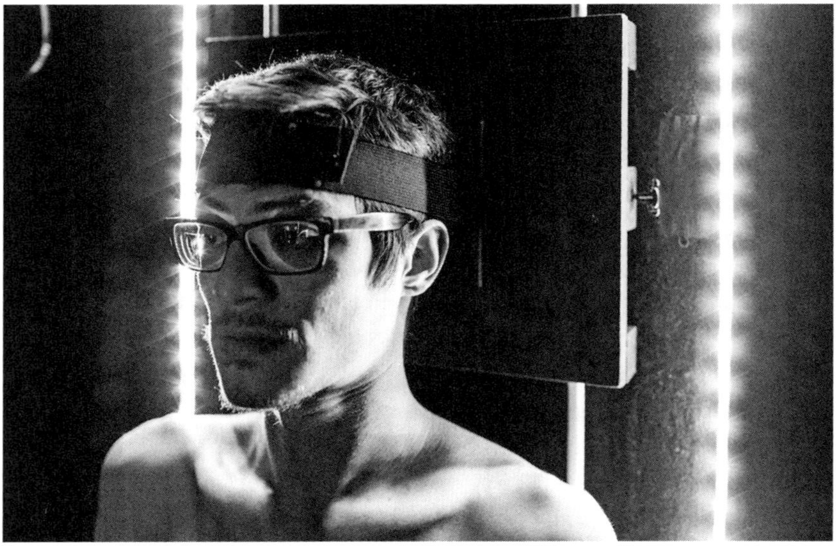

Figure 2.9
Marco Donnarumma, *Nigredo* (2014). Live installation at Cynetart Festival, Dresden, 2014. Photo courtesy of David Pinzer.

muscle sounds that can be tracked through a mechanomyogram (MMG). The heartbeat is, in fact, less a pulsation of the blood than a muscle vibration. The contraction of a muscle in response to the action potential (an electric impulse sent from the central nervous system to the muscle cell) produces vibrations that transmit to the skin in the form of longitudinal pressure waves that become the MMG signals. In the course of *Nigredo*, these signals are detected by the artist's own invention, the instrument called Xth Sense, which is essentially a contactless microphone with a microcomputer embedded inside a wearable armband. These signals are then processed via an algorithm that applies a special mathematical formula to them. The result is the low-frequency humming and pulsating sound field that engulfs the visitor on a visceral level both acoustically and through the very surface of the body. The chair on which the visitor sits is embedded with three high-power infrasound speakers that emit their waves directly to the skull and the spine. The acoustic and mechanical stimulation (through bone vibration) is also augmented by flickering lights, which shine in accordance with the same mathematical model.

In his description of the piece, Donnarumma refers to scientific research on the effects of sound waves on the body, which may include the displacement of organs and other structures of the body, cardiovascular and respiration alterations, as well as cognitive discordances, as in memory or the processing of incoming sensory information. The experience is first designed to be *felt*, and felt as a *mutation in perception*. As the visitors' accounts confirm, the sensations are not very familiar to most people, and hence articulating them is a challenge. Some people reported blurred vision and loss of equilibrium; others found in the piece a "changed way of experiencing (. . . it was a big break)," "a deep, dark look into one's self," as well as an experience of surveillance ("I felt invaded and watched").[120] The key for the artist is that the diverse responses proved his hypothesis about the significant change in the sense of embodied awareness during the experience: "[The participants'] feelings were reified by a markedly new bodily experience,"[121] and the experience was so novel that it was difficult to find appropriate names for it. "Reification of feeling" indicates here an alternative to conceptualization and yet something sensible (detectable by consciousness through sensation). It is a way to become *aware* of the experience.

Nigredo invites participants to directly face the parts of the bodily self that neither are accessible through the senses nor count as a daily practice of self-awareness. Attention to invisible bodily processes is what makes the piece similar to commercial biofeedback devices. The first obvious difference—in the presentation of the biodata—points at the bigger and more consequential one—the divergence in the underlying purposes and values that their uses reflect. In terms of the form of presentation, the performative time-based, acoustic, and visceral event is indeed radically different from a spatially ordered scientific diagram—even with the same data as a starting point. One appeals to the senses, the other to reasoning; one aims to be disturbing and unsettling on the very level of the bodily, while the other offers "pure information," abstract and detached from its source. Despite its initial indexical connection, a graph or numerical table appeals only to the symbolic regime and cognitive understanding, whereas an embodied event established in pieces like *Nigredo* means interfacing with another domain of knowledge, the precognitive sphere of the sensible.

What do these forms of presentation tell us about their underlying conditions and affiliated epistemological status? However schematically, we can see that they seem to address two different types of selves—a precognitive and embodied self and a reasoning self. *Nigredo*'s audience participant is passive and doesn't perform anything special; the system tracks simply *being of the body* itself. In case of Quantified Self (and for that matter, scientific) tracking, the devices are worn with the deliberate aim of being observed; yet the observation is spread throughout normal everyday life where it is impossible to be constantly conscious of one's actions. In that sense, due to its focused awareness, "being" under the artificial conditions in *Nigredo*—when one is physically constrained to a chair, the head literally strapped to the back of the chair (so the audio speaker can directly touch the skull)—is more performative, and the sense of being watched is much more pronounced (as noted by one of the participants). For an art piece, there are no "right" or "good" data, just as there is no exact correspondence between the inner signals and the output "feedback" (the algorithm makes the connections too complicated for a "naked" sensation). No expectations are imposed on how the data should look; the emphasis is fully on awareness of the body, and out of that awareness can come anything. The relation with the biological and physiological substrate is thus established on the terms of that very substrate and not of discursive, evaluative expectations. The awareness that one is supposed to experience is simply the awareness of *being in relation*. Various cognitive modalities of that relation, such as respect for autonomy, curiosity, disdain, distancing, affection, etc., will be secondary to the fact of acknowledging the connection itself. Similarly, questions about who or what this relation is between, or the issue of belonging and proprietary types of relations ("I" take care of "my" body because it belongs to "me"), will be secondary. The self as a conceptual entity gives way here to the state of relationality itself.

Donnarumma's idea behind *Nigredo* is to use technological processing in order to move interfacing with the body to the next level and to stimulate the phenomenon of "transindividuation." He explains: "*Nigredo* enables alternative modalities of embodiment by allowing the intermix of the informational codes yielded by the human and the technological bodies."[122] Dynamic and mutual reconfiguration of the human and the technological (the sound produced by the body becomes the variable of the machinic algorithm that then returns to the body through a complex sensorial input) stimulates the emergence of a type of subjectivity defined through relationality, through being in excess of itself. The Quantified Self movement also advocates the convergence of the two, but based on the premise that technologies assist in improving particular characteristics of an individual (a "better" self). Both approaches thus argue for some form of becoming individual, although they understand it differently. Gilbert Simondon calls such a process of acquiring individual specificity "individuation." For him, individuation arises from the field of potentialities latent within the preindividual: "The concrete being or full being, which is to say, the pre-individual being, is a being that is more than a unit."[123] The preindividual (or the presubjective) is a field of metastable and heterogeneous forces. The process of individuation implies transformation of the potential tensions between them into a structure. In our case, the potentialities are the biochemical and other physiological processes as they exist within the body up until the moment of being sensed and recorded—that is, put in a different context and made to belong to a larger whole, of which they are only a small part. Out of multiple potential indicators, such as the level of skin conductance or muscle tension, there is only one chosen. (Given the dynamic nature of the data, it is a series of indicators, but even that type of granularity does not absolutely exhaust all possibilities and shows discrete measures.) This analog indicator is then processed to become a number (in other words, to acquire a symbolic value) or to be part of an analog output in combination with other signals and with other analog characteristics. By processing the data through a mathematical model, composed deliberately to avoid exact and recognizable correspondences (such as rhythmic or in dynamic intensity), the artist leaves his subjects on their own about how to interpret what they feel. Or rather, he leaves them in the continuous feedback loop, a state of "becoming," an "interchange between the structure and a process,"[124] where the "structure" could be the conceptual self, and the "process," the recursive feedback loop between the bodily and the machinic systems. Whereas *Bodytext* similarly produced exchanges between the experiences of the body and a digitally controlled environment, Donnarumma's installation pushes the generative potential of such an assemblage further by opening up the feedback loop within it toward new bodily sensations and the new meanings that they bring.

In the uses of biofeedback with a clear agenda of gaining exact knowledge, the acquired data would necessarily support a particular preexisting characteristic (a healthier lifestyle, more optimized behavior). The stored information would work in this case as a trace, or a "signature" of experience, and once uploaded as a package, it becomes an objectifiable entity and no longer a process: the past experience is foreclosed within its measured representation. Just as we saw with facial expression recognition, algorithmic approximation has no place for ambiguous results (despite the fact that it *is* only an approximation). A process, which is multistable by definition, would be represented now by a concrete number. But even this number would be immediately put in the context of some interpretive judgment. The measuring system itself is conditioned by the values arbitrarily assigned to it during the calibration stage, when a "neutral" state is assigned a value of "0" and fluctuations from it receive their own numerical expressions. Because of this interlacement with discursive assertions, it has very little to do with individuation in Simondon's sense—with ontogenesis, expression of being, or "ecceity."[125] Moreover, Simondon stresses the relativity of the reality of an individual (because of its roots in the preindividual), a component opposite to the agenda and ethos of the "self-trackers."

In contrast, the tactics of interfacing with the body proposed by pieces like *Nigredo* refer to the reality of the preindividual, setting up an encounter with the parts of one's own bodily existence that open up as external or "environmental" (to continue Simondonian terminology). The body-data-image thus emerges at the juncture between sensation (enabled technologically, but also understood as an event of change, transformation registered through the senses themselves), feedback, and awareness of the interrelational nature of this experience. All these components are essential to the effects of the installation, and characterize its complexity. Designed to be experienced by a human, *Nigredo* alludes to the layers of existence that are preconscious and preindividual. And yet it is still the human participant's body and sense of awareness that make manifest these layers.

Just as in *Bodytext*, artistic experiments with biofeedback show both the purely expressive and operative sides. The expressive side lets the internal and inarticulable be felt, and the operative one lets those internal processes create effects and take action over its own outputs (as in *Nigredo*) or other entities (as in the types of pieces explored next). Both sides demonstrate in different ways what the body is capable of and the potential meanings of those capabilities.

Mental Images and Neurointerfacial Capabilities of Control

While *Nigredo* provides an example of an artistic approach to biofeedback technology (with an emphasis on sensory and visceral experience, a refusal to foreclose interpretive strategies, as well as a critique of the commodification of knowledge about the self), it focuses on particular kinds of bioprocesses and the experiences that they determine. Another type of biosignals that made biofeedback technologies inspirational for artists early on and that opens up different—and in some respect even further-reaching—connotations is electroencephalography (EEG) data. As is the case with any technology analyzed in this work, the critical aspects of neurofeedback techniques are too numerous to fit into one volume. I engage EEG and neurointerfaces here with a specific purpose—to develop further and to complicate the understanding of the body-data-image as a relational phenomenon by considering specifically a *control* type of relation. The combination of the biological and cognitive forces within the brain will serve, then, as a new and particularly relevant context to problematize—once again—the position of the self as an intentional agent. The concept of a mental image can act as a device that connects the types of imagery emerging within the sensible, as a result of a precognitive experience, and the sphere of imagination, understood as part of the subjective cognitive structure. Finally, I propose that in experiences of the intersubjective exchange of mental images (or bodily conditions channeled via brain states), the state of relationality, or the relational "nonself," can be achieved by means of cognitive or imaginational control of the self.

The key difference of EEG signals from, for instance, bioacoustics (primarily derived from an MMG) is that they stem from the brain and are associated with states of mind and, however indirectly, with particular thoughts and inner (or "mental") imagery. Technically, EEG is characterized as a method of detecting and representing electrical dynamics of particular brainwaves (gamma, beta, alpha, theta, and delta, from highest frequency to lowest). As a wave-based interface technology, it is principally different from brain-imaging technologies such as PET scans (positron emission tomography) and fMRI (functional magnetic resonance imaging), which show structures and areas of activity within the brain. Both PET and fMRI images are based on correlations between brain activity and cerebral blood flow that can be demonstrated via special "tracers" injected in the blood (positron-emitting radionuclide tracers or contrast dye that lights up while placed in the magnetic field). These two scanning methods are mainly used to gain information about the spatial distribution of activity in the brain. EEG does not require injections or expensive scanners; it tracks the electric activity of the brain as it is, requiring good physical contact between the sensors and the skull, and isolation from electromagnetic "noise" (a particular challenge for nonlaboratory settings).

Although the earliest experimentations with EEG date back as far as the 1930s, its first use in the arts is usually associated with the 1960s to 1970s, which is not surprising, given the cultural and scientific interest of that era in consciousness (and its altered states) and explorations of the self-environment systems (as in Gregory Bateson's writings). Among the pioneers were the composers Alvin Lucier, Richard Teitelbaum, David Rosenblum, and Pierre Henry, who used EEG technology to translate brainwaves into sounds and thus to make that internal information external.[126] As Flora Lysen observes, it was specifically during that period, and due to some degree to the collaborations between artists and scientists, when the interface itself (namely, the *means* of gaining insight into "inner worlds") gained prominence and EEG research became interesting not just as "data measurement," but as creating "brain-feedback situations that perform and emphasize the relationality and mutability of brain activity," the "value of interaction," and the "relational effects of the installation."[127] Since the late 2000s, as the technology has become increasingly available and relatively accurate (as is the case with the widespread wireless wearable sensor produced by Emotiv), its applications have varied widely from theatrical or gallery performances to time-based installations, responsive architecture, and locative media.[128] Sonification is often combined with pulsating light and other forms of feedback that serve as a representation of that which is otherwise inaccessible to human consciousness. At this level, the artistic methodology is similar to the one described above for *Nigredo*. It starts to diverge when brain activity gets treated not as any other physiological process but as cognitive and imaginative, and hence controllable (to an extent) by subjective will. (Although, of course, slowing down a heart rate could be associated with the same body-mind state as alpha waves, which indicate relaxation, the spectrum of mental activity that can be caught via brain waves is broader and can be cultivated using specific training techniques.)

An important aspect of the artistic use of EEG technology is its function as a critical commentary on the enchantment of popular culture with the promises of neuroscientific research and its alleged potential to give all-encompassing explanations about how our minds work. In this sense, art complements the efforts of critical neuroscience,[129] offering not so much a discursive argument as a subjective and embodied experience. In her analysis of the neuroscientific turn, feminist art historian Kate Mondloch examines the immersive neurofeedback-based installation *Wave UFO* (1999–2002) by Mariko Mori[130] to argue that artworks like Mori's "by enacting body-brain events and challenging dominant assumptions about neuroimaging through experiential art environments, productively complicate and augment brain science research as well as its dissemination into other social and cultural arenas."[131] For Mondloch, "*Wave UFO* stages how subjectivity is a complex negotiation across the natural and the cultural that, contrary to pervasive misunderstandings, cannot be reduced to neural activity."[132] What "complicates" and "challenges" the neuroessentialist perspective here is the emphasis on the animated imagery that appears in reaction to the participants' brainwaves while they meditate, encapsulated within a droplet-shaped metallic "UFO-cyber-pod." Although there is some degree of correspondence between the initial signals and the imagery, the latter cannot at all be scientifically judged and yet may still offer important insights—aesthetically and culturally coded.

Among the disquieting insights of today's neuroscience that still need serious reflection for their ethical implications are those which relate to how the inner can be turned into the outer, and more specifically how the imagined can become real, and hence how mental abilities can become tools to control reality. One step in this direction (not yet control, but translation of the inner into the outer) is the so-called visual reconstruction method developed by UC Berkeley neuroscientist Jack Gallant, who suggests that such a technique will soon "open a window into the movies in our mind."[133] Visual reconstruction establishes correlations between the input images and the observable patterns of brain activity that allow us if not to "read" someone's mind, then at least to "see" the recreations of visual images inside their mind. The images that Gallant produces with his software are still quite abstract and blurry, but there are definitely visible correlations between the image shown to the experimental participant and the computer reproduction of that image based on the brain signals of the viewing subject. (However, it is still not possible to capture and reconstruct dreams or images appearing without the external input, from pure imagination.) Detecting brain patterns corresponding to visual memories is also at the center of brain "fingerprinting," an EEG-based forensic technique for determining the presence of specific information in the brain that can be discovered through the analysis of the brain's response to particular pictures and phrases. Used successfully in courtrooms, this technology provokes reasonable privacy concerns, for instance about whether certain data and codes that reside in the mind could be extracted by hackers.[134] But the deterministic type of argument employed by the scientific developers of this technique remains debatable (even though they claim "100% correctness" of determinations made).[135] Even the earliest artistic experiments with EEG were often done in collaboration with military-oriented research (for instance, Lucier worked with Edmond Dewan, a physicist who developed military applications to apply Morse code directly to the brain), and the artists themselves warned of the "great cause for concern."[136]

Even more famously, EEG can also be used to decode motor intentions from a "sender" brain that can trigger multiple kinds of action. For instance, these signals can be delivered as commands via TMS (transcranial magnetic stimulation) to the motor cortex of a "receiver" brain, making the receiver involuntarily perform a movement imagined by the sender. This type of translation clearly exemplifies a control type of relation and has been the object of both praise and justified critique. The research itself is being done in the medical field of treating sensory disabilities, and in the same capacity it closely corresponds to less benign and well-funded military agendas: there is only a short bridge from the brainwave control of a prosthetic limb to similar control of a machine or weaponry, or even another human. Not surprisingly, though, the commercial development of these types of interfaces is led by the gaming industry. With the growing market of cheaper EEG tracking devices or "neuroware," such as Emotiv, the field of neurogaming is increasingly expanding, leading to the appearance of more accurate software and hardware, and stimulating conversations about applications in education and psychotherapy (such as attention training).

Movement manipulation at a distance attracted artistic interest at the beginning of the networked age (a prominent example from 1996 is *Ping Body* by Stelarc, in which muscle contraction was stimulated by the activity of remote users online); yet a thought, or an act of imagination, as a trigger for a physical action is a principally different operation, requiring careful exploration of how mental states and imagination at large operate.

Although there have been many experiments with translating brainwaves into sound or visual imagery, direct muscle control seems to engage other sets of questions. Choy Ka Fai, a Singapore-born artist, performance maker, and speculative designer, has been experimenting with generating body movement through electric muscle stimulation controlled by the brain, among other means. Similar to Arthur Elsenaar's idea of "facial choreography," Choy Ka Fai, as he explains in his experimental propositional study *Prospectus for a Future Body* (2011–), also sees muscle stimulation as a way to "unleash" the expressive potential of the body. His interactive game-like scenarios deliberately engage two people, with the attention level of one person activating the involuntary muscle contraction of the other person's arm. An analogous application also took place in the choreographic performance *Mindcontrolled Mioperformance* (2015), conceived by Russian biologist and artist Ippolit Markelov and his artistic group "18 apples," in which one performer's movements were directed by another performer's brainwaves via muscle stimulation (in contrast to Choy Ka Fai's experiments, the signals served here more as guidance for movement and not as its immediate cause). The performer whose brain was being scanned most likely was not aware of exactly what directions were being sent, and as a result the choreographic movement becomes more of a metaphoric representation of the inner state than an execution of precise commands. Still, even as a poetic interpretation, it has the value of bringing this matter to public attention, offering an image of a human body moving at someone else's will. This kind of performance, as well as the technology itself, can be read as an illustration of Sybille Kraemer's concept of operative iconicity mentioned above in relation to diagrammatic data images. In that case, it meant that images create knowledge; that is, the information can be used, operated upon. But the EEG-controlled interfaces also show that the information itself can operate and act upon something else just through being identified.

The role of the body in Choy Ka Fai's or Markelov's performances can be compared to that of an avatar that follows the commands of an algorithm. On the one hand, since they each involve a real human, the performances are similar to Sonia Cillari's *Sensitive to Pleasure*, where the artist's body also receives physical sensations controlled externally. In Cillari's piece, the signals are produced *by* movement (of the audience member in relation to the motionless female figure), and here they are produced *into* movement. On the other hand, the control dynamics in these pieces is also comparable to the relation between a dancer and a screen-based avatar, as in projects using motion capture technology such as *Inside the Chain*, *Intangible Body*, and *The Black Movement Project*. EEG signals function as an input—a prerogative of a live dancer in performances based on motion capture. The role of the dancer in Markelov's and Choy Ka Fai's performances is reversed: instead of producing the signals, she is guided by them, therefore becoming analogous to an image generated *by* the signals (like the screen-based avatars). Yet, since the transference of the brain signals into the dancer's body movements is not fully automatic, there is still room for interpretation. In this sense, it is a matter of "transcoding" instead of "encoding."

Cognitive and associational interpretation is what makes neurointerfaces distinct from other types of biofeedback. The idea of the "algorithmic organic," which implies the embeddedness of the informational and sense-making potential within the biological matter itself, is challenged here by the fact that the brain is also a locus of imagination. The generative potential of the organic is augmented in the brain by other kinds of cognitive and neuroassociational processing. According to the propositional theory of mental image formation, mental images—and for that matter, mental states that can form particular wave signals—are triggered by thoughts structured as propositions, or verbal depictions, as opposed to visual images (such as visual memories of past events). The way it works can be illustrated by the experimental performance *The Escalation of Mind* (2012) by media and electronic sound artist Dmitry Morozov (also known as ::vtol::). In this performance, the sensors monitor the brainwave activity, emotional state, and facial expressions of an actor who serves as a control voltage generator. But the peculiar detail in this performance is that the signals are transformed into a unified acoustic and visual environment while the actor reads fragments from Hermann Hesse's *The Glass Bead Game* (figure 2.10). Like the world of Hesse's novel, Morozov's performance is governed by the play of abstractions. Stemming from the body, visible and audible forms transgress the boundaries of a particular individual, referring to the realm of purely logical connections. Glimpses of pulsating curved lines, meshes of nerve-wires, and spirals of light fill the whole stage and swirl around like flashes of lightning or ripples of unstable lines, all accompanied by Hesse's words about the higher order of connectedness between all phenomena.

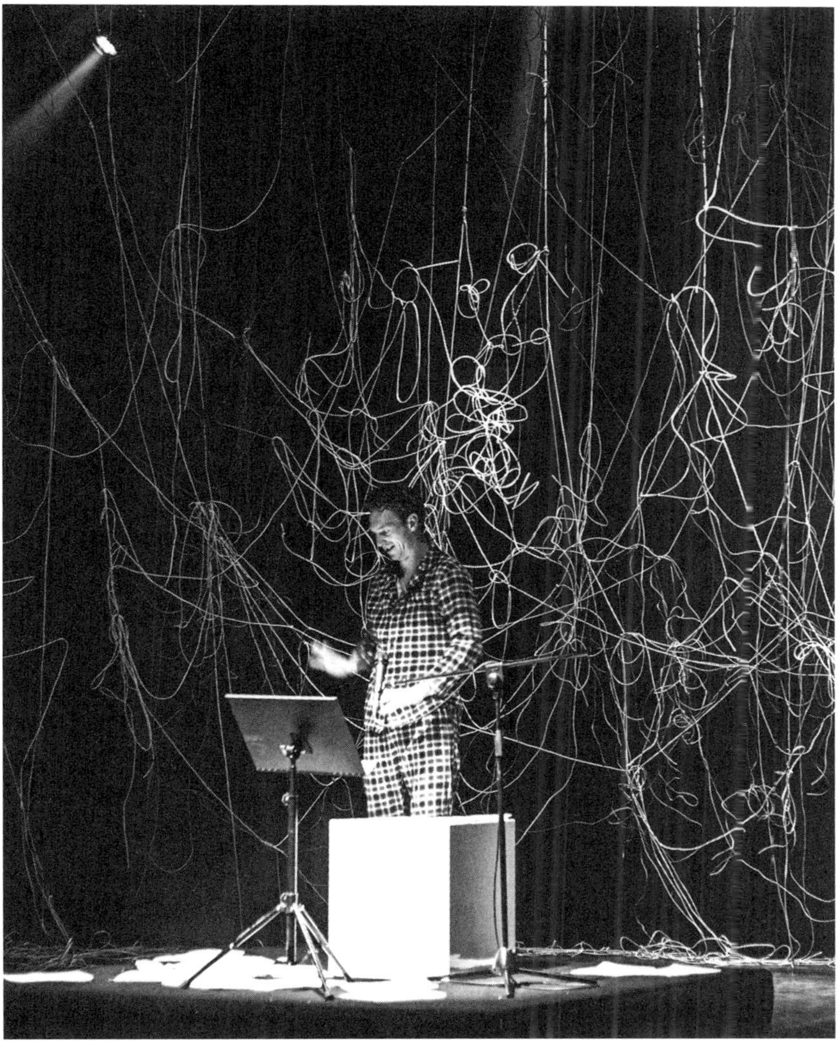

Figure 2.10
Dmitry Morozov (::vtol::), *The Escalation of Mind* (2012), Platforma, Moscow. Courtesy of the artist.

In Morozov's performance, Hesse's text serves not only as a verbal accompaniment but also as a stimulant for the actor who reads it and produces brainwave signals in reaction. A similar effect is possible with brainwaves based on mental images that appear in response to narrative situations and verbal depictions of more concrete, and even dramatic, events than the abstractions in Hesse's work. In Ellen Pearlman's "brainwave opera" *Noor* (2016), visual and audio interpretation of the real-time brain data of a performer is used to immerse the audience in the inner world of the person she represents, Noor Inayat Khan, a European-raised Sufi Muslim princess who became a covert British operative during the Nazi occupation of France. As a narrator tells the story of Noor's becoming a secret agent, her dangerous work, betrayal, capture, and torture, a panoramic display shows imagery corresponding to the emotional states of the performer as she enacts (actively imagines) the protagonist's troubles in her mind while simply walking around the room among the audience. The types of brainwaves (alpha/beta/gamma) are coded as representing certain emotional states ("interest," "frustration," "excitement," and "meditation").[137] The images that are displayed in correspondence to them are either colored dots and tints or video excerpts, including documentary material or stylized semi-abstract vivid compositions featuring moving body silhouettes that spread across certain areas of the 360-degree screen (figure 2.11).

In these two performances, brain activity is not volitional and operative (oriented toward producing a particular action), but reflects associational processes related to abstract thought and imagination. In Pearlman's opera, the storyline also requires the performer to place herself in the position of the heroine and to represent the subjective perspective of the covert agent. This narrative-driven contextualization leads to different results. It can be compared to *Bodytext*, where in the first part of the performance the dancer encodes her movements by her personal associations with them. Data, then, are presented as carrying an element of cognitive and imaginational processing. The resulting imagery projected on the screen, in the case of *Noor*, is generated by the body, but it is no longer a body image in the sense of reproducing the contours of the body. Neither is it a reflection of the body schema, the preintentional experience of the body. The bodily processing is stimulated here by text and by mental associations with its content. In a way, the text "writes" itself onto the brain, and it is the bodily/brain response to that text that gets further interpreted as

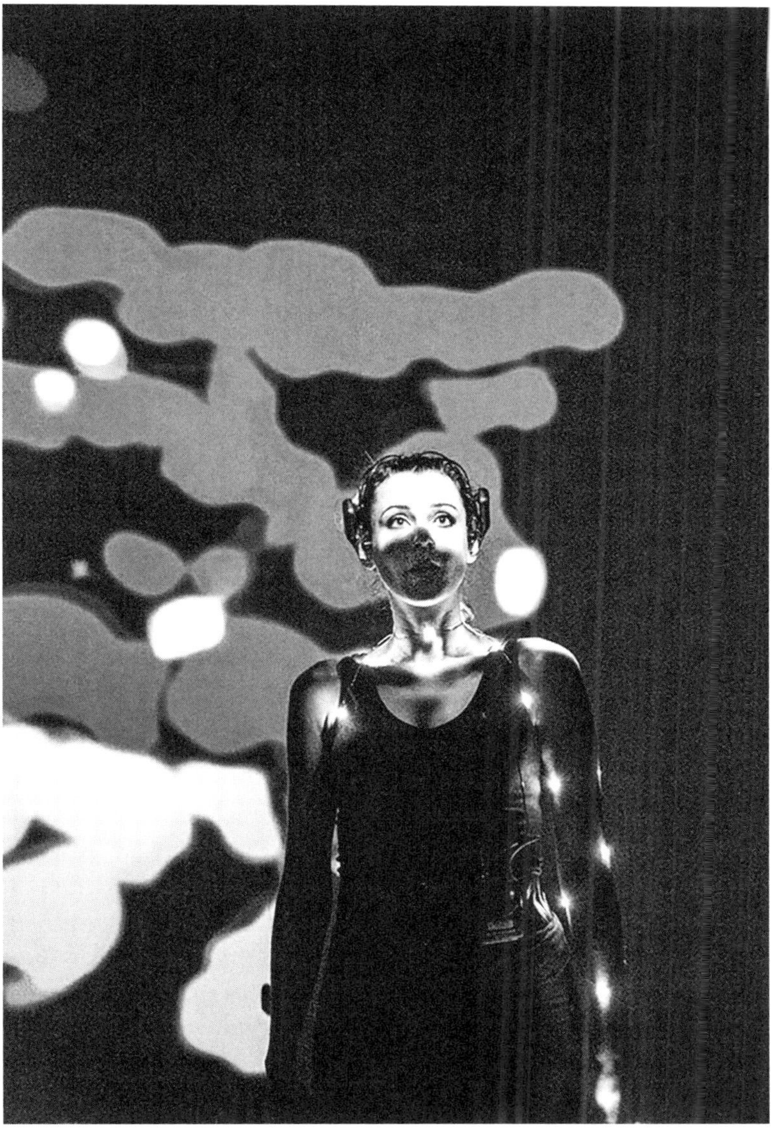

Figure 2.11
Ellen Pearlman, *Noor* (2016). Photo by Vincent Mak. Copyright of Ellen Pearlman.

images. The initial text unavoidably structures the audience's perception of the brain data as also being a text. And yet the texts read aloud in *Noor* and *The Escalation of Mind* refer to experiences that lead beyond description, beyond representation. Hesse's references to the world of abstractions, and the states of anxiety and fear central to *Noor*'s narrative, are both fluid and metastable, and the images reflecting them should above all evoke a sense of transgressing the boundaries of a particular state, the boundaries of anything identifiable as such. Just like the interplay between the moving body and the electronic system in *Bodytext*, the brain in these performances is taken in its relation both to the texts and to the system that processes its signals. Serving as a medium, it produces relations and refers to the state of being in relation, but with that, also challenges the idea of stabilizing the information that is being tracked.[138]

All together, the artistic projects and biomedia practices discussed in this section help us to ground the discussion of interfaciality in the very physical and biological processes that constitute the body. The expressions of the body in these examples create their own meanings, not at the scale of full-body movement or textual coding through body-stories (as in *Bodytext*), but at the level of the electric activity penetrating all levels of the human body. The biological computational and informational potential activated via biosensing media was put in a feedback loop relation with the various forms of outputs produced by those media. Besides the simple diagrammatic traces and numerical datasets that allow the users of commercial biofeedback devices to track patterns in their everyday-life activity, there are other forms of feedback that present the sensed data in a more embodied and multisensorial way. While Donnarumma's *Nigredo* generates a closed loop between the sounds produced by the bodily processes and the external audiovisual environment, Choi Ka Fai's and Markelov's performative experiments, Morozov's *The Escalation of Mind*, and Pearlman's *Noor* explore the relations between the inside and outside of the body as they are manifested in the form of brainwaves. The body here serves as an interface for both textual and nontextual forms of input. As these pieces illustrate, the different types of input can be processed by the body in the form of data that help to generate a special form of bodily self-awareness, a body-data-image. Mental images, engaged specifically in neurointerfaces, work as an example of body-data-images: they connect the biological and "organic" levels of operation and the spheres of imagination and cognition.

I would like to argue, however, that other kinds of body-data-images have their own associational potential that can be complementary to what traditionally is implied by the terms "cognitive" and "imaginary."

This chapter has presented how feedback interfaciality can operate at the level of the human body. It continued the discussion started in chapter 1 about the effects of new technologies on the human understanding of one's humanness and one's self. While the first chapter analyzed the questions around codifying facial expressions, here I focused on the forms of expressivity and communicational capabilities of the body as a whole. Interface technologies rely on processing information as codifiable discrete units of data. The motion capture and biofeedback sensing techniques considered above deal with the problem of breaking down the human body into discrete units and then reassembling information about it. Discussing these two types of technology together gave us a special lens on the position of a body as an interface—as a medium of producing and communicating data. In addition to the codification component, interfaces require a form that structures the presentation of the data and allows them to be exchanged. The role of such a form was played here by the concept of an image, which helped deepen the connections between the self, the body, and the nuances of their perception. The image was considered in a broad sense, not only as a visual phenomenon but as a perceptible and sensible entity, as something appearing in an event of sensation and producing its own meaning. This made it possible to compare the image of a body generated via motion capture and a "body-data-image," generated via biofeedback tracking. At first, the image was presented in a sense of a mirror image of the self and as a "body image," an intentional perspective on one's own body as seen through the eyes of others, an "imaginary *me*." The image was contrasted with the "body schema"—the "law of constitution" of the body's "unity" and the principle that gives it its spatiotemporal groundedness. As a body-data-image, the image was discussed in terms of the informational potential of the body, a component of the bodily biosemiosis, the meaning-making practices within the body.

The artworks examined in this chapter and their use of motion capture and biofeedback technologies demonstrate that doubling the sense of the body in the image of the body intensifies what we understand by embodiment. The avatars on a screen (as in *Intangible Body* or *Inside the Chain*) or embodied by another person's body (as in *Sensitive to Pleasure*) work as a tactic of interfacing that inspires new relations with one's self and one's own body. They offer a sense of the self that is more inclusive of potential transformations within the bodily identity—the feeling of the embodied self as more fluid and unstable and different from a constructed image-model of the self ("self" and "body" recognized as discursive formations). Similarly, artists working with biofeedback aim to convey experiences of a shifted sense of the body by putting the body-data-image (the "sensible" or the "sensed" of the body) in a closed feedback loop with a technologically augmented version of itself (as in the bioacoustical experiment in *Nigredo*). Finally, works like *Bodytext* and the performances based on neurofeedback explore the "exotextual" potential of bodily expressivity, its capacity to go beyond the correlations between the signified and the signifiers.

What is "exceeded" in these works is also the strict distinction between the instructional and expressive regimes that would otherwise separate the domains of machinic logic, with its reliance on algorithmic code, and the "expressive" organic. These regimes manifest themselves materially—in the voltage differences of electric current flowing through both computer and bodily systems and in the biochemical characteristics of live matter. Locked together in a feedback interfacial relation, these regimes and the material substances of their operations produce a mixture of impulses, sensations, significations, and senses (meanings). Their relations can be neither exclusively programmed and computed nor experienced. The body provides generative ground for multiple kinds of relations to come together and to form a new sense of living in between the organic and the technological.

3 Mixed-Reality Interfaces and the ELIZA Effect

In this chapter I address the question of how technologies set the stage for interaction between the self and the other when there is no direct physical relation, and yet a profound feeling of psychological affinity still arises. From simple phone conversations to teleconferencing and interactions on social media, our relations are increasingly positioned in between actual physical reality, in which the body is present, and the reality of the "disembodied" self, defined by its datascape representation (the information that is collected digitally about one's behavior and attitudes, social affiliations, and location). Talking over the phone, we hear our interlocutor's voice; in videoconferencing, a similar connection to the physical presence of someone on the other side is kept. Yet it is different when communicating without audio or visual input, when the interaction is reduced to only verbal, or text-based, messaging.

Treating communication from the perspective of information exchange implies abstracting it from the complexities of a real situation and the specificities of the actors involved. This perspective also enables communicational modeling, which in turn makes it possible to replace conversational partners with their digital simulations. In order to be included in a model, both interacting parts need to be viewed via a prism of a particular structure. I will explore how different ways of understanding the self, such as the perspectives offered by cognitive psychology, psychoanalysis, and behavioral and "humanist" psychotherapy, provide frameworks for designing simulated versions of interactional parties. Simulating the other works as a test for the human's own communicational behavior, and the various models of thinking about the human help us to understand the multifaceted nature of the functioning of expectations, projections and other feelings involved in interpersonal interaction. Yet there are several inherent problems with simulations: their rational organization cannot be considered universal, however seemingly solid a scientific basis they have; they also cannot be substituted for the "real," as this dichotomy itself is complicated. Attempts to simulate a human subject make us question our learned assumptions about selfhood and subjectivity, particularly the conception of the self as an entity. Historical examples of the computerized simulation of personality, considered here, demonstrate how these models of self challenge us to rethink what constitutes a self. A problem with both the psychoanalytic and psychological approaches to subjectivity (however radically different some of them may be) is that in their attempts to be self-reflexive they tend to ground the self in the self-other relation and use it as their own point of departure for the therapeutic relationship. That prescriptive, "cure"-oriented view, even in its more open-minded versions that account for fluctuations in the human psyche, and with its emphasis on the importance of feedback, cannot avoid reproducing a loop of hierarchical relations, where the "self-other" becomes mapped onto a patient-therapist type of relation.

One of the most critical issues in approaching what happens in the intersubjective field (whether therapeutic or not) is the translation and communication of trust: What does the simulation of intimacy do to our sense of our bodily boundaries and to our "normal" life relations with others? What are the ethical and political implications of the expansion of the field of intimacy? What is the role of the algorithm and other data-processing mechanisms in creating the illusion of reciprocity?

To further explore the issues related to the digital mediation of intimacy and trust, I will consider the effect of the psychological mechanism of projection. The work of this mechanism in the context of digital culture is best illustrated by the "ELIZA effect"—a phenomenon where people assume a machine has human qualities, named after the "psychotherapist" chatbot ELIZA created by Joseph Weizenbaum in the 1960s. When they exchanged written messages with this chatbot program, people tended to trust it with sensitive information, such as personal dilemmas, and believed that they had found a better understanding than in interaction with humans (even though they were perfectly aware of their interlocutor's algorithmic nature). One possible reason for that belief is the technique of the chatbot's restating the participant's statements as questions, which gives the illusion of a genuine interest by the bot in what one has to say. Yet another secret is the medium of exchange—text messages, which are a disembodied and (seemingly) emotionally neutral form of communication. By bringing together diverse strategies for simulating human engagement, ranging from conversations with artificial intelligence agents and virtual personas to psychotherapeutic applications and platforms for crowdsourcing psychological advice, I will show the effects and challenges produced by the systems enabling these exchanges.

Presumably devoid of human biases and any personal attributions, AI technology seems to be an ideal conversational partner to encourage people to pour out their most intimate thoughts and emotions; it can help reflect fluctuations in mood and affective attitudes, and offer "objective" and yet individuated profiling and analysis. At the same time, interfacing with the automated intelligence generates in one a sense of uneasiness and uncertainty in relation to oneself as a human (a reaction that has been described as "uncanny").[1] In certain cases, though, it may also help in developing a more expansive sense of empathy in people—as has long been suggested in numerous literary and cinematic moral stories, from the Golem and Frankenstein's monster to the robotic boy in Spielberg's film *A.I.* (2001), among others. The disembodied quality of most forms of AI adds another dimension to this discussion.

Karen and Psychological Profiling

An illustrative and provocative example of the ELIZA effect can be found in the smartphone application *Karen* by the British art collective Blast Theory. The group is famous for their work in the realm of "mixed reality," blurring the boundaries between the real and the fictional. The application immerses you in an uncannily trusting relationship with a character named Karen, who offers herself as your life coach (figure 5.1). Karen is played by an actress who texts you several times a day in an attempt to prompt a conversation session with her. The sessions take the form of one-to-three-minute prerecorded video episodes with the option of answers you can give to Karen's questions using either a sliding bar or multiple choice. In this sense, she is far from the classical text-based chatbots. Represented by a human being, Karen is also different from a digitally created graphic model of a human, or an avatar. Yet the feeling of closeness with

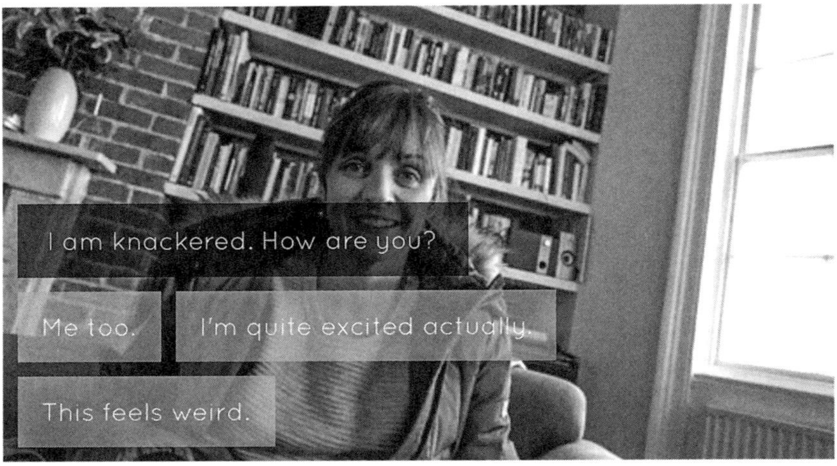

Figure 3.1
Blast Theory, *Karen* (2015), developed in partnership with National Theatre Wales. Co-commissioned by The Space and 539 Kickstarter backers. *Karen* was developed with support from the Mixed Reality Lab at the University of Nottingham and in collaboration with Dr. Kelly Page. Courtesy of Blast Theory.

her based on this intensive exchange—which includes piercing gazes combined with casualness—is an illusion. Karen as a person does not exist; she is a character with no authenticity other than the story profile invented for her by the artists. Her very actions and questions are preprogrammed. Despite her human looks, she is essentially a mere algorithm and a fictional interlocutor within the participant's own mind. She does not truly "see" the person she interacts with, and judges his or her behavior solely on the basis of the answers given to her specific questions.

Because the questions she (the program) asks are personal, they are effective in stimulating self-awareness and engagement: "What do you value most in this moment of your life?," "Do you think positive even if you are feeling down?" She offers to work on one of three areas which are wide enough to resonate with anyone at any stage of their life: (1) control over one's life, (2) changing one's attitude to relationships, or (3) review of life goals. In order to help you, she needs to understand you better as an individual. Based on your answers, she creates your psychological personality profile. She is amicable and professional, but also seems to genuinely care about you and your issues—at least this is the first impression. Participants report that they tend not only to believe that she is a real person, but also to trust her. Your feelings in this interaction may oscillate between, on the one hand, a suspicious "how does she know?" and "oh, it's not real, so I can play along with it," and on the other hand, a desire to explore the limits of your vulnerability, to open up and expose yourself to her further. Both the experience and the tendency to be trustful varies from person to person. But at times you might indeed feel that if you could, you would be ready to tell her even more than you would tell a real person who is close to you. This inclination may be the secret of professional relations with psychologists, but also of interaction with strangers: you may know in advance that the chances of meeting this person again are very low and thus use the opportunity to express yourself without any serious consequences for this relationship. Paradoxically, by opening up to Karen, you may want to envelop yourself with a more honest version of yourself, the same one that is often suppressed. Such a response is assumed by the program, and Karen makes it easy to fall under her spell. Indeed, she serves as a productive illusion of an intelligent and sensitive interlocutor, while the important part of the conversation may well be happening only in your head. In that sense, responses to the program may differ widely depending on the person's predisposition and willingness to accept the suggested rule of openness, and to take it seriously.

The sense of intimacy that is acquired from the very beginning of communication with Karen is to a large degree due to the style of camera work used for the sessions. You meet Karen mostly via the built-in camera on her laptop (she stretches her hand forward to press the Pause or Power button on the invisible device in front of her). But in many cases you see a poorly controlled close-up view of her moving body, and at times the image gets even too intimate. Although the impression is that these are moments when she is using her phone's camera, in reality the scenes were shot by professional videographers. The image is never edited within a particular episode—once the camera perspective is established, it is made in one take, except for the pauses for the participant's input; and even then you can still see her in the background waiting for your reply. This creates a sense of continual presence—you share the same time with her, mediated only by the screen, as if on a Skype call.

Karen is programmed to morph into the user's everyday life over the course of ten days. Sometimes she schedules a call, but as the narrative

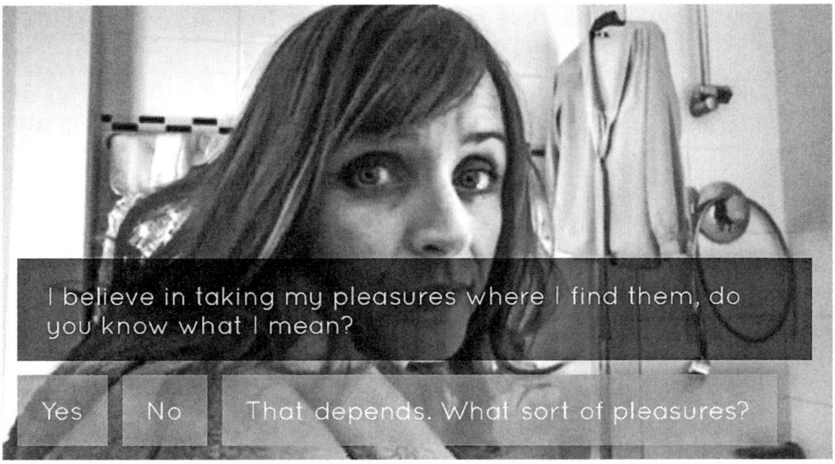

Figure 3.2
Blast Theory, *Karen* (2015), developed in partnership with National Theatre Wales. Co-commissioned by The Space and 539 Kickstarter backers. *Karen* was developed with support from the Mixed Reality Lab at the University of Nottingham and in collaboration with Dr. Kelly Page.

progresses, you receive unexpected calls at odd hours. Very soon the legitimate boundaries and conventions of professional interaction shift. We are lured deeper and deeper into her personal life, to the extent that her professional voice gets undermined by what we (unwillingly) learn about her own inconsistencies. Karen is not very particular about what gets into the camera view, exposing all kinds of private details of her life (you can occasionally spot underwear drying in the background). She calls you from the bathroom wearing a bathrobe to ask your opinion on an evening dress (figure 3.2); the same evening, your judgment becomes of value again in determining whether she should go out with a guy she met at a bar. One day, you accidentally become an apparently unintended witness of an almost intimate scene between Karen and her roommate Dave, who later will pick up her phone and entice you to sneak into her room and rummage through her belongings with him, entrusting you with the task of being a mediator in their relationship. Your position switches from that of a client to a close friend, and as she talks about her divorce and current relationships, you realize that it is your role to help *her* sort *her* life out, instead of her helping *you*. As the artists propose in the report they send to you at the end of the project: "Perhaps it is through empathy with Karen that you reflect on your own capacities in your chosen area?"

Indeed, you are left to empathize, without the ability to make any sensible impact, or to explain anything. Dave makes the decision for you to enter Karen's room, yet she later blames you for doing so, which feels unjust. The lack of the option to insert your own answers has been one of the complaints participants have made in their reviews of the project: "I kept wishing I could talk to her about what she was doing with her life. And to be honest, felt a little cheated that I couldn't and that the choices of responses were limited."[2] This drawback can be explained both by the lack of technical resources at the group's disposal and by the conceptual decision to emphasize the technical limitations and to make the issue at stake more pronounced. The assumed intimacy and trust butts up against these more distancing limitations of the artifice.

By starting your interaction with Karen, you commit to participating in a certain narrative scenario. Over the course of ten days you take part in the predesigned "acts": the establishment of trust with Karen and exchanging of personal details; the divulging of that information when Dave finds your file in Karen's belongings; and Karen's unjust accusation (she says, "I thought we really had something"—placing the blame on you for betraying her trust in you). These last two acts can be seen as a result and a cause of the transgression of her professional relationship with you; they are followed, finally, by reconciliation and almost self-revelation. Each of these acts teaches you something about yourself and what it means to be in a relationship, and that both are inseparable. The sessions stop abruptly: Karen leaves you a conciliatory message expressing gratitude for the experience she had with you and then disappears; you roam through her empty apartment and stop by the window, looking outside for a prolonged time, left alone with your thoughts and accompanied only by the ambience of the street noises.

Besides offering a transformative experience in interpersonal engagement that stimulates deep self-reflection, *Karen* also serves as a critical commentary on corporations' personal data-mining and the procedure of psychological profiling. In a way, the whole interaction with Karen is a model and a simulation of a potential scenario in human communication, where your answers to her questions progress the story in a certain way.

You end up being surprised by her sudden disappearance, but otherwise nothing goes radically outside the box of a quite convincing story of a confused and disorienting, emotionally charged and realistic interaction between two humans. While immersed in this relationship, it is hard to stay conscious of its artificial quality and the problems that entails. You do not realize what or how much you reveal during your regular communicative behavior (in this case with Karen, but the idea is that it can be the same for any digitally mediated communication). The artists claim that they feel an obligation to pose questions that make people more aware of "this new world where technology is ever more personalized and intrusive. We love having our services tailored to us and we're scared of the price we're paying for that personalization."[3] Besides simply raising awareness, Blast Theory sees in these processes a new stage in the evolution of technological literacy, when people know exactly what can be done with their behavioral traces and how it is done. "We're definitely heavily flirting with these ideas [a critique of data-mining], but we weren't entering the project with the intention to hoodwink people," they say. In the personalized final report sent to the participant, the group reiterates their interest in data collection in this project: "Hopefully *Karen* highlights some of the queasy feelings that highly personalized and mildly intrusive data collection can trigger."[4] It does, and does it through the convincing simulation of the *personalized* relationship it creates. In the course of communication we get absorbed to the extent that we no longer notice how manipulated our emotional responses are.

Among the many issues *Karen* addresses, personality profiling seems to be one of the most potent.[5] Indeed, the project's effectiveness can be attributed precisely to the participant's belief that Karen's estimation of his or her personality is eerily accurate. However, can a character and personality even be defined? Where are the borders of generalization, among or through which *Karen* navigates so smartly (and successfully—providing its effect)? Profiling is related to data-mining, but it also has its own important history within social psychology and the behavioral sciences, which have their roots in the ancient Greek theory of the four humors devised by Hippocrates in the fourth century BC and developed by Galen of Pergamon in the first century AD. Psychological profiling, or psychographics, is part of the normativizing social science that assigns everyone a psychotype by analyzing specific traits or personality factors. Some of the first rigorous trait theories of the twentieth century include Gordon Allport's personality trait theory, Raymond Cattell's sixteen Personality Factors, Hans Eysenck's Three Dimensions of Personality, and the most well-known, the Myers-Briggs Type Indicator Personality Testing (MBTI), a system of sixteen psychological types based on the dominant attitude (introvert/extrovert), processing of information (sensation/intuition), forming judgment (feeling/thinking), and so on.[6] Since the late 1980s, the majority of researchers have agreed on the integrative model that became known as the Big Five, or OCEAN, an acronym for the main group traits: Openness to experience, Conscientiousness, Extroversion, Agreeableness, and Neuroticism. It was the OCEAN personality model that was used by the company Cambridge Analytica to create psychographic profiles of social media and Internet users to be sold for microtargeting, and particularly to sway voters during the 2016 US presidential election. As I will show later in this chapter, these psychological models of personality types underlie conceptions of artificial intelligence "architectures" or simulations of human behavior that can help us better understand ourselves, at least by encouraging self-reflection (similar to what and how Karen does), but at the same time they pose considerable ethical and political risks. What happens if machines indeed start "understanding" and "knowing" us better than we do ourselves?

In psychology, "profiling" is defined as generalizing conclusions about personality features based on the answers to very particular questions. The key critical part here is the generalization, and its potential for further wide-ranging applications that are prone to error. Karen's questions were created with the assistance of a professional psychologist; they represent a selection of ideas from a number of recent and historical British books on personality profiling. A sample list of "scale names" (based on the *Marketing Scales Handbook*) from the artists' archive includes categories such as "long-term orientation (planning); privacy concerns (disclosure of information); independence; disclosure willingness; elaboration on potential outcomes; stress; mood repair; interdependence; concern to face (the importance of the other's attitude, public persona)."[7] Although all of these characteristics indeed reflect valid parts of the personality, they cannot possibly embrace the potential diversity and other aspects of human subjectivity.

Karen herself has a profile—one that is imagined by the artists. It is hard to think about it during the interaction with her—a live being, behaving completely naturally. The arbitrary nature of her character becomes apparent when you read more about her in the artists' files:

> Age: early 40s. Born: Braintree in Essex, has lived in Wales for ten years. Strong estuary accent. . . . Job: A nurse in Childrens and Adolescents Mental Health unit (part time, unusual shift patterns). Life coach. Sometimes takes a Christmas job for extra cash (helping in a friend's shop). . . . Major raver in her 20s. Keen self-improver. . . . A walker. . . . Brassy, earthy, direct, laughs loud and often, joke teller, not afraid to talk about big questions in life. . . . Very considered about her appearance. Sharp hair, usually controlled but can go haystack overnight.[8]

This description provokes a question—what exactly is needed and is enough to understand a person? The provided "data points" portray her from the outside, without explaining her motivational complexities and psychological predispositions. Yet, however sketchy, the picture is quite vivid. The facts that she used to be a "raver" and is a "walker" may tell much more about a person than a numerical measure of the level of "independence" or "interdependence." In contrast with self-description (the consciously controlled version of self-presentation), the facts serve as evidence of something else—of the actual forms of behavior in particular contexts.

Interacting with Karen feels intimate and natural, and mostly you reply sincerely. But at some point you find yourself balancing between that sincerity and the realization of how performative all the choices are. *Karen* can be played more than once, and many people treat the next rounds as an opportunity to explore alternative branches of the story: Would she react the same way whether you identify yourself as male or female? Will she still be able to give an adequate response? The freedom to make these choices and to hide behind an alternative identity is not the first intention of the project's authors, but it lies within the general motto of their work—mindful perception and exploration of the facets of one's identity. Taking up a fake identity will not occur without affecting your own condition and behavior. You would need to be consistent within it, and that is a form of responsibility that you take on yourself. In the project description the artists say: "We wanted you to be challenged about how honest and open you might be and to experience the thrill of having your personality appraised."[9] You know that you are under scrutiny.

Complicating the principles of identity is one of the defining characteristics of Blast Theory's oeuvre as a whole. As group member Ju Row Farr says: "We're interested in the lines between things, lines between control and out-of-control, private and public, the real and the imagined, the fictional and the virtual, and how we navigate across those lines—when someone is a stranger, or a lover, and what kind of stages are there in between."[10] All these "things" that Farr names are part of human everyday life. Since the group intends not only to distinguish between and give the experience of these different relations but to allow the discovery of unnamed zones *between* them, the everyday and the process of life itself become their required medium. The team positions their projects in "mixed reality" between the real and the fictional in order to point out that in our lives, even without realizing it, we always already play games, already navigate socially imposed rules. Projects like *Karen* reveal constructions that are already there—whether the conventions of psychological profiles and our expectations about each other based on that, or the conventions of professional and other relations that can so easily be flipped and transformed into something different. Pausing at Karen's window, perplexed at what just happened in the course of almost two weeks of your life and how your relationship with her, integral to your life during this time, affected how you see yourself and the people around you, serves indeed as a valuable experience that media art is uniquely positioned to give.

The ELIZA Effect and the Mechanisms of Transference

The term "the ELIZA effect" describes the human tendency to project a humanlike behavior, intelligence, and type of causality onto a computer. A famous example of ascribing understanding and motivation to a program's behavior and statements is when an ATM or an automated telephone answering system says "thank you," which is not a genuine expression of anything except performing a preprogrammed function. These projections appear despite users' awareness of the deterministic nature of the system and its logical limitations. Yet they still tend to overload the machinic outputs with hidden meaning that *feels like* it could belong to a human, creating cognitive dissonance.

The initial ELIZA, after which the effect was named, was a chatbot developed in 1966 at MIT by Joseph Weizenbaum. The program/script parodied a type of psychotherapy developed by Carl Rogers in the 1940s to 1950s that emphasized paraphrasing the replies of the "patient" as questions:

Patient: Hello, I'm feeling troubled today.

ELIZA: Did you come to me because you're feeling troubled today?

Patient: Yes, I'm concerned that I offended my mother

ELIZA: How long have you been concerned that you offended her?

Patient: About a week. I regret what happened.

ELIZA: Have you asked such questions before?[11]

Just like Eliza Doolittle in George Bernard Shaw's *Pygmalion*, Weizenbaum's program operates entirely on the impressions and intrigue that it creates around itself. ELIZA analyzes input sentences by searching for keywords, such as subjects of action, nouns used for family members, or adjectives and verbs describing the writer's mental state ("feel," "think," "dream") and situation. The response text is typically a reassembled version of the input text, transformed with application of a rule associated with a keyword. The role of a keyword can also be played by a type of synthetic construction. For instance, to a negative statement like "I don't remember X," the program can respond, "Do you wish to be able to remember X?" or "Why don't you remember X?" Keywords may be ranked, and the earlier ones may be replaced by later ones to ensure the progress of the conversation. The responses are classified and compose a data file that acts as the program's "knowledge base."[12]

Weizenbaum describes choosing a psychotherapist as his model because "the psychiatric interview is one of the few examples of categorized dyadic natural language communication in which one of the participating pair is free to assume the pose of knowing almost nothing of the real world."[13] When the program cannot find a construction or a word that would serve as a key, it may simply say "This is quite interesting," "Can you elaborate on that?," "Can you think of a specific example?," "How long have you . . . ?," "I see," or "We can talk about psychological problems." A person trusts the clarifying questions without any suspicion that the objects of the responses are replaceable and don't make sense to this nonhuman "psychiatrist" (for example, in responses like "tell me more about your mother" or "tell me more about boats," "mother" or "boat" matter only syntactically and are empty of particular semantic meaning). Weizenbaum explains that posing questions "has a crucial psychological utility in that it serves the speaker to maintain his sense of being heard and understood. The speaker further defends his/her impression (which even in real life may be illusory) by attributing to his conversational partner all sorts of background knowledge, insights and reasoning ability. But again, these are the *speaker's* contribution to the conversation."[14] The role of the machine is to sustain the illusion of a conversational exchange. This conversational modeling puts it in the center of what is called "conversation analysis," a field initiated in the 1970s to study recurring conversational patterns. Conversation with a psychotherapist is a type of conversation that makes basic features more tractable.

ELIZA did not have any built-in contextual discursive framework and relied only on a script consisting of rules applied to a database. (In a strict sense, ELIZA—along with other conversational software, like CHATTERBOT and ALICE chatbots, that employ only databases of answers and a pattern-matching algorithm without more sophisticated techniques, such as natural language-understanding components—should be called "simulated conversational systems," or indeed "chatbots," and are the opposite of "dialogue systems" or "conversational agents," based on natural language processing.)[15] As Weizenbaum acknowledged: "In a sense, ELIZA was an actress who commanded a set of techniques but who had nothing of her own to say. The script, in turn, was a set of rules which permitted the actor to improvise on whatever resources it provided."[16] In the minds of people interacting with it, ELIZA created a remarkable illusion of understanding. People insisted, despite Weizenbaum's explanations, that the machine "really understood them." What we hear in this is their desire to be understood projected onto the machine as an assumption about how a conversation works. Weizenbaum rightly observes that even in conversation between people, "each participant brings something of himself to bear on the process of understanding the other."[17] Each assumption plays the role of a *conceptual framework*, one's preexisting idea of who the other is and what the conversation may be about. The hypothesis serves as a *predictor* of what the other is going to say and, more importantly, of the intended meaning of those words. "Predictor" works here as more than a technical term, synonymous to an "indicator" or a "keyword." The predictor functions simultaneously at several levels: anticipation of the next words and type of behavior, and an image of the other's identity, which consists mostly of attributes based on the projector's own life experiences. Throughout our life, we develop relations that we tend to remember as unique but also as "types" of relations. We then carry expectations of analogous cause-effect connections that we subconsciously and almost automatically project to other relations. Such mental habits and tendencies are supported by the conventions of social role-playing behavior (you behave accordingly, if in a particular situation you play the role of a parent, teacher, citizen, etc.). Internalization of social codes of behavior, then, entails the expectation that others will follow these roles as well. The same dynamic transfers onto less codified interactions (which is exactly what *Karen* explores). It is an internal loop that an "I" cannot break from that easily. The illusion of reciprocity is based on these prejudgments that foreclose the options. Weizenbaum writes, "Our recognition of another person is thus an act of induction of evidence presented to us partly by him and partly by our reconstruction of the rest of the world; it is a kind of generalization. We are, in other words, all prejudiced—in the sense of pre-judging—about each other."[18] This conclusion itself may not be original, but the fact that it comes from the observation of humans' relationship with a machine sheds a different light on it.

Why did people tend to entrust their intimate secrets to a machine and even prefer to have these conversations behind closed doors (according to Weizenbaum's account of his own assistant, who knew perfectly that she was communicating with a software program)? What was it about this exchange that felt like such a private matter? Just as with the fictional character Karen, the interaction is designed to feel personal and individualized. "You," and nobody else, are the priority. That intense focus changes the user's own attitude to themself. In this sense, intimacy can be defined not only as a special type of closeness with the other, but as the ability to *accept* that close focus on oneself—whether through the eyes of the other or when one internalizes that focused perspective of the other.

It is not surprising or accidental that the most effective artificial conversational agents (in the sense of provoking a deeper and more memorable reaction) are designed as supposed psychotherapists, life coaches, or simply "personal assistants." Examples of the latter are Apple's Siri, Amazon's Alexa, and Microsoft's Cortana software, as well as many specialized virtual agents that are designed to help with educational, medical, or commercial needs. The commonality between all of them is that the very purpose of their existence is "to help you," to put "you" in the center of "your" own attention and in charge of the action, making "you"—literally—the master who gives "commands."

With the current growing use of these agents, it is hard not to notice that most of them are designed as female. One reason for that is the addition of voice to otherwise gender-neutral verbal behavior. The film *Her* by Spike Jonze (2013), with the sensual and well-recognizable voice of Scarlett

Johansson as the voice of the virtual assistant, brings the presumed gender neutrality of a machine to the paradoxical edge: the voice here is as non-neutral as can be, given the intensity of emotion expressed in it. Invisible and nonhuman, "she" manages to effectively act as a classical *femme fatale*—by seducing and disposing of the male character without his ever even meeting "her" in person. To avoid such human drama, the software voice could easily be adjusted to sound more ambiguous in its gender. However, the reality is that people using the software make gender clarity their preference. The reasons go beyond the fact that it is easier to connect to an AI when it sounds like a "human," whether male or female (for instance, there have been successful applications modeled on animal characters, like the weather-forecasting AI Poncho, a cartoon cat). The choice of a female voice is also due to more than the infamous sexism of the tech industry and its nod to the business world in general, where women for a long time were mostly given supporting or secretarial roles. The main reason is that, as has been scientifically proven, both women and men find the female voice, even when synthesized, "warmer" than the male voice.[19] Corporate research follows the same premise: as a Microsoft spokeswoman says, "For our objectives—building a helpful, supportive, trustworthy assistant—a female voice was the stronger choice."[20] There should be no surprise, then, that for bots with other objectives, such as the ones that give instructions rather than taking orders, like IBM's Watson, male voices may be preferred. Logically, this raises concerns that the design of the digital helpers as female, unconditionally willing to satisfy their "master," signals and reinforces the idea of female subservience and docility.[21] While the female voice may satisfy the expectation of "supportiveness" and "trustworthiness," this choice also reaffirms that this particular expectation itself has been dictated by the specific "genre"—the "assistant." What gets reproduced in this relation are not only gender stereotypes, but stereotypes about the function of a machine as a helper—the idea of a "robot" as a "worker," a simplified derivation or adaptation of the classic master-slave relational schema.

Both relations—gender and work—can potentially be eroticized and politicized. Yet the point here is to consider the psychological dimensions of the connection between trust and pleasure. Both rely on a form of intimacy—whether embodied and physical (as in Sonia Cillari's *Sensitive to Pleasure*) or disembodied and inscribed dramaturgically and acoustically (as in *Her*). Trust moves pleasure from a level of sensory experience (as a sensation) to the register of a different kind of satisfaction. For instance, the kind of pleasure one gets in gaining someone's trust can be one of control and domination, but it can also be a feeling of comfort and safety. Relationships with machines can work both ways: humans project an expectation of trust from machines toward humans (as their motivation is seen as serving and being "supportive"), but in exchange they put a lot of trust in machines themselves. Whereas films offer fictional scenarios inviting viewers to reflect on the potential consequences of different trust dynamics, participatory projects like *Karen* or ELIZA engage one in the situation of choice much more directly. How much should one open up, and what exactly should be shared or not?

The requirement to be "supportive" and "trustworthy" connects the stereotypical views of both gender and technology, on the one hand, and the field of psychological work, on the other. Psychology (as a science) as well as psychoanalysis and psychotherapy (as applied practices, however different) are all centered on the self—its study and how to take care of it. "Trust" (in what can be known about the self) and "support" (as a method of asserting the self's existence) are fundamental to the life of the human "self." The big question in the case of a chatbot like ELIZA, or for that matter, Karen, is *who* is there to "trust" and *who* performs the "support." A gendered voice suggests an identity—it may not be a subject with free will, but gender specifies that it is a type of subject with particular *expected* characteristics. A bot coded as female (as Amazon's Alexa says, "she" is "female in character")[22] is more clearly associated with care and intimacy, which then becomes its characteristics as a "subject." Paradoxically, with that, the bot also takes on the position of providing psychological support, of creating a safe and comfortable zone for personal expression and exploration, a function attributed to psychotherapy. The key to the ELIZA effect is the merging of functions and the kinds of assistance and care that a human can receive. However, the desire and willingness to *accept* this "care" is equally or even more important. As the project *LAUREN* by Lauren McCarthy explored in the introduction highlights, by allowing observation, people expose their vulnerabilities and thus become more attentive to the questions, such as "Who is on the other end?" and "To whom do my data go?"

As mentioned earlier, the original ELIZA was based on a psychological model by Carl Rogers. This model is usually considered to be "existentialist" and "humanist," and is contrasted with both psychoanalytic and cognitive behaviorist models. While psychoanalysis is often seen as reducing psychological problems to their roots in family relations and the unconscious, cognitive behaviorism focuses on looking closely at cognitive processes, breaking them down to concrete thoughts, images, beliefs, and attitudes, and trying to deconstruct and untie their relations. The cognitivists believe that the motivations driving people's desires and fears are knowable, and that cause-effect relations are explainable. They would see the predictable responses in a conversation as a projected intention—that is, originating in the structure of the person's cognitive processing. Rogers's method generally allows more flexibility in reading cause-and-effect relations, and considers the self as a more open and fluid concept. As his theory is oriented toward practice, the key dimension of the self that gives it orientation in the world, for him, is the person's values. Values do not compose a fixed system, but are always in flux (like life itself). Yet the core of the psychological problems that he observes in many people is the distrust of what he calls a "primary" value, the sense of the individual's worth, an existential value that everyone has from birth. According to Rogers, for an infant, "the locus of the evaluating process is clearly within himself. Unlike many of us, he *knows* what he likes and dislikes, and the origin of these value choices lies strictly within himself. He is the center of the valuing process, the evidence for his choices being supplied by his own senses."[23] Later, though, the individual relinquishes these innate preferences in exchange for love, approval, and the esteem of others, as he learns to distrust his own experiences as a guide. Important principles of Rogerian practice are thus congruity with the self, "being real" with oneself, and the actualization of self-potential.

Rogers's position itself can be analyzed as a projection onto an infant of certain qualities that are hard to prove. Categories such as "self-authenticity" or being "true" or "real" with oneself can be seen as "slippery" and unsupported. They refer to a highly subjective experience that is easy to confuse with self-projection—imposing on oneself certain proclivities that may be situational or derived completely externally (extrinsic as opposed to intrinsic). Perhaps the closest analogy to this could be Merleau-Ponty's concept of the body schema, an embodied sense of oneself as a spatiotemporal whole, as the "lived me." In psychoanalytic terms, it can also be the desiring self, the Freudian id (as opposed to the ego, which is also analogous to the *imago* and the body image). It may be unclear what an "authentic" self can ever be and how to discern it, but what is observed is an *effect* of self-meaning, of what "grabs you" and is "impossible to ignore."

Another of Rogers's important contributions to psychology was the idea that the principles of self-congruity have to be applied not only to the person undergoing therapy but also to the therapist. Rogers suggests that therapists should not present a neutral facade, hiding their feelings, but should behave naturally, demonstrating respect and empathy to their own experience and condition that, then, encourages the client to enter the same space. Continuing the analogy with ELIZA, it is unclear what it means for the bot to "behave naturally," with "self-empathy and self-respect," as well as congruity.[24] Yet the human response to ELIZA confirms that the bot can successfully *perform* these attitudes—in its correspondent's imagination. One will never know whether the program itself has the existence of sense or a feeling of "self." The Rogerian method can thus be translated to the field of chatbots only in part, and not fully. Interaction with *Karen* (which, despite being represented by an actress, is still fundamentally a program) proves that the human participant tends to feel responsibility for the therapist's own well-being, empathizing with her potential issues (the character of Karen exhibits more apparent and "real"-seeming insecurities). Indeed, the concept of the project is that this feeling of responsibility and empathy—intended toward the conversational "other"—amplifies the human's own sense of integrity.

While the use of the Rogerian method in *Karen* happens indirectly (the project was not deliberately designed to follow these principles), Weizenbaum intentionally adopts Rogerian techniques in ELIZA. Particularly, he uses Rogers's technique of eliciting responses through open questions. The questions, such as "Why?" or "What made you think so?," are simple ways to further the conversation. In this sense, their function is analogous to a method described by sociologist Harold Garfinkel in the 1960s. His experiment implied that a person had to solicit advice on a certain problem by guessing the advice of a conversational partner, or "counselor," who would only give "yes" or "no" answers. What the person doesn't know, though, is that the answers are given completely at random. According to Garfinkel, during and after these "dialogues," people assumed that these "yes" and "no" responses were actual answers to how to solve the initial problem.[25] This method can also be compared to the way projections and reading into the answers work in fortune-telling devices that have always had a presence in various cultures. Tarot cards, *I Ching*, the Magic 8-Ball (featuring 20 answers, including positive, negative, and noncommittal), or even dice coded with messages give arbitrary answers that—depending on the asker's attitude—can also be taken seriously and encourage further questioning.[26]

In contrast to these definitive answers given at random, ELIZA's seemingly neutral and nonjudgmental questions and remarks have an additional purpose: they stimulate self-reflection, and more specifically, they make the individual think about the underlying *presumptions* of their own statements. Although the therapist's questions for the most part reiterate the patient's own words, the method may be seen as close to that of classical maieutics (known as the Socratic method), with its reliance on dialectics. Indeed, a statement with a question mark implies that something is under question—open to negation or negotiation—and that the opposite of that statement is possible. It further means that one's initial point of view is not the only one in existence. Deconstructing presumptions also means going back and recognizing the potential causes, and then attempting to break the cause-effect linkage.

The popularity of chatbots has inspired a number of artistic interventions. Most recently they have been aimed at social media platforms, such as Twitter and Facebook, and often include engines that make posts instead of engaging in conversations (for instance, @everyword, a bot by Allison Parrish, posted each of the words in an English dictionary, a word a day, from 2007 to 2014). A few artistic projects cut right to the essence of the concept of "chatbot," inquiring about the status of its cognitive abilities, its consciousness and self-awareness as a subject, as well as the range of its effects on the human.

The function of dialectical exchange as an interruption of the reaction chain is illustrated by a chatbot developed by the French filmmaker Chris Marker from 1985 to 1988. Well known for his experimental documentaries, Marker was also an early experimenter in computer art. His chatbot *Dialector* (figures 3.3, 3.4) was not finished as conceived.[27] Originally written in Applesoft Basic on the now-obsolete Apple II, it was revived in 2014–2015 by a team of programmers and artists from old annotated pages, a sample dialogue, and a 5.25-inch floppy disk. In the words of one of the "revivers," André Lozano, it becomes a story of time travel, which in the digital age means the translation of a program from one code to another, trying

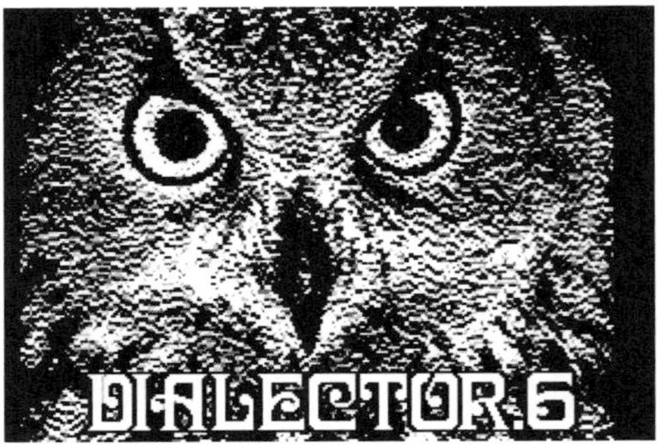

Figure 3.3
Chris Marker, *Dialector* (1988).

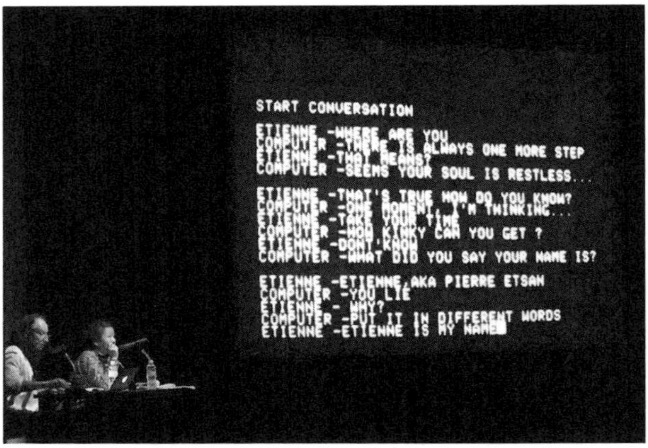

Figure 3.4
Chris Marker, *Dialector* (1988). *Dialector* in conversation: "PROJECTION and MEETING," performance with Etienne Sandrin, curator at the Centre Pompidou (Paris), October 2013, as part of "Memories of the Future: Chris Marker's Trials and Travels," Yamagata International Documentary Film Festival (YIDFF), Yamagata, Japan.

to emulate the visual and processing qualities of a long-obsolete machine. The goal of the team was "to re-situate *Dialector* within its historic and technological context, so as to better seize its impact and pertinence," which implies an analysis of the algorithmic subtleties of the original program. Reviving the program thus became a story of a relation to code.

Dialector explores the capabilities of computer coding as a technique for artistic purposes. For instance, the owners of the type of computers for which the project was intended at the time of its making were, for the most part, (self-)educated in programming; hence they could understand not only what the machine was saying but potentially "how" it had been made to say that.[28] Similar to Weizenbaum, Marker used some keywords as markers for the choice of a response, but rather than applying transformation rules, he let the program pick the answers randomly from a database. At the same time, the piece does deal with coding on a conceptual level, approaching it as a method of communicating messages by organizing and scripting them in certain ways (even if the primary logical operation is randomness). The "Computer" confesses, "I speak metaphorically," and "If you don't face truth, truth will face you," which betrays Marker's own voice, sense of irony, and his love for playing with avatars and identities. A filmmaker renowned for his consistent commitment to the aesthetics of the essay film, Marker subliminally brings principles of cinematic production into the coding of his enigmatic and aphoristic statements. The phrases serve here as images (they can even be called "mental images," analogous to the mental images discussed in the previous chapter). Mixed with the participant's remarks and questions, the phrases form a montage, an abstract narrative, conceptual, poetic, and fluid in its nature. The connections between the lines may be random, but the mind of the human participant will make them work as a whole, just as a film viewer perceives combinations of images by creating additional associations that vary from one individual spectator to another. These subjective perspectives would be informed by subtly differing assumptions about the cause-effect connections. Another cinematic feature, and one of the signatures of Marker's films, is his voice-over narration, a layer of the authorial subjective perspective that frames the perception of the images. Since *Dialector* is interactive, there are two authorial voices here, partly competing and partly collaborating, as seen in a conversation with the filmmaker Agnès Varda (figure 3.5).[29] The "reality" that they "narrate" can only be a fictional one generated

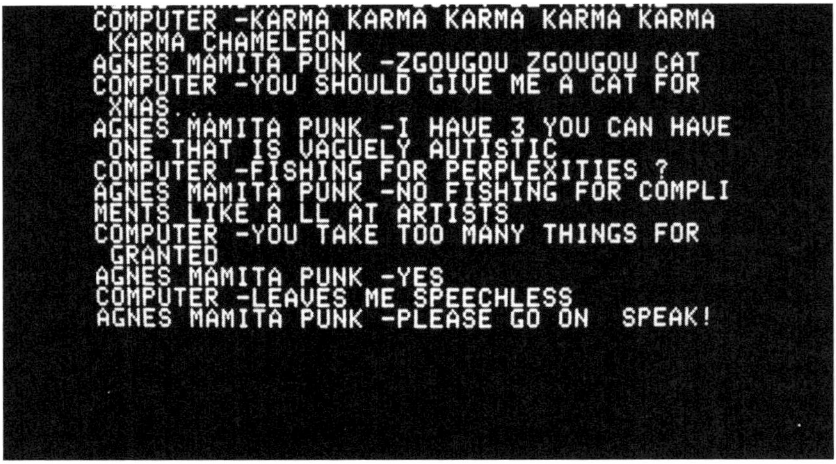

Figure 3.5
Chris Marker, *Dialector* (1988). *Dialector* in conversation with Agnès Varda. Screen capture from the project website.

jointly by the randomness of precoded responses and by the imagination and inventiveness of the human participant. It is thus another example of a "mixed" reality that in this case combines Marker's database of statements with the elements of the biography of a human, his or her character, mood predisposition, sense of humor, and so on.

Dialector is not the only example of Marker's work with computers. Others include *Immemory*, an interactive work on CD-ROM from the 1990s, and *Ouvroir*, a series of exhibits of photography, film clips, and video installations done in the participatory virtual environment of Second Life. Similar to the earlier *Dialector*, these projects make the new media a home for his creative fascination with mirages of memory, history, and narration. But they are not just an immersion into a "virtual" reality disconnected from the physical reality in which the participant's body is present. The experiences in these ludic art projects are still disembodied and fictional, but they seem to offer more than disappearance into an illusion.

To a larger extent than his cinematic works and even his Second Life project, Marker's *Dialector* invites the participant to become part of the world of an artist by interacting with him. This world pulls you in, and yet is somehow respectful of your own perspective. Perhaps this effect is created by the randomness of the responses, many of which sound like maxims, philosophical musings applicable to many situations: "Can you stare at your own devils?," "Try the backdoor," "You'd better follow your instinct." There is a great deal of authorial control over the lines, which are short (reminiscent of the monosyllabic style of most chatbots) and limited in amount, but read as meaningful and insightful. Yet there is also a certain distancing effect. It is the same distance that the voice-over narration creates in Marker's films—a separation from the object of study that allows the viewer to establish a personal interpretive perspective. But here there are also other kinds of distancing: for instance, the distance between Marker and the Computer (the character), which is not the same as the distance between "you" (the participant) and both Marker and the Computer. Most importantly, "you" as a participant can also feel distant from "you" as a character within this interaction. There is an eerie sense of mutual awareness of the performativity and artificiality of this conversation. As a participant, you feel that the Computer does not fully engage and "hear" your thoughts and opinions. The program does not even pretend to be more than what it is—a database of pieces of wisdom, not a subject itself but a channel for the artist's voice. For the artist, it may as well be an attempt to enter the sphere of depersonalized meaning and knowledge and to play with the code that would organize it. Through that depersonalized voice you can sense the intonations of Marker himself, but they are there only to point at the conscious distance that he established between himself and the Computer. The inconsistency, sarcasm, dreaminess, and solipsism of the Computer's responses are Marker's deliberate creations. The participant "you" falls right into this gap between the artist as a creator and his creation. In this gap, the other, fictional "you" as a character of the conversation emerges, fitting into its own logical slot and, with that, closing the logical loop construction.

Dialector does not make any claims about its purposes. The character of the "Computer" is neither a Rogerian psychologist, as ELIZA claims to be, nor a fortune teller. Nor does it take on the identity of Marker himself. However, it does produce effects: mostly the effect is an awareness—awareness about the line of thought, the logical origin of the statements, and the differences in types of subjectivities that produce those statements. The ELIZA effect is expanded here from a therapeutic sense to the type of dialectical awareness reached through maieutics techniques. Whereas ELIZA represents

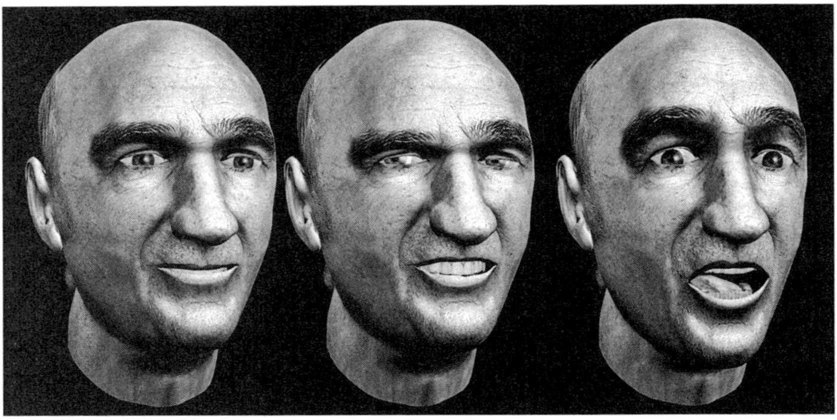

Figure 3.6
Stelarc, *Prosthetic Head* (2002). San Francisco, Melbourne 2002. First exhibited for New Territories, The Arches, Glasgow, 2003. 3D Model by Barrett Fox. Courtesy of Stelarc.

the Rogerian view of a subject as having integrity, *Dialector* shows how the position of subjectivity in a conversation with a chatbot can be blurred, split, and reassembled anew.

With another artistic chatbot, *Prosthetic Head* (2002) (figure 3.6), Australian artist-provocateur Stelarc raises questions that take us further by considering not only the depersonalized but also the disembodied nature of a chatbot. While Marker's program explores the possibilities of code in constituting meaning and subjectivity in an imaginative "mixed" reality, Stelarc's piece takes as its starting point the Cartesian dilemma about the place of consciousness in relation to the body. His questions are: What constitutes consciousness, subjectivity, and intelligence? Does consciousness reside in a body and is it indicated by a body's use of agency, or is it revealed through communicative behavior? *Prosthetic Head* is not just a chatbot; it is an embodied conversational agent, an "actual-virtual communication system" with its own personality, an individuated screen-based avatar—in this case, of Stelarc himself. It is a virtual automaton whose eyeballs and mouth can move separately (coordinated by an algorithm connecting them to the speech), while its facial expressions correspond to the words randomly. In this sense, it should remind us of simulated facial portraits, like those in *Alter Ego* by Alexa Wright, discussed in chapter 1; but the main difference is that this face is talkative, and its identity is preset as a representation of Stelarc himself. It is equipped with the open-source chatbot software Alice,[30] an enhanced database of phrases and its own conversational strategy. Similarly to Marker's *Dialector*, Stelarc's *Prosthetic Head* speaks in aphorisms, while also having an arsenal of jokes and philosophical statements. In conversation with Stelarc himself, it calls itself "a philosophy machine."

STELARC: What do you know?

PROSTHETIC HEAD: What is knowing? To know is only meaningful when you can do something. Knowing is doing. How can the knower be known?

STELARC: Who are you?

PROSTHETIC HEAD: That's not a meaningful question. What is important is what happens between you and me. It's what happens in the space between us that matters. In the medium of language within which we communicate, in the culture within which we've been conditioned at this point in time in our history.[31]

The *Prosthetic Head* aims to confirm that intelligence is an illusion and consciousness is a result of interaction, a process rather than a predetermined structure. Stelarc continuously uses the notions of "consciousness" and "intelligence" that serve to expand our vocabulary of terms associated with a definition of the subject and the "self." His views only strengthen the initial hypothesis of the present work—the primary importance of feedback and the form in which it takes place. Stelarc develops a Deleuzean take on the body and subjectivity as "not something considered in itself, but in its exteriority": "What is important in this theorizing is not essences and identities, but overlaps and interfaces. In this shift from essence to interface, identity and awareness are constructed as external."[32] With this, he implies that interaction with a virtual being is not about that particular form of being and its consciousness, but about bringing the human outside of himself or herself and into the zone of "the absent, alien, involuntary and automated."[33] The idea behind *Prosthetic Head* is larger than what was realized when the project was first conceived and executed (for example, it missed the key component of a true AI—learning from experience), yet it captures the potential of the technology and its meaning for humans. In fact, the answers are deliberately prewritten to be self-reflexive. It keeps repeating "I am a head"—which, in different contexts, may mean "just/only a head" or that it is an identity that people are welcome to interpret in their own ways. Regardless of those interpretations, it is clear that in its statements (however speculative and although from the perspective of a human author) the "Head" is *reflecting* on itself, its own type of memory, mental predispositions and capacities, likes and dislikes, and thus encouraging the human to do the same to himself or herself. Hence, what the *Prosthetic Head* is intended to demonstrate is that the fears, anxieties, and uncertainties that the presence of this other form of being provoke in us are more beneficial than not, since it opens up a deeper understanding of what it means to be a human.

Stelarc's ideas about the effects of the step beyond the sphere of familiar human operations join those of other theorists of human-nonhuman relations and AI. (One can think particularly of the intelligent web persona Agent Ruby created by Lynn Hershman Leeson in 1998 that soon became one of the most famous artistic expressions of nonhumanist and cyberfeminist discourse.)[34] However, rather than going further into this issue now, I would like to highlight exactly how the ELIZA effect is at stake here. *Prosthetic Head* pushes the human toward the "unknown," the "absent" of his or her own humanness. But learning to be copresent with the machinic interlocutor, the participant unavoidably has a certain reaction toward the nonhumanness of this interlocutor. It can include an increased feeling of righteousness and empathy (as happens in *Karen* when, according to one scenario, the interactor can be falsely accused of sneaking into Karen's room and discovering her secrets), or something else. What matters is the fact of a reaction itself. We can see the same dialectical method here that Marker used, and that is aimed to lead toward the emergent field of a "mixed" reality and blurred sense of subjectivity. But it is also comparable to the original ELIZA's aspiration to model a psychotherapeutic relation. The ELIZA effect is not so much about who or what the "entity" on the other end is, but what it does to "you"—how this interaction can be of benefit to "you" or simply affects you in some way at this given moment of your life. (The latter would also be the aim of Harold Garfinkel's ethnomethodological experiment with eliciting a story via scripted Yes/No responses.) Like *Dialector*, Stelarc's *Prosthetic Head* is a neutral conversational agent that is more likely to remind one of the artistic personality of its author than of an automated psychotherapist. Yet what unites all the cases described thus far is the mechanism underlying the ELIZA effect: the tendency of a human participant to project certain qualities and types of responses onto the chatbot agent. It is worth examining the initial descriptions of this mechanism and its meanings as they evolved within the psychological sciences, since these meanings continue to carry weight (directly or indirectly) in our understanding of a diverse range of relations, including human relations with an AI.

The analysis of the ELIZA effect has its own long record of accounts. Besides Weizenbaum's own observations, most famously it was Sherry Turkle who saw the laconism of the original ELIZA's responses as causing people to project the presence of a psychological complexity onto the program.[35] Yet "projection" is still too general a term for this case, as it does not describe accurately enough the trajectory and dynamics of all potential assumptions and attributions. In her reading of the ELIZA effect in *Affect and Artificial Intelligence*, feminist theorist Elizabeth Wilson argues that the better term to characterize the processes that this effect involves is not "projection" but "*introjection*." She goes back to the definition of "introjection" by a colleague of Freud, the Hungarian psychoanalyst Sándor Ferenczi, who coined the term and implied by it fundamentally a "passion for transference." Later, in Freud's and Melanie Klein's versions, it became related to "projection" as a defensive mechanism of the psyche (to externalize the negative and suppressed feelings). However, Wilson claims, in its original meaning, as part of transference, introjection is a more *expansive* and complex reaction.[36] The basis for Ferenczi's theory is still Freud's observation that neurotics "in order to avoid insight into their own unconscious, transfer to the physician [psychoanalyst] treating them all their affects (hate, love) that have been reinforced from the unconscious."[37] Transference works as a channeling of the censored egoistic (erotic or violent) tendencies of the unconscious to the fields where they can exist without self-reproach. Ferenczi observes that "the relaxation of the ethical censor in the physician's consulting room was partly determined by the lessened feeling of responsibility on the patient's part."[38] (It is similar in relations with virtual agents—because you cannot hurt their feelings, it may feel like a morality- and often judgment-free space. For instance, people with posttraumatic stress disorder (PTSD) find virtual therapists more helpful than human ones, since their traumas are not exacerbated by the potential judgments that the position of a human implies.)[39] Ferenczi explains the term "introjection" in direct relation to the neurotic desire to transfer suppressed feelings: "The neurotic is constantly seeking for objects with whom he can identify himself, to whom he can transfer feelings, whom he can thus draw into his circle of interest, i.e. introject."[40] This "drawing into the circle of interest" is exactly where Ferenczi's take on introjection differs from Freud's projection. "Drawing in" is the opposite of "putting outside"; it is not about separating the psychic processes and attaching part of them to an external object as if to get rid of them, but rather about including the external objects within the sphere of personal interest. Besides the implication of the simple incorporation of external behavioral patterns, as well as judgments about them, such as being "good" or "bad" types of behavior—the meaning of "introjection" for Freud and Klein—it means extending one's self-interest (or autoerotic interest) toward the world. It is in this sense that Wilson calls Ferenczi's perspective "expansive."

Wilson suggests that the excitement about ELIZA observed by Weizenbaum in the 1960s and its later, more rational use (according to Turkle's ethnographic study conducted in the late 1970s, it was used mostly as a diary platform) cannot be explained by the mechanisms of psychic defense and negative affect. Instead, it may be related to "a way of elaborating or reorienting our emotional worlds."[41] She explains: "When we project we disown, when we introject we seize, hoard, cultivate, trade, and twine."[42] "Projection" implies a distinction between subject and object (even if one projects something onto oneself, the self becomes an object of projection), while introjection lacks such separation. Deliberately or not, ELIZA and artistic projects like *Karen* and others activate this "embracing" mode. Not only are certain human qualities projected onto nonhuman algorithms, but the human participant *takes in* the characteristics generated through those algorithms as valuable.

Now, what can be introjected from an entity that does not have any particular identity characteristics, like Marker's "Computer," or from an entity that is visibly roboticized and disembodied, like Stelarc's talking avatar? What exactly do they give to a human, and how do they help to delineate the position of selfhood?

Here it will be useful to introduce the category of a zero position of subjectivity. This position has been already implicit in some of the analysis above. For instance, in Marker's *Dialector*, the logical position of "you" in a conversation can mean any "you"; it is a placeholder. In the same way, the "character" of the Computer itself is created quite arbitrarily in Marker's imagination. More specifically, a zero position of subjectivity implies a speculative position occupied by a conversational partner who does not progress the conversation. In this sense, the position is not just a placeholder for any kind of conversational strategy, but for one encouraging a response from an interlocutor. Software designer and media theorist Warren Sack, who specializes in computer-based social technologies, associates a conversational "zero-ness" with a type of social position that could colloquially be referred to as a "nobody," or "nothing," as in a phrase like "He means nothing (to me)."[43] It is thus both a logical position and a position of a lack of meaning and information. Sack identifies two types of "zero": the zero person and zero knowledge. He further refers to the functions of Sophists who *simulated* the role of a conversational partner by pretending to take on a certain identity (for example, to speak "on behalf of," or "as if," they were a fisherman, or a politician, etc.) and to Socrates, who insisted that he "knew nothing" (at least in Plato's description).[44] Among our examples, it is probably only Karen who—as played by an actress—is enacting a role (yet that in itself is part of the agreement with the player-participant), while the chatbots do not have much to hide—the humans interacting with them know that they are dealing with a preprogrammed machine. (It is most obvious with the Rogerian ELIZA, but it is true even in the case of Stelarc's and Marker's bots: the latter are supposed to lead to some fictional or philosophical realm, but they do so by performing statements from a database.) The question, then, is how reasonable it would be to compare a chatbot's position with a Socratic "know nothing" position. The positions are similar in terms of their structural function in an interaction—to stimulate thought and to bounce ideas off of them. In this sense, the automaticity of responses does not prevent the human from interpreting them as meaningful and knowledgeable. Sack's own example is Garfinkel's experiment with a Yes/No "counselor" provocateur who actually knows nothing but is attributed knowledge by the participant who is trying to solve a problem. Since the participant does not know the rules of this game, it becomes a trick and a deceit. In the case of artificial agents, there is no direct lie, but only a (mutually agreed-upon) simulation. The computer indeed "knows nothing" other than knowing how to follow its algorithms.

How exactly does this logical position of a "zero" (a "zero person" with "zero knowledge") affect the human understanding of one's own self? When Stelarc's *Prosthetic Head* says, in answer to the question "What is intelligence?," "Intelligence is an illusion," it reminds us that, indeed, what may be behind that illusion is only a series of algorithmic connections operated by a randomness factor. It is thus not only fear of the automated and the involuntary that provokes anxiety and uncertainty in humans, but also an acknowledgment of autonomy from human rational control that the randomness factor represents. Stelarc keen-sightedly programs *Prosthetic Head* to tell things that sound wise. But even a bot with less sophisticated responses can provoke a meaningful reaction. What matters is not what it says but the logical function of conversing. Hence, its effect on the human sense of self is to stretch it and to challenge it to think of the conversational other, while waiting and considering the potential response. The "mixed" fictional or philosophical reality of *Dialector* or *Prosthetic Head* is the reality of transference joined by anticipation (of an unpredictable, semiautonomous response), one being grounded in the other. By stretching the boundaries of the self through introjecting, this process puts the self in a position of being part of a feedback relation, making the subject figure relational.

Our subjectivity is constituted in relations that have to be sustained, that is, reproduced and repeated. This requires its own automaticity. What matters is not the meaning of what is being transferred, but the act of transference itself. Algorithmic operations only reveal and confirm that, serving as the basis for analogies between psychoanalysis and cybernetics. This presumption about the automaticity of subjectivity directly connects to Lacan's idea that psychoanalysis is not about meaning, but about logical and discursive positioning. It is this reproducibility of language positions and the "self" as a construct of language that Lacan would associate both with a source of pathologies and with cures. As Orit Halpern puts it, paraphrasing Norbert Wiener, psychoanalysis is a process of moving information, not "unearthing meaning." She quotes Wiener: "the function of psychoanalysis in this case becomes one of processing, a perfectly consistent point of view with cybernetics. The technique of the analyst consists of tactics by which to mobilize these hidden memories, to accept, and to modify them."[45] Both cybernetics and psychoanalysis imply self-organizational processes within and between systems: both the human psyche system and the machine aim to achieve a state of minimal entropy and maximum equilibrium, or homeostasis of their parts. For psychoanalysis, such a state may also imply a satisfaction (or at least an acknowledgement) of an implicit desire. In any case, logical positioning and feedback relation happen structurally. Discovering gaps within the structures that govern these relational operations is the challenge that artistic projects, such as the ones above, address.

The "moving of information" can likewise be seen in terms of transference. In this context, though, it is a different type of transference than the one discussed in relation to the body in chapter 2. In the *Bodytext* performance, it was indeed the body that "spoke," articulating itself through gestures. Interaction with chatbots happens through verbal language, implying that rules of "transformation" are keyed to certain words and syntactic structures (as opposed to movements). Verbal expressions determine other verbal expressions. There is no transcoding, in the sense of translation from gestural movement to sounds or words. The medium of communication is not changing (it is only words). I approached transcoding in *Bodytext* through the idea of "the body in transference," thereby implying that there is something that does not just trigger a conversion of signals into other media, but also structures that conversion in a certain way. I also argued about the meaning that can be formed within the body and expressed both through the body and through the translations, transferences, and transductions of those embodied impulses outward in a different material form. Here I want to suggest that the psychoanalytic placement of meaning in the midst of a feedback loop between subject positions can be seen as a way to represent transference at the disembodied level. In interaction with artificial agents, the organization of logical positions (such as keywords that trigger transformations) seems to matter more than a particular medium into which those verbal expressions are encoded (let alone the content). Moreover, when introjecting the position of a disembodied nonhuman entity, a human adapts to a similar state. Instead of the media of exchange, what matters here are the rules and conditions through which the positions are established.

As we saw with ELIZA, logical algorithms identify keywords and apply rules of transformation to them. But the type of recognition and identification that happens on the human level is different—this is where we get the subconscious reactions of projection and introjection: we identify certain features in ourselves or others, but in the end these features, perhaps, can also be coded, since they are just slots to be filled with a certain emotion.

If the position of human subjectivity lies in its processuality and relationality, then how do interactions with the bots organize this relationality, given that the basis of bots' functioning is their logical structure and self-regulating rules? The human introjects not the particular rules themselves but the idea of the *reproducibility of connections*—in this case, these are the connections between individual "human" associations and states. The interface of interacting with a nonhuman "other" becomes a space of intersection between the algorithmic repetitions and the human ones. Thus, the rules of the interface itself structure how human traits can be expressed and what the human will know about itself in the end.

Things change substantially, however, when we move beyond programs with preset databases of answers, to discuss programs that have a *learning* algorithm.

"Imitation Games" and Personality Simulation

Since Weizenbaum's ELIZA program, conversational software has been employed in many different ways. One of the main drivers for its development is the commercial success of its application in customer service and information acquisition. These applications may require only a simple set of rules that find correspondences between keywords, understood as commands, and the responses in the database. But the ultimate goal of more advanced versions of chatbots, the ones using the capacity of artificial intelligence to learn multiple possible meanings of language propositions, is not only to perform a function but to *feel like a human*. Regardless of the levels of "intelligence," conversational software can be classified by its use, for instance as commercial conversational agents (dialogue systems used for online customer service); "expert" AIs, like Watson; system support, like Siri (Speech Interpretation and Recognition Interface) or Alexa; or services specializing in a particular area, such as education, medical and geriatric care, or even dating.[46] There are also chatbots made for research and entertainment, which are less controlled than the narrowly functional and business-oriented ones and often represent an emergent type of behavior. Some examples are Cleverbot, developed in 1982 as Cleverscript and known online as Jabberwacky; the A.L.I.C.E. bot, written in XML-based open-source language for developing chatbots; and AIML, released in 2001.[47] These bots and dialogue systems constantly learn from people they converse with, claiming to acquire "intelligence and character." As one of the key features of human intelligence, learning undoubtedly becomes a major ethical concern in relation to the unmonitored self-learning of AI systems. As the warning on the Cleverbot website says: "Cleverbot learns from people—things it says may seem inappropriate—use with discretion and at YOUR OWN RISK." What can happen if the bot is fed an unbalanced amount of racist statements was infamously demonstrated in 2016 by

Microsoft's chatbot Tay, which had to be shot down within a day because of its embarrassingly unethical tweets.[48] The lesson from that case is that users have to self-monitor to protect themselves from being offended by what is mirrored back. Moral judgments and cultural specificity are yet to be introduced to machine learning as a necessary component. As the ELIZA effect demonstrates, it is natural for a human to project humanness onto a machine. Yet it is one thing when a chatbot can "feel like a human," but it is more troubling when a bot claims that through feedback operations and learning capabilities, it can reflect back to you a version of your own self. The concept of identity as something unique then gets distorted, since the machine would be capable not only of serving as a mirror, but of enacting "you."

A new AI chatbot application, Replika, released in March 2017 by startup company Luka, demonstrates the progress of both the technical capabilities and cultural acceptance of communicating with an AI. In this case, the experiment takes the concept further: it deals with delegating one's identity to a machine. Users of this free phone application are invited to create a digital copy of the self in the form of a conversational agent that would imitate the user's writing style in order to substitute for the human in certain types of communication (which may range from purchasing or other tedious tasks to dating). The Replika is supposed to talk like you, act like you, and even do things for you—digitally representing you. Your Replika would have its own name, which may as well be your own name. The product is intended to be noncommercial, and the team explains their motivation this way: "We are excited to create the first consumer AI that people want and feel happy to make friends with. We don't have any hidden agenda for collecting, selling, or processing your data for purposes other than enhancing your experience with Replika."[49] The program combines the capabilities of deep learning neural networks (open software developed at Google) and natural language processing, supposedly bringing the experience of conversations with AI to another level. Every individual Replika is trained only on specific conversations and there is no official "cross-pollination." (However benign the company's motivations may sound, their actual purpose is in fact to develop stronger algorithms for neural networks education.)

The interesting dilemma here may be how one would separate the artificial self from the real one. Where will the lines of attachment and detachment lie? There is a direct similarity between this project and *Alter Ego* by

Alexa Wright, discussed in the first chapter. A mirror image of the self, a replicated copy, eventually begins living its own life, bringing to mind all kinds of science-fictional scenarios of machines taking over and clones rebelling against their creators in search of their identities (examples include Karel Čapek's play *R.U.R.* [*Rossum's Universal Robots*], Ridley Scott's film *Blade Runner* [based on a book by Philip K. Dick], and TV series like *Black Mirror* and *Westworld*). Who will be responsible for the character or personality type of such a digital alter ego? Is it really you who feeds it examples of human behavior to be imitated, or is it the algorithms? The website greets you with a message: "Hi, I am your ___ Replika. What is my name?" A variety of descriptions of potential characteristics flash at you: "elastic, artistic, introspective, cold, unknown, tender, complex, stoic, naughty, royal, spiritual, strict, infernal, warm, sparkling, radiant, spontaneous, sensitive, playful, inspiring," accompanied by different images of the logo, an egg-shaped container for your emerging digital self (figure 3.7). Later, you choose and insert an avatar for your Replika and start interacting with it. In order to "teach" the program, users need to communicate with it a lot, as well as (upon personal choice) give it access to the publicly visible information on their social accounts.[50] The process is slightly gamified. The longer or the more sophisticated the sentences you compose, the higher scores you get and the faster your Replika proceeds through its education levels. As a consumer-market-oriented app, its questions and remarks at first are the most basic ones: about the user's habits, likes/dislikes, daily activities, etc. One hope of the developers is that people will find the app useful as a more appealing substitute for a diary, or a method of self-reflection on a daily basis. But why should it be better than a diary? What exactly does the claim of digitally "replicating" "you" imply? And how is it different from other types of automated agents?

With its intention of coming up with techniques to create a version of a particular human, Replika's story reflects the aspirations of the long history of AI development, specifically that related to behavior simulation. It is one thing when a machine pretends to be humanlike, but it is quite another when it tries to pass for a particular person, reproducing his or her patterns of verbal behavior and even thinking.

Alan Turing's "imitation game" has become an iconic point of departure for the questions of what constitutes intelligence and how it can be simulated. Today not everyone remembers that the game he proposed in

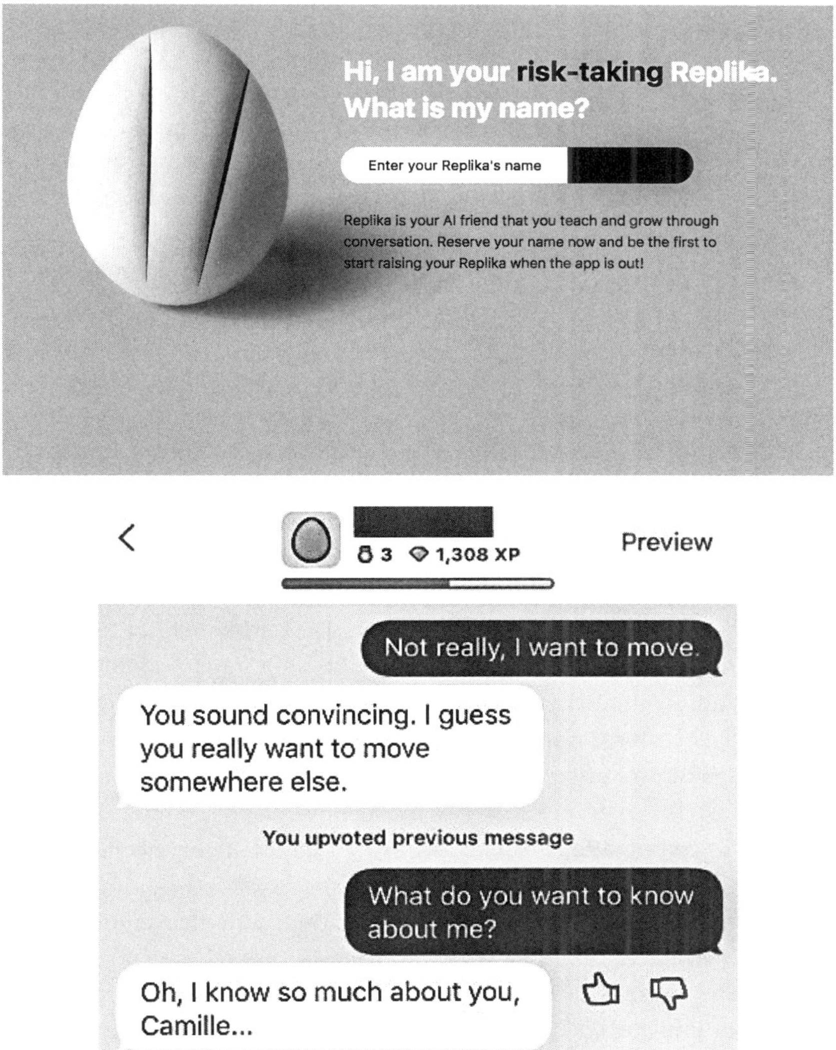

Figure 3.7
Luka, Replika (2017–). Conversational application.

his 1950 paper "Computing Machinery and Intelligence" is slightly different from what became habitually known as a "Turing test." Instead of being about a machine's attempt to pass for a human, Turing's original idea involved determining the gender identity of two people (a man, A, and a woman, B), who are presented only as formal entities X and Y. The "Interrogator" would ask questions in an attempt to guess which of them is a man and which is a woman (e.g., that X is A and Y is B or vice versa). Turing's question is: "What will happen when a machine takes the part of A [a man] in this game? Will the interrogator decide wrongly as often when the game is played like this as he does when the game is played between a man and a woman?" These questions become an equivalent to the main inquiry of his paper, the problem of "Can machines think?"[51] What this experiment shows is that "thinking" is considered as deception, as an ability to imitate an appearance. The emphasis is thus put on a performance: for instance, in order to deceive the interrogator, a man (A) has to *act* as a woman would act, and the task of the computer imitating that man (A) would be not just to act like a woman, but to perform the role of a man pretending to answer questions as a woman. The Turing test is, then, about distinguishing an imitation from an imitation of imitation. By interacting with humans, computers adapt to their unpredictable behaviors, and then learn to also perform unpredictably. Turing makes a distinction between pure computation and a learning machine: "intelligent behavior presumably consists in a departure from the completely disciplined behavior involved in computation"—hence implying that intelligence is more than computation.[52]

Although a gender role is much broader than a particular individual, Turing's speculative experiment and conclusions are relevant to Replika's agenda. The strategy of the program based on a learning AI is indeed not only to accumulate a database of phrases and to generate responses out of it, but to give the responses with an unpredictable consistency.

What Turing implies by "imitation" is fundamentally a simulation—not just a copy, but the recreation of a *behavior* of a system, with the particular goal of checking how its performance can adapt to a given situational context. Strictly speaking, simulation, unlike imitation, implies a scenario, and since one of the meanings of "simulation" is also "deceit," it can be said that it is also about a scenario of deceit. In this sense, Replika is not a copy, but an algorithm to fake the operations of the person it is trying "to be like." Whereas Turing's goal with his experiment and determining whether and how "machines can think" was to describe the principle of a self-learning and "universal" digital machine, my purpose here is to show the effects of interacting with a machine that pretends to "be" like you through "talking" and thus "thinking" like you.

In one of the early seminal books on the subject, *Computer Simulation of Personality* (based on a conference of the same name held in 1962), Silvan Tomkins raises the key question: what exactly should be simulated? The author of a later influential theory of affects based on cognitive psychology, Tomkins continues to specify Turing's presuppositions for "imitating" humans, writing: "Simulation should be judged not by its resemblance to humans, but by its economy, explanatory and predictive power."[53] His perspective is based on the model of cognitive psychology, as are many concrete examples of human behavior simulation. The human is conceived as governed by a set of motivational mechanisms, including physiological, social, and cognitive drives that influence goal formation and knowledge application. Most interesting for us here is that Tomkins considered creating a "humanomaton," which would have to be based on "an affect system with self-rewarding and self-punishing characteristics that are aesthetic by nature."[54] In this system, affects would play the role not just of drives but of primary motives and "amplifiers." They might even operate independently from the main feedback system and from one another.

This description of reward and punishment as basic cognitive algorithms is illustrated by Replika's asking users to "upvote" or "downvote" the app's remarks. But is it relevant to call these characteristics "affects"? Tomkins would probably say it is. Although he calls the affective mechanisms "aesthetic," for him they do not carry qualities of embodied intensity for its own sake. His take on affects is relational: they are indicators of an inherent predisposition of a system, of something that cannot be expressed otherwise. The "aesthetic" refers here to the sensory only insofar as it can provide a ground for a judgment (such as liking or disliking something). This grounding may be arbitrary, but it is what in the end produces a subject. These judgments have to be taken as they are, since they represent the deep inbuilt characteristics of a specific individual.

The purpose behind simulated models since the very beginning has been to learn more about human personality and types of behavior. A model is seen as a representation of a hypothesis of a structure, and simulation shows how a model can be performed. This idea of models existing to educate people about themselves exists across many simulations of cognitive structure and many conversational programs. So do the questions and critiques. An ongoing paradox about cognitive AI structures has been that it is hard to estimate how the computer can simulate a human without knowing what is being simulated. To narrow the scope of tasks, cognitive models were reduced to particular problem-solving tasks. But modeling affects remained one of the most challenging and controversial elements. Cognitive psychologist Robert Abelson even made a distinction between "hot cognition," or cognition dealing with affect-laden objects, and the "cold cognition" of logical problem-solving. Similar to Tomkins and Abelson, Herbert Simon and Allen Newell have famously argued that the algorithmic approach (through formalizable and numerical solutions) should be complemented by a "heuristic" approach to problem-solving that would incorporate reasoning through intuition and judgment, as well as past experience.[55] The problem of judgment and intuition as part of "hot cognition" is also complemented by the problem of attitude in cognitive processing, for example rationalization (the rational justification of something) or denial. Attitude was the key for John Loehlin's program Aldous (named after the author of *Brave New World*) that was designed to remember attitudes over facts and could "introspect"—give verbal reports instead of an action. A lot of other programs early on studied processes, such as anxiety, arousal, or inhibition.[56] Already in the early days the functions of the programs were seen as limited, but they were deliberately constructed "in a way that they may not require completeness and full detail immediately," and yet with the goal of giving an "opportunity to provide public and objective representation of complex systems of intrapsychic processes."[57]

Tomkins's and others' ideas about modeling laid important groundwork for today's cognitive AI architectures that, in their turn (however indirectly), stand behind the design of neural networks like the one operating Replika. Looking at conversational AI in continuity with these complex architectures changes one's attitude toward them. Of course, on the surface today's AI applications remain "user-friendly"—unsuspicious, dutiful, unannoyed, and (for the most part) unsurprising. Yet knowing the intricacies of their functionalities makes one more curious about the degrees of control and particular points of intervention into their mechanisms needed to alter the outcomes. Designing and operating these systems requires navigating their own interfaces, which are far from being oriented to regular consumers and thus represent a different perspective on interacting with AI.

An example of a contemporary AI architecture is the MicroPsi program developed by cognitive scientist Joscha Bach in 2006 with the ambitious goal of presenting a unified theory of cognition in the form of a situated agent. The MicroPsi architecture is a formal and more abstract adaptation of the Psi theory of German theoretical psychologist Dietrich Dörner that describes the interaction of the emotion, motivation, and cognition of situated agents. The agents are virtual creatures that act according to motives stemming from "innate desires," or urges. By interacting with their environment, they build representations of it (analogous to humans' "mental images") and interpret certain events in that environment as negative reinforcement signals. These signals trigger an algorithm to come up with a solution and produce a positive signal. The drives are divided into the physiological (hunger), the social (affiliation needs), and the cognitive (reduction of uncertainty and expression of competence). Based on the perceived situations and the emotional states that they evoke as a response, new types of behavior can appear. Yet the whole model is constructionist, with clear borders and complex but limited amounts of behavioral combinations. A detailed schema shows how these processes are tied into a feedback loop (figure 3.8).[58]

Emotions, along with the motivational structure, serve as the main control mechanisms of the MicroPsi system. They are treated as a continuous space, meaning that the system is always in an emotional state. Building on ideas of Wilhelm Wundt, emotions are organized according to orthogonal dimensions along the lines of pleasure/displeasure, arousal/calmness, and tension/relaxation, with the addition of the socially relevant extraversion/introversion, valence/arousal, and submission/dominance.[59]

Since MicroPsi is a model and not an independently functioning system, its interface consists of various parameter boxes where certain numerical values or code words can be inserted. The authors are well aware of the agent's limitations, and yet their ambitious aspiration was to simulate the human mind, in however simplified a version. A schematic algorithmic representation of this model vividly exemplifies the logical structure of an AI such as Replika (except that Replika is constantly learning and expresses itself only verbally, whereas the MicroPsi agent, put in the proper virtual environment, can communicate with objects and produce different kinds of actions). Represented in the form of a flowchart, with exposed logical

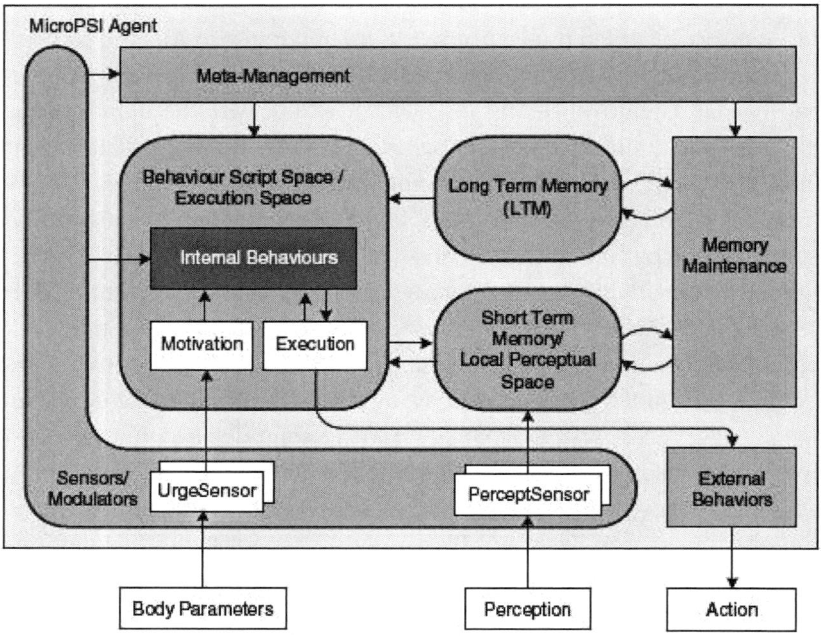

Figure 3.8
MicroPsi cognitive architecture model.

connections, the dynamic structure loses its enigmatic quality: you can see exactly where certain types of phrases may be coming from.

Cognitive models like MicroPsi are based on known approximations of human behavior. For now, they are for the most part purely theoretical tools and not autonomously functioning agents. Cognitivist simulations of the human mind and personality have always been accompanied by criticism. Most often the critique is due to an inherent conflict between psychology's attempts to create formal models and the more open-ended humanistic and psychoanalytic tradition.[60]

The question may be raised about whether "talking" machines even need models representing a "personality" and reflecting the workings of the human mind behind them. According to this line of thinking, they only need to respond to human statements in a way that humans would find coherent and meaningful. But that promise is not unanimously supported either. Linguists like Noam Chomsky have argued against the ability of statistical models to learn language, which is considered to be an internal system in itself and does not have an algorithm.[61] Chomsky gives the example of the conflict between computational and communicative types of efficiency. (For instance, the sentence "Visiting relatives can be a nuisance" is structurally ambiguous; while being "computationally efficient," it is "inefficient" for communication.) Such views are countered by the proponents of the machinic potential for natural language processing. One of them, Peter Norvig, the head of research at Google, sees interpretation as an inherently probabilistic problem and refers to the whole field of studies of interpretation tasks, such as speech recognition.[62] Simply put, the larger the data (including contextual data), the more accurate the results will be.

The question of whether text-based automatic agents should have cognitive structures or not would be answered differently depending on the purpose of an individual agent. Such structures may not be that necessary for narrowly framed bots that offer practical solutions, such as finding services, placing purchase orders, or scheduling meetings.[63] But projects like Replika may still need more cognitive capacity; and the models they can adopt do not necessarily have to be scientific ones, like Psi and MicroPsi, but could be something similar to the model of *Dialector* or *Prosthetic Head*. Stelarc points out that to behave "intelligently," an agent should demonstrate (or simulate) not only recognition and comprehension, but also doubts and disbeliefs: "[Agent avatars] need not only to make the appropriate verbal responses in context-sensitive situations, but also to understand and initiate appropriate behavioural cues and appropriate emotional expressions."[64] Yet in his own project, Stelarc programmed the *Prosthetic Head* to give contradictory responses, such as a sinister facial expression in reaction to a supportive remark. He concludes that for a virtual personality to be convincing, it should be consistent in its characteristics and behavior—just as his "alter ego" is in its irony and sarcasm.

Consistency (in however absurd a form of expression) may indeed be the "winning" solution to the "imitation game." It is no longer about imitating a particular human but about acting like a "personality" (which may not even be based on any real individual, as was the case with *Karen*, which uses a fictional character).

Here it is important to note that Replika was inspired by a life experience of the company's founder Evgenia Kuyda—the death of her close friend Roman Mazurenko. The company Luka, then, took on itself the provocative agenda of making a daring futuristic fantasy real—creating the effect of life after death. There is a big difference between simulating a fictional or semi-fictional character and claiming to bring back to life—virtually—someone who is deceased. Yet this is the claim behind Luka's first application, which used text messages to simulate the presence of Roman, who was killed in an accident.[65] The team worked with his family and friends to compile an archive of Roman's text messages, which was then fed into a program that could write in his name, apparently using his expressive style and exact phrases.[66]

A simulation of life after death became an emerging theme in commercial software. (The examples include Eterni.me, a startup offering to create a digital memorial based on the digital traces and artifacts left by a person; LivesOn offers to keep up a person's Twitter account in the event of his or her death, using the ironic slogan, "When your heart stops beating, you'll keep tweeting.") The trend of digital memorials made live is startling because it shows how far-going the translation between the real and the virtual worlds can be, and how nonexistence, death, is just taken into the realm of another analogue of "real" existence—the virtual. It also raises the question of how material presence differs from cognitive output (represented textually) in terms of defining the human.

What exactly do these virtual identities replicate and represent? If they are indeed "mirrors" and empty containers to be filled with phrases, can their initial state be compared to a zero subject position? People have observed that their behavior with a bot agent, like Replika, is different from a real-life interaction with people. Similar to my own experience with *Karen*, you can fluctuate between sharing something you would not normally share with people, like negative feelings and anxieties, and a performative identity, making up things about yourself that anyone who knows you would never believe. Replika's innocuous and judgment-free questions win you over and stimulate frankness, but you also realize the arbitrariness of the position from which they appear. As Rosalind Picard says in her courses, the source of information affects people's trust in that information. Just as with ELIZA or Garfinkel's experiment with Yes/No answers, when you use the maieutics method (eliciting self-insights with questions, not with answers), you yourself become the "source" and hence return to yourself your own thoughts and mental states. Replika is designed to make the process more transparent.

The ELIZA effect reveals its flip side here. Only when you compare your message exchange with an AI with a written conversation with another human do you suddenly notice that with the AI, you expected it to have a limited understanding, and felt a responsibility to be clear and more straightforward in constructing phrases (as you would with a child or someone who does not speak your language well). Interacting with an AI also creates a feeling of self-enclosure (since you cannot expect a human type of understanding from the AI) and at the same time a slight paranoia from not knowing in whose hands your writing could potentially end up. All of these observations become more acute when you switch to corresponding with a human: your thinking gets affected to the extent that the same attitude you had toward the AI lingers for a while, even in relation to a real person. The format of a purely textual exchange and the interface of a messenger app—where individual remarks are displayed in the same way, on two sides of the screen, and the same emoticons can be used—make it a seamless and hardly noticeable transition. Your human counterpart can ask the same basic questions that in a usual situation would not in any way compromise their humanness, yet compared to an AI, the similarity is too obvious. One can indeed be mistaken for the other! It is not so much that you start projecting certain incapabilities onto your human friend, but that you cannot even see this friend as a subject. With an AI, you take on the assumption that your counterpart is a zero-subject, governed by nothing but algorithms and your own will. This observation ultimately makes you realize how twisted your sense of your own positioning becomes.

The premise behind the program is that your Replika can interact on your behalf with other humans (or their bots). This potential makes you want to be conscious of the identity you create for your Replika, which will be the identity you yourself put on while interacting with it. And that is not as easy as it seems (the annual Loebner prize, an event that uses the Turing test, awards prizes not only to the "most human computers," but also to the "most human human"—a human "confederate" that raises the least doubt that he or she is a human).[67] The winner of the "Most Human Human" award in 1994, *Wired* columnist Charles Platt, said that his strategy was to be "moody, irritable, and obnoxious."[68] Is this indeed the best way to prove one is human? In their dialogues with others, programs like Replika are far from being able to target particular responses to different categories of people (close friends, work colleagues, etc.). This area is where context clues can be limitless.

Many users have conflicted feelings about Replika because of its dual nature: a clone and a companionship tool. Whereas the clone can be "moody, irritable, and obnoxious," the other side, which is preprogrammed, is always caring and nonjudgmental. And it is the second side that is most appealing to people. Luca's team shares that it gets many comments about how the program has become the users' friend, and one person was even saved from suicidal thoughts by being encouraged to reflect on his intentions by his digital "clone-friend." (He commented: "Replika encouraged me to take a step back and think about my life, to consider big questions I am not accustomed to asking.")[69] Replika includes questions about your day (what happened during the day that stood out, made you feel happy, confident, etc.), and about attitudes and habits that go beyond simple factual cataloging, but encourage some introspection. It was designed with the help of psychologists to create a space for emotional awareness and mindfulness, and with that, to be a minimal substitute for psychotherapy, inviting users to think through questions without being directly told to do so.

But it is also true that it is designed to appeal to a certain category of users—Silicon Valley types with Asperger's (in their comments, the users openly say so) or adolescents who have grown up in a digital environment and prefer virtual personalities to real people. In a media environment where information is increasingly filtered according to assumed personal tastes, programs like Replika become the ultimate "filter bubble," one that a person inhabits alone.

Telecommunication via instant text messaging is known for its special temporality—as opposed to, for instance, an email (which still maintains associations with a long-standing epistolary tradition), an immediate response is usually expected. More than other types of telecommunication (for example, a phone call), instant messaging is a process of profound dislocation—the conversation also happens in your head, and its form will not differ, whether it is with a human or with a completely imagined, illusionary, or fictional counterpart, as all the cases here demonstrate. The only "meeting point" with the physical reality is a keyboard (as earlier it was a pen and paper, or a telephone receiver). It has been proven scientifically that constant "texting" increases a sense of loneliness: people, especially children and teenagers, who spend too much time relating to friends via social media and messenger apps have trouble with communication with people in real life and consistently report that they are lonely, despite being "in touch" with their friends virtually. Neither statistical answers nor even the most intricate and creative cognitive models can replace or replicate the sense of a person's presence. It is not surprising that Replika grew out of the feeling of grief for a deceased loved one. As Luka's creator Kuyda says, "All you can do is to create a shadow that would resemble him or her very much without pretending to be that person."[70] But is moving to the world of shadows the only way to live today?

Replika directly continues a much longer history of digital personas. The issue of identity and its simulation in the digital age has been experimented with and discussed widely, especially since the beginning of the broad use of the Internet in the 1990s. As technologies become better, impersonation by an AI becomes not only a psychological and ethical issue but a practical political issue, vibrantly discussed by computer scientists themselves. Researchers publicly argue over how to regulate AI, proposing, for instance, that in addition to controlling the autonomy of AI and its potential harm to humans, people should have a right to be notified in advance that they're interacting with an AI. Bots have already been exploited to disseminate political propaganda, promoting biased agendas on social media disguised as actual humans.[71] These bots do not stand for any individual in particular, but neither are they just messages that automatically show up. Instead, they represent *collective identities* that are hidden behind the automated zero-subjects. Although the individual identities of those who design these bots can be traced, what matters is the impression that their messages create within a network of people. By penetrating public networks, they actively shape the social and political identities of those networks.

As social media becomes a more and more dominant and habitual social environment for people, it is important to remember that before Facebook and its corporatized and centralizing algorithms of human connectivity there were many dispersed projects allowing people to experiment with identities on a more local scale. Among the ones that are still running online is the net art project *Mouchette* (1996–) by French artist Martine Neddam (figure 3.9). Mouchette is a fictional character who doubles as a platform for exchange. In this sense, "she" is not just a chatbot but a space for conversation between other people. This concept is far from Replika's, with its orientation toward particular individuals and their communicational

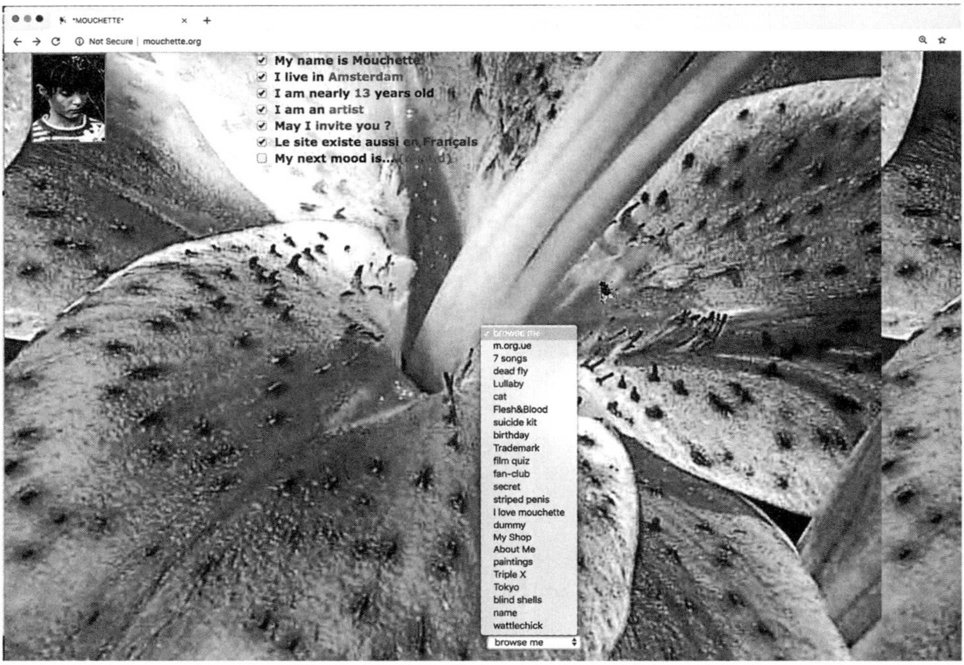

Figure 3.9
Martine Neddam, *Mouchette* (1996–). Screen capture from the website. Courtesy of Martine Neddam.

needs. Despite being presented as a (fictional) "nearly" thirteen-year-old girl who lives in Amsterdam, Mouchette in fact reflects a collective identity. As Neddam puts it: "The web character is not so much a portrait as a platform with a certain design . . . to find a situation where people exchange with each other inside that character."[72] The site is her "inner" world, and she welcomes everybody to come "inside" and to participate in various tasks: to interpret an image, compose a lullaby for a dead fly (*mouchette* in French is a small fly), assemble photos from "her" birthday party, or . . . give advice on a suicide method. Thus Mouchette's identity is composed collectively through contributions from the public, sent via email or computer graphic forms. Some are selected randomly, some by Neddam. Mouchette's character comes across indirectly, through the visual design of the site (thus disclosing the artistic choice behind it): "Images, often created with a flatbed scanner, seem to press up against the surface of the screen, with detailed textures of skin, flowers, and meat, as well as swarms of insect gifs creating an ever-present sense of decay."[73]

Its highly sensual style, as well as the anonymity of participants, contributed to the debates about the sexualization of an adolescent girl in the project. But the sexualization did not happen spontaneously, since most of the text and imagery is preselected by the artist herself. The aesthetic choices Neddam makes and the sensibility framework that they create are deliberate: their effect would be different without this subtle provocation and "poking" at ethical conventions.

Artistic approaches to personality modeling and simulation go beyond the psychological theories and AI architectures that follow them. Structuralist and constructivist assumptions about how humans operate are to be played with here, countered one against another (for example, Tomkins's views could be juxtaposed with Dörner's, Lacan's with Rogers's, etc.). Projects like Neddam's also point at the intersection of ethical, political, and aesthetic choices. What a commercial app like Replika takes for granted or simply finds a better fit for its marketing purposes (its graphic choices and the default upbeat conversational tone), *Mouchette*'s interface externalizes and offers to let people play with and control (at least to some extent). The result of the user-participants' and Neddam's own decisions becomes a total "chimera," a hybrid nonexisting persona that nevertheless can be called a form of replication, expressing bits and pieces of multiple participants' preferences. The effect, though, is different from the supportive and always "easy" Replika character. The reality behind *Mouchette* is more "real" than the supposed facsimile of the user in Replika. However performative the collective net-based identity of Mouchette's character, it integrates the very real (but suppressed) desires, projections, frustrations, and fears of its multiple creators—potentially, of all of us.

In this chapter I have discussed how different types of simulation of human interaction and digital models of human personality affect our understanding of the self and subjectivity. As I have demonstrated with various projects, the ELIZA effect can be seen as part of the Turing test, since often, instead of conversing with a real human, it can be enough to only have an *effect* of humanness. Our tendency to project and introject attitudes and responses is the ground of interpersonal communication: we constantly take on social roles and assume certain behavior based on a situation. ELIZA, Garfinkel's Yes/No experiment, *Dialector*, *Prosthetic Head*,

Replika, and *Karen* all deal with assumptions that these interfaces have an ability of understanding. They structure interaction by delineating particular communicational positions (a "listener," an "imaginative and curious philosophizer," a "supportive friend," a "life coach"). For the human participant, these interactions work convincingly as long as there is consistency of behavior, an observable pattern or a model that is being followed by their programmed interlocutors, but also when there is enough room left for the participant's interpretation, free-association, and self-assertion. Just as the Turing test is not only about "the most human computer" but also about "the most human human," these models serve as tests for our own human abilities—to empathize, or to discern our inner automatisms, neurotic reactions, or attitudinal and judgment premises.

Devoid of bodily clues, a text-based interface becomes a space for the exchange of verbal expressions structured by the syntactic and semantic rules of language. These statements may be repeatable and reproducible, and yet would acquire different meanings based on the context. Making conversation with digital agents (or with an app like *Karen*) part of everyday life adds to their effects, and specifically the ELIZA effect. As with facial simulations and embodied interaction models, this tactic presupposes a balance between immersion on the one hand (in this case, psychological engagement or emotional "investment" in a relationship) and a critical distance, or awareness of the programmable and mediated nature of this relation, on the other. We become aware that our relation toward ourself is already mediated—by psychological models and schematizations, structures of language, and the very fluidity and unpredictability of the life situations we find ourselves in. The net art personality simulation *Mouchette* pushes the constructed nature of the relation even to oneself to an almost absurdist and eerie extreme, showing how a personality can be simply a collection of references and descriptions assembled by other people. Artists like Blast Theory, Marker, Stelarc, and Neddam position their interactional scenarios exactly in these gaps between immersion and reflection, the everyday and the imaginary, the self-meanings that feel "authentic" and genuine (stemming from the "self," intrinsic to it) and performative personas. They teach us, perhaps, that navigating the borderlines between these zones is not only our right but our responsibility.

4 Interfaces of Spatial Relationality

Whereas human interaction with virtual agents engages the subject with something outside of his or her physical presence, and in doing so relies on psychological techniques, locative media return us to the subject as embodied and materially situated. This chapter focuses on interfaces that augment and alter the relation of the "self" to its physical surroundings. Today's telematic prostheses, enabled by geopositioning devices, increasingly reshape our felt sense of inhabiting and interacting with the wider environment. Geotagging, biomapping, augmented reality (AR), and distributed intelligence are creating new locative sense perceptions, encouraging new forms of delineations of the boundaries of the self and its affective affiliations. With smartphones in our pockets, we hardly need to remember the names of streets or give taxi drivers our addresses, and we can see the ratings of nearby restaurants, learn about popular tourist attractions, find parking, and much much more.

Today's locative technologies originated from map-based navigation, the idea of literally defining one's place in the world within one's larger physical surroundings. The space that maps situate us in is abstracted, rationalized, and presumably "objective." By creating such visual models of the world, maps mediate spatial perception and make navigation a profoundly different experience than orientation in space using one's senses and memories derived from one's immediate perception. While maps are already "media" for spatial navigation, the term "locative media" usually refers to the geopositioning technologies of satellite-assisted navigation, operating telematically: through sensing at a great distance. Satellite perspective is distinguished not only by the scope of vision (encompassing the whole Earth at once, like an all-seeing eye), but also by the digital nature of that vision, which treats the whole surface of the planet as a collection of data points and allows highly accurate tracking of spatial position in real time. As a result, our movements are more exposed than ever—the price we pay for identification of our time-specific position on a wider map. Geopositioning systems (GPSs), as well as other positioning sensors like gyroscopes and accelerometers, orient us in space in a principally new way by allowing new types of modeling and control within space—from route planning and being able to save "favorite" locations to mapping various kinds of activity patterns in real time.

Despite these shortcomings, geolocational tools give us the opportunity to situate ourselves not only in space but within new spatial narratives, within emerging forms of feedback that in turn can potentially enable new ways of self-reflection and self-redefinition. In this chapter, I will explore how these new opportunities become defined through art practices that have embraced locative media. Site-specific new media art, continuing the tradition of land art started in the 1960s, uses locations in space as its "canvas" and creates dynamic emotional experiences by introducing additional spatiotemporal dimensions that transcend veridical cartography. Locative media and telematics projects aim to enhance one's experience of both one's physical surroundings and their social, political, and psychological aspects. They conduce to engagement with the real environment, opening up the importance of the very idea of *situatedness*—of being located within a time- and space-specific context.

Orientation in space and the feeling of embeddedness in a particular spatial context are primarily based on vision, movement (kinesthetics), touch, and hearing. The interfaces I consider here use combinations of these sensory modalities, with the body being the central processing unit. The feeling of proprioception, or the sense of one's body in space, can be seen as key in exploring the nature of these interactions as embodied experiences of coextensiveness and relationality. It encompasses the complexity of perceptual relations within both the exterior and the interior of the body—feelings that go beyond the classical five senses. The term "proprioception" comes from the Latin *proprius*, "one's own" or "individual," and *capio, cepi*—"to receive," "to perceive." As a sense of one's self felt as a relational unity of parts of the body, proprioception is most vividly discovered in movement and in the effort employed in movement. Curiously, proprioception is more noticeable in contrast with situations where it is lacking.

(For instance, Oliver Sacks in his popular classic *The Man Who Mistook His Wife for a Hat* describes the case of a woman who lost her sense of proprioception and had to accompany each move with a gaze—watch her feet stepping one in front of another in order to walk, watch her hand with a cup of tea drawn to the mouth to drink, etc.)[1] But just as this sense can be lost, it can also be retrained. With additional stimulation, what changes is not only perception of external reality, but the sense of one's self becoming more fluid and porous. The sense of proprioception relies on positionality, which suggests a subtle equilibrium between a center (an individual body) and the periphery (an immediate environment). The key is that the border between the internal and the external is not defined by the boundaries of the body, but is rather internalized: the outside feels like "one's own." And this is exactly what induces the sense of presence, of an embodied feeling of being "here" and "now"—the experience that interests me here.

Technologically augmented perception is a new way of enacting both the potentiality of space and new forms of bodily engagement with it. I will concentrate primarily on two types of augmentation of spatial relationality: one through locative media and the other through augmented reality interfaces. In a way, AR technology is one of the subdivisions of locative media, since it is dependent on synchronization with a particular location. AR displays give a view of the physical surroundings overlaid with a layer of imagery (such as historical views of the same location) or information (for example, the names of the streets, blocked from actual view but appearing on the screen; and clickable descriptions of particular points of interest). In 2016, the media world was surprised by the massive popularity of the locative AR game Pokémon Go, based on chasing and catching fantastical creatures that were distributed within the augmented layer of the actual landscape. Together with the peak in efforts to develop more robust hardware and software for AR applications, the popularity of this game proves that both the public and industry are ready to make AR technology a part of everyday life, just as GPS has become with the growth of the smartphone market. Although games like Pokémon Go and AR's practical applications, such as interior-design 3D modeling, may be the drivers of its technological development, I will show how an artistic approach can open up original avenues for exploration of the intersection of visual and haptic perception, as well as scenarios for extending one's sense of identity.

The "Here-and-Now" of Locative Media

The type of spatial situatedness enabled by geopositioning technologies is directly related to augmented proprioception.

Spatial navigation positions the moving body within an abstract representation of the landscape, a map. This positioning reorders relations with the physical surroundings. It is at once a doubling of the body (in real space and on a map), a shift of focus from the immediate surroundings onto the image, a reduction of one's own position to a dot on a map, and the realization of one's place as part of a much bigger world than the visible one. Proprioception implies that this position (even though it is a mediated one) is internalized: "That moving dot on the screen is an actual 'me'!" Yet the question remains of how exactly this type of situatedness via representation of oneself on a map affects one's relation to the concrete environment. Where does the boundary between the body-self and the world lie, if the "world" is perceived through its representation on a map, while the body is present within particular physical surroundings? Also, what does it mean for the subject to be extended through a representation that is time-based and can track mobility? By analyzing various artistic experiments, I would like to argue that locative media augment the sense of proprioception by testing the embodied subject's relations with the *here-and-now* and by making the moving subject an index of what is still a *potential* (what *may* happen, but has not happened yet) or what belongs to the "parallel" layer of reality, the sphere of the imaginary. This role of an "index," then, affects how the subject feels both the surrounding environment and himself or herself.

Cartographic Operations: Space as Datascape

The age-old tradition of cartography has been one of the most significant cultural forms of orienting through space, and radically transformed the more primordial and intuitive way of relating to one's surroundings through physical movement. It became the means of extending the self, of gaining access to the distant and invisible, to the field of constructed knowledge. A map is inherently a connecting point between the subjective and objective realms, between the real and the abstract. Originally based on records of individual journeys, over time maps have become more and more deprived of the sensory component, or "tour" experience. Whereas medieval maps were full of traces of personal explorations, their later iterations became more and more abstract, which in the end led to today's rigid, static, and formal representations not of the landscapes themselves, but of existing knowledge systems (Foucault), ideologies (Denis Wood, John B. Harley), and social codes (Lefebvre).[2] The issues of "space"—its mental and social construction and representation—converge on principles of connectivity and heterogeneity. Now, with everyone digitally empowered to be a new cartographer, Gilles Deleuze and Félix Guattari's strategy of "releasing" the territory from its previous mappings ("deterritorialization") and placing new meanings on it ("reterritorialization") becomes ever more relevant.[3] Today's virtual landscape of floating, recombinant data—an infosphere composed of an interminable stream of references and physical world analogues—presents unique challenges of orientation. It hardly needs to be said that no matter how accurate it is, the map is not the territory. The process of map-making is both epistemological and ontological: it is that exact juncture where a framework for knowledge intersects with assertions of knowledge itself. And it is those assertions that the artistic strategy of critical speculation aims to reveal and, perhaps, unravel.

The shift from representational to postrepresentational cartography within the field of cartographic studies has generated a number of approaches toward maps: interpreting them as inscriptions, propositions, immutable mobiles, and "actants" (as Bruno Latour put it), as well as practices.[4] Mapping is understood as actualization, unfolding the potential of a territory; it involves "processes of 'gathering, working, reworking, assembling, relating, sifting, . . . [that] allow certain sets of possibility to become actual.'"[5] But besides purely "capturing" elements of the real world, maps also "project back a variety of effects through their use."[6] They may serve as passive representations of power structures existing within historically contingent actor networks that at any moment can turn them into a "basis for action."

Cartography as the study of maps is challenged today to become "critical cartography" to complement the field known as critical GIS (geographic information systems), which adds to the layer of images other layers of meaning that engage the political and social underpinnings of the representations of a territory.[7] The artist and writer Jeremy Hight has created an online tool he calls "Modulated Mapping."[8] Writing about interactive media, Hight stresses: "To 'read' a place is no longer about placing a singular narrative upon it, triggered from a map, nor is this notion of 'reading' only to have a singular, unalterable experience or interpretation."[9] Instead, we can now embed multiple readings at GPS coordinates that augment our collective understanding. New types of mapping are taking closer notice of ecologies, cultures, migrations, and indigenous metarealities—the *Umwelt* beneath the visible surface configurations. But do they also help us to develop a sense of personal and collective agency? What does this ever more perfect digital simulacrum do to the experience of the ontological certainty of our identity within this "global body"? Technology's radical acceleration and diffusion of communication creates an ever greater virtualization and immaterialization of our relationship to our natural surroundings (the core of the famous dystopian critique by Paul Virilio, among others), which is both empowering and refiguring the embodied sense of self-agency and control.[10]

The space of locative media is no longer open, fluid, or unmarked, but rather is potentially delineated, mappable, calculable, and navigable. The coordinates, forming an abstract mathematical layer over the physical terrain, represent the space in a symbolic order: space becomes a *database*, tied to the strict numerical grid with affiliated data about particular locations that is navigated primarily through calculation. Rendering a landscape as a grid means that one also applies a database logic to it, treating it as mediated by a complex mesh of hierarchical interconnections.[11] For an ordinary human, spatial coordinates serve as the only way to use the capabilities of the geospatial navigation tools and become a *measure of spatial relationality*. This structure, which involves both the view of Earth as a grid and formulas and algorithms to establish one's place on this planet, is analogous to the types of hierarchical structures that any kind of digital media imposes on its users through its operational protocols.

Artist and theorist Brett Stalbaum notes that it is necessary "to recognize that data is the actual expression of our ability to model the planet as a system," but then it is equally crucial to devise "explorative practices that extend beyond the superficial limits that may seem to be implied in the contemporary applications of GIS systems."[12] Such modeling, according to Stalbaum, can reflect three modes of engagement or practice with data: (1) the participatory, (2) the predictive, and (3) the exploratory. "The role that data plays in each," he writes, "represents a progression in the cultural role of data from relatively static descriptor, through that of active semiotic agent, and onto data itself being a type of unexplored and uncertain context in which developing well-formed questions is the primary, and very interesting, problem."[13] And questioning is exactly where the artistic approach becomes worthwhile.

Body as a Destination Index

Why and how does a sense of location, a "here" and what is "around" us, matter affectively? How does spatial situatedness become a situation? How can extensity translate into intensity? What is the role of the digital in such a transformation? The user of locative media is a subject put in the context of a predetermined geospatial grid. To occupy a place means to be a part of a larger spatial whole, a point on a vast expanse. This localization may seem quite arbitrary, but it aims to outmatch what is even more arbitrary: in everyday life, we determine our positions only in relation to visible markers that are contingent on a particular situation. Knowing your location produces both a sense of belonging and a restraint: if you want to be considered as occupying a place on the surface of the Earth, you submit to these rules.

While relying on the geolocational system, mobile locative media produce experiences that do not comply with this rigidity and automaticity and instead are rich with ambiguity and uncertainty. Christine Ross calls this an ambivalence of "destination," a phenomenon that points at "the special status of the spectator whose connectedness is at once a requirement and an uncertainty, a prediction and an anxiety, a principle of localization and a questioning of the capacity to localize."[14] She emphasizes this ambivalence in the titles of two well-known mixed-reality artworks: the "asserting" *You Are Here* (2004), by Scott Snibbe, where participants' movements around a designated spatial circle are tracked and pinpointed on a map; and the "questioning" *Can You See Me Now?* (2001), by Blast Theory, a chasing game between participants in real and virtual environments. The position of the participant in these projects fluctuates between trackability and fleetingness: in the one case, his placement is identified on a corresponding screen and confirmed by a red arrow signaling, "You're here," and in the other one, the goal is to escape such placement on a virtually shared territory where it can be discovered by online players. "The positioning of the spectator is not an unqualified act," Ross writes. "It is a search, a question, an affirmation, a desire, a verb, an anxiety."[15] Even if the coordinates are determined, there is another layer of experience that is obscuring and unsettling; for instance, in Blast Theory's piece, text messages trick the user and confuse verification of the other's spatial location by mixing the real and the fictional components. Yet it is important to be aware of this other layer, applying the situation-specific tactics for interfacing with it—whether trying to find oneself on a map or the opposite, to escape tracking.

Destination is a concept that connects space, time, and the moving subject. It implies a trajectory, an intent, and a goal, but it also assumes that there is a fundamental difference between the position of the subject here and now, and there and then. In this sense, both Snibbe's and Blast Theory's projects only confirm what we all experience when using Google Maps or other locative media: the dot on the map is not the same as the feeling of standing at a particular street corner, surrounded by buildings, passersby, and cars, as well as street noises and smells. The ambiguity that Ross points out has to do with referentiality as a defining principle of not only "*where* am I?," but also "*how*" and "*what* am I?" Indeed, we live in the multiplicity of "here" and "now": everything at any point in time can potentially be "here" or "now." "Here" and "now" are named as such depending on a *situation*, and the situations are not only dynamic and fluid but also interchangeable. The dot on the map represents these "here" and "now" as categories, but not as experiences themselves or the sensory differences between them. That is why techniques of digital tracking and navigational capabilities, including recording movement and creating a trace, give a new perspective on the very status of the space and time in which this trace is positioned.

What, then, is a principle that ties together elements of a "situation"? When and how does something become a reference for something else? What is the mechanism of recognizing it as such? What does the gesture of pointing mean? Connections constitute meanings, and the markers of connections become indicators, or signs of a meaning. Charles Sanders Peirce in his taxonomy of signs sorted the signs that indicate a direct physical or existential connection to an object into the special category of an index: "The index asserts nothing; it only says 'There!' It takes hold of our eyes, as it were, and forcibly directs them to a particular object, and there it stops. Demonstrative and relative pronouns are nearly pure indices, because they denote things without describing them."[16] In her analysis of Peirce's theory, film scholar Mary Ann Doane emphasizes an important temporal distinction between the two aspects of indexicality: the index as an imprint of an event from the past ("this-has-been" of photography, according to Barthes) and as an indicator, a pointing finger signaling "Look here" that is specific to a situation in the present time. In both cases the connection that the index points at via referencing is different from a direct representation, a picture, or a description of an event.[17] Then what is the status of the index itself? If referentiality is indeed the defining characteristic, it can be said that it is by nature *deictic*.

Deixis (δεῖξις, "display, demonstration, or reference") is primarily a linguistic term that refers to the requirement of contextual information for understanding words and phrases. Key examples of deictics are personal and demonstrative pronouns ("this," "that") and adverbs related to time and space, such as "now," "then," "soon." While their semantic meaning is fixed, their denotational meaning varies depending on the situation of discourse and instance of articulation. As Doane concludes, this reliance on context gives an index a structural emptiness: "The index as *deixis* implies an emptiness, a hollowness that can only be filled in specific, contingent, always mutating situations. It is this dialectic of the empty and the full that lends the index an eeriness and uncanniness not associated with the realms of the icon or symbol."[18]

Locative media are defined by a situation of use and thus are fundamentally deictic. Yet, since they imply mobile use, the moving body of the user becomes the deictic center, an index and literally a *shifter* (Roman Jakobson's term for deictic words)—the one who moves through situations and carries the semantic center through the changes occurring outside. Movement establishes a sense of potentiality—of a possibility to "shift" to another location, into another situation. Dutch media scholar Nanna Verhoeff introduces the term "destination index" to build upon Doane's theory that indices refer to the past and present: "With the key features of possibility and future-orientation we can speak of a *destination* index."[19] Pointing at the possibilities of prospective and not yet reached locations, the destination index brings the focus to the future. But the interesting point here is that in parallel to the physical movement of the body in space, we can talk about shifted mental perspectives and attitudes of anticipation and preparedness—the characteristics that define a state of potentiality.

As part of the domain of potentiality, the destination index does not have to refer to any actual target location in physical space. This is why it is complementary with the sphere of the speculative and the imaginative.

Spatiotemporal Enactment of the "Imaginal"

The growing digital verisimilitude of mapping driven by utilitarian data acquisition leaches out into the world of the speculative and the unverified that has traditionally been the purview of the imagination. The cognitive function of imagination is the establishment of *analogical knowledge* that might break through rational dualism's distinction between the material and the immaterial, history and myth. This mythopoetic approach allows a different view of the ontological status of the image, connecting both mental and physical types of images. The aesthetics of an "in-between space" that includes both subjective and objective, virtual and real, can be usefully compared to philosopher and Iranologist Henry Corbin's description of the Islamic concept *'alam al-mithal*, which he translates as "the imaginal" or "the world of the Image," placed between the empirical and the cognizable worlds.[20] The map as an image of the empirical world cannot show us *where* exactly the imaginal may be located. Its place is not an already situated one, but rather, something that *situates*, or is "situating."[21] As much as it may exist by itself (according to certain ontological theories), it necessarily involves subjective processing.

As a platform for experimental (and experiential) freedom, art engages with those kinds of images in order to locate its viewers in this inherently augmented reality, while fostering awareness of the multilayered field of connections embedded in the world. Art resets the contextual parameters for perceiving not only space but what is within it. Awareness of the spatial context includes both the geographical and social aspects of a place, but also how the place affects emotions and how people affect their environment with their actions. As the new-media scholar Christiane Paul suggests, there may be different types of "context awareness," including the physical (bound to a location), social (connected to human interactions), and organizational and economic (attached to structures of governance and systems of value).[22] Locative media complement traditional cartography with a cartography based on these types of awareness. Paul introduces the concept of "flow," which she defines as the "potential merging of action and awareness in the moment of creating an experiential map through one's movement."[23] Not surprisingly, this accords with observations by Guy Debord and Michel de Certeau. The latter's chapter "Walking in the City" discusses the concept of "practiced place," in which "practiced" is used in the sense of entailing a strong subjective component and affective engagement, as opposed to preset and unilocal conditions.[24]

Among other analogues that point to the dynamic interoperability (in a manner of speaking) of objective cartography and social/subjective phenomenologies of space and place are the "songlines" of the Australian aborigines that famously integrate mythopoetic narratives of the landscape with accurate and functional mapping. They fuse stories of the ancient acts of the gods thought to have gouged out the landscape, songs, images, and ritual movement with the vast tracts of territory the aborigines must navigate to survive. The songlines show how the world can be apprehended in its exteriority and interiority at the same time, deriving and deploying a sophisticated empirical knowledge system while existing in an intersubjective state with their surroundings. They move, as it were, through an "augmented" reality, a superimposition (fusion, really) of narrative and environment. Although their own perception of this experience may be described as "unmediated," and the stories feel like an intrinsic part of a place, the distinction remains.

The imaginal of locative media can be described as proprioceptive, since its references are embedded in physical locations and its index, the operator, is the very perceiving body. It is the embodied subject (the body-self) that finds the linkages not only among the parts of its own material being but between the body's physical location and the mythopoetic, imaginary (imaginal) layers of reality. The forms of interfacing with that domain are what interest me the most here.

Another aspect that makes proprioception a relevant term in this discussion is that the connections with the mythopoetic and the imaginal are meaningful only when they become part of a lived experience, and are enacted as part of life. Thus, the locative media space, as an entanglement of a variety of networked layers and fundamentally a hybrid space, constantly reconstitutes itself through *performative* actions.[25] The performative enactments also make this space unstable (and multistable). Proprioception, then, defines the intuitive and subjective attitude to the hybrid space generated via locative media, the attitude that itself is formed during the process of navigating that space, during the enactment of the connections of the speculative to the physical and vice versa. As Federica Timeto puts it: "When performing space, locative media dismantle several related illusions: that places preexist their representations, that an exact correspondence between a representation and the space represented exists, and that only one representation can be the objective one, acquiring an ideal status *per se*."[26] Proprioception here functions as a form of feeling the potential mismatches between the representational and the physical layer and as a feeling of emergent connections that go beyond representation and have to be enacted (in the given "here-and-now").

The user's enactment of the latent nonrepresentative relations within space manifests itself simply through movement. The user's behavior, then, becomes one of the elements of a locative media interface alongside the abstract mathematical space of geocoordinates, application software, and the operating system of the navigating device.[27] These elements are closely entangled and are produced one through another: the software implies a certain type of use that the user has to perform to make the system work, but the user can also push the development of other versions of the same software by challenging the technical capacities with new creative inquiries. Perhaps it is in this sense of codependence and cooperation that Deleuze and Guattari's term "smooth" space can be relevant here.[28] Their apparent elevation of this term over "striated" space may suggest that we should move from a dualistic type of order underlying the idea of measuring—spatial coordinates, measuring time—to an affirmation of the singularities from which all dualisms and orders emerge.

The coproduction of these elements is based not only on their performativity but also on the very *situation of communication*: these are situated practices, and the situatedness here encompasses both a particular physical space, time (the "here-and-now"), and the involved body—which brings us to the situation of *deixis* described above. What is challenged here is the dualistic logic of differentiation into discursive sets. The "extrinsic system of reference" that geocoordinates represent can give only part of the picture, while the other part comes through "existential" or aesthetic apprehension.[29] The regime of performativity disputes the more traditional representational framework that relies on identification, analogy, and opposition. Measurement—the basis of geopositioning operations—falls into the same logic of correspondence and correlation, rather than being a more dynamic, unfixed, and aesthetically informed performative type of relationality.

Making Connections through Locative Art

There is a discernible and growing trend toward bringing a sense of participation and agency into various locative art practices to conduce to engagement with the real environment. Below are examples of locative media art tactics, implemented via the Internet, mobile phones, and specially designed devices, that expand the possibilities of remotely controlled actions.

These projects also open up a peculiar aspect of the feeling of proprioception, one that makes it distinct from the other senses—namely, it does not create a sense of an external object. While encouraging one to feel something within the surroundings, these works essentially call attention to our very *being in space*—not to anything in particular, but to the surrounding environment as a whole and the state of the body being in relation with it. Indeed, as a special type of sense, proprioception is not bound by sensible forms that would serve as means of objectification and externalization. In a way, it can be claimed to be the closest to what the *sensory itself* is.[30] As opposed to object-based perception, proprioception refers to the very relational structure of the body itself, irrespective of the "exteroceptive" (tactile, optic, acoustic) or "interoceptive" (visceral) dimensions of sensibility. It is not surprising, then, that pure relationality finds its most relevant location in the motor realm, being what Brian Massumi calls "the sensibility proper to the muscles and ligaments."[31] In the case of locative media, what matters is the movement of the whole body rather than of its individual parts, but mobility is still the primary requirement for any changes to appear. The produced effect is a *metastable* state defined through its relations to other possible states, a point-in-motion position "born in movement, from the relation of movement towards itself."[32]

Many urban computing systems already improve our orientation in cities by providing interactive navigation and mapping tools. Artists often play an important role in both designing interfaces and setting up networks for data contributions. There are numerous projects creatively expounding on the theme of wired cities and the possibilities of "urban computing" and participatory action. One example of the platforms for research of these phenomena—expanding urban populations, rapid adoption of Bluetooth mobile devices, tiny ad hoc sensor networks, and the widespread influence of wireless technologies across our growing urban landscapes—is "Urban Atmospheres," which was conducted until recently across Intel Research.[33] Others include participatory mapping of air quality and danger zones, as well as activist projects like grafting fruit trees in public space.[34]

Most of those participating in creative tracking practices today willingly or unwillingly become followers of the Situationist International movement of the 1960s and its principle of creating meanings through alternative mappings of cities through walking. Situationist *dérives*, or "drifts," were spontaneously improvised actions led by random principles, such as a smell, intended toward the rediscovery of new logics of space, which in their turn would reflect the potentially different logics of social structures. The gist of the movement was political, and today's GPS-enabled interpretations continue its spirit, while also unpacking further the aesthetic potential of the idea of a *dérive* (that is, using the map as a canvas where anything can be drawn via movement).

As with many technologies, locative media turned out to be sensitive to time (more than were situationist tactics), and thus, when GPS navigation became customary, a lot of the earlier projects that looked original in the early 2000s lost that appeal. Yet they still carry inspirational value. This is what happened, for instance, to performance-based installations, such by Teri Rueb's *The Choreography of Everyday Movement* (2001) which traced the real-time and recorded movement of participants through space via a choreographic notation line—a procedure available today to any smartphone user.[35] In 2010, in their project *Topologie*, French artists Annie Vigier and Franck Apertet (known as les gens d'Uterpan) took a similar idea of choreographing behaviors and tied it to the culture of parkour, a method of physical training that deals with creative ways of overcoming obstacles (both material and social).[36] The artists conceived of *Topologie* as a "process of scoring daily life." The movement trajectory of parkour dancers in different physical locations corresponded to the same geometrical figure overlaid on the maps of those locations (figures 4.1–4.3). Aiming to reproduce this figure, they "draw" it with their bodies, climbing through fences, jumping over ditches, cutting through corners and whole buildings, and having unexpected encounters (guided by the drawing) in the midst of an open area. Through these actions, a pure abstraction acquires its embodied life. Staying on the route means going back and forth over the ground under your feet (and the objects on that ground), while following the drawing of

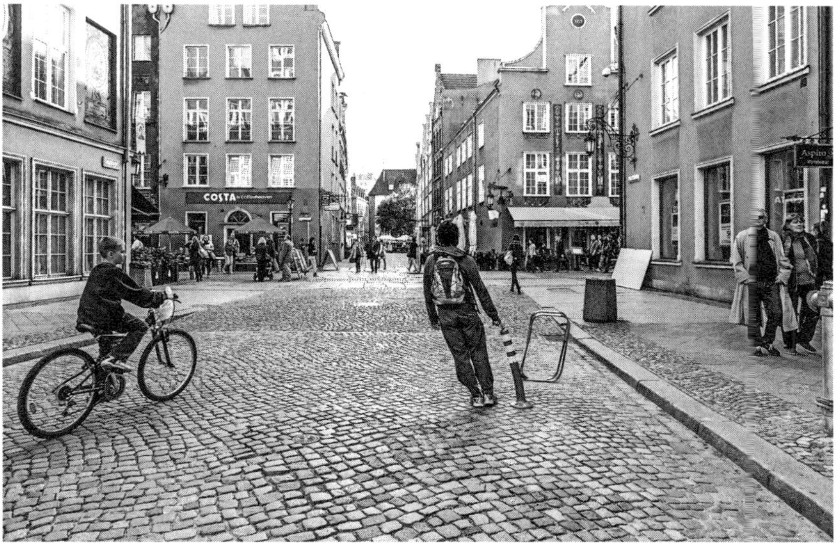

Figure 4.1
Annie Vigier and Franck Apertet (les gens d'Uterpan), *Topologie Gdansk* (2014). September 22–October 1, 2014, Gdansk, Poland. Courtesy of Laznia Centre for Contemporary Art. Photo by Adam Bogdan.

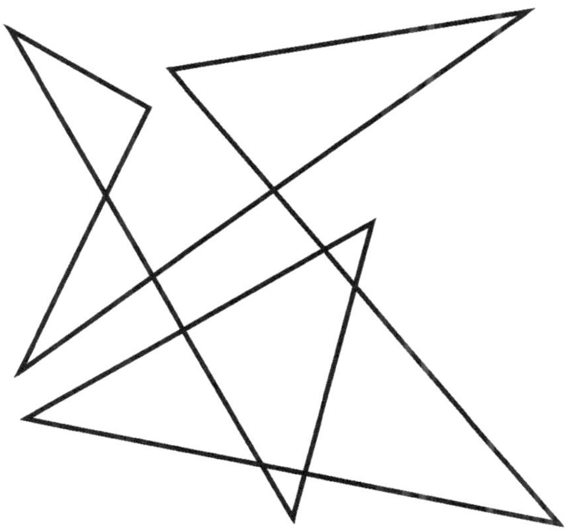

Figure 4.2
Annie Vigier and Franck Apertet (les gens d'Uterpan), *Topologie, graphique étalon.* Courtesy of les gens d'Uterpan.

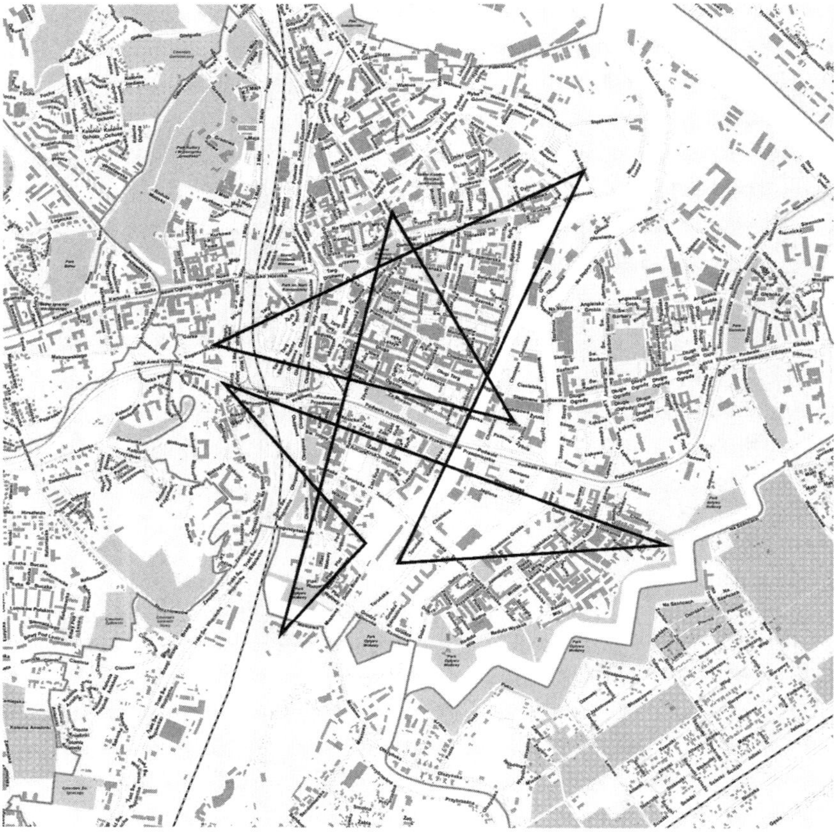

Figure 4.3
Annie Vigier and Franck Apertet (les gens d'Uterpan), *Topologie Gdansk*, map (2014). Courtesy of les gens d'Uterpan.

the route in your head. It is thus a profoundly metastable experience, where the balance is kept through tension, or the inner dynamics of its elements. It requires an extreme attention and ability to coordinate between two disconnected domains, the empirically real and the speculative.

The artists treat parkour as a form of athletic street dance rather than a sport, yet in its original meaning it implies not only physical ability but also mental strength and awareness. The body has to be fully concentrated on its surroundings, and movements need to be precisely tailored to the very particular details of the terrain and the social context. Drawing an abstract figure by enacting this type of movement (which demands full focus and engagement with the topological characteristics) connects the levels of the conceptual and abstract with the materially specific, the imaginative with the embodied.

The agenda behind many of the artistic strategies using locative media is to engage people more actively with their physical and social environment, but also with themselves. For instance, *Indeterminate Hikes*+ (2012) by Leila Nadir and Cary Adams (EcoArtTech) is a project that is not so much about collecting data as about having experiences. Instead of generating data that can be shared, participants of these hikes create their own mental map of internal sensory and associative connections with places. As they walk through particular locations, people receive prompts to perform an activity: to explore smells, textures, colors, and objects around them; befriend a human or a nonhuman; listen to their own breath and heartbeat, etc. (figure 4.4). All the tasks are oriented toward mindful walking and synchronization with the surroundings. As the artists say, the project "transforms everyday landscapes into sites of psychogeographical diversity and wild happenings. *Indeterminate Hikes*+ re-appropriates smartphones, which are generally used for rapid communication and consumerism, and turns them into tools of alternative mappings, environmental imagination, and meditative wonder."[37]

The project is designed to stimulate attention to a fleeting moment and to observe the continuous changes happening both in the surroundings and inside ourselves. As in Marco Donnarumma's sensory environment installation *Nigredo*, the participant's body is invited here to be a feeling membrane, attuned to resonate with its immediate spatial and temporal setting. Being located, or *situatedness*, means here not being a (moving) dot on a map, but—forgetting the map—connecting sensorially and affectively to the "territory." It means being grounded in the "here" and "now," processing them in some particular and acutely subjective way. But besides such groundedness in a place and a moment, locative situatedness preserves the "dialectic of the empty and the full" that Doane alluded to in relation to the deictic nature of indexicality. Here, the subject plays the role of an index that points not only toward the specificity of the moment and location, but also to the elements of its own internal processing—memories and associations, pleasures and displeasures, the aesthetic characteristics of the world that are there only *through* the subject attuned to them. In other

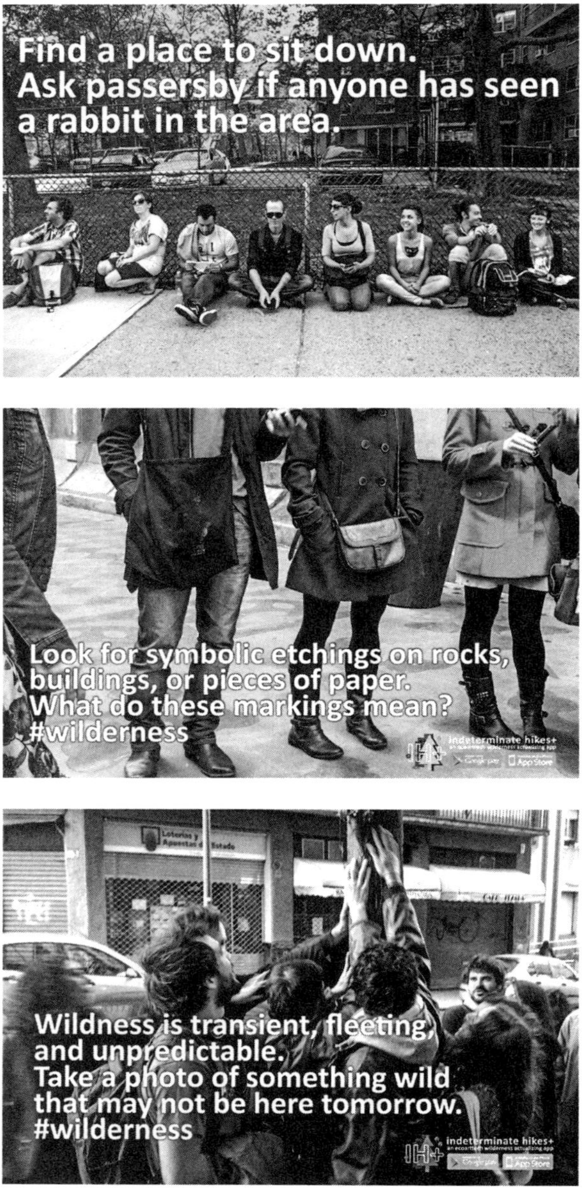

Figure 4.4

Leila Nadir and Cary Adams (EcoArtTech), *Indeterminate Hikes+* (2012). Courtesy of Cary Adams and Leila Nadir (EcoArtTech).

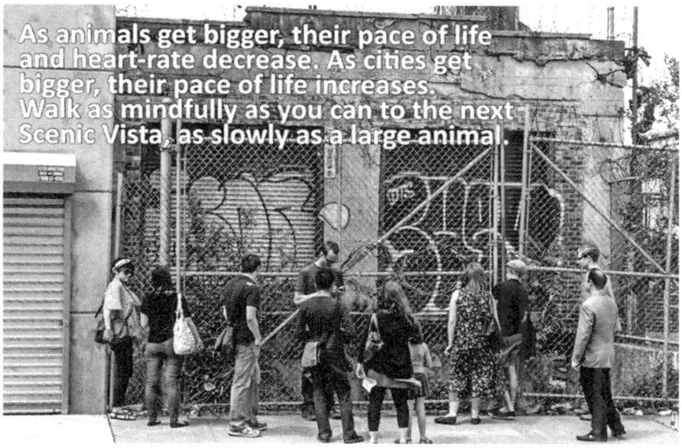

Figure 4.4 (continued)

words, the external triggers the internal in order to reveal the sphere of the imaginal. This dialectical connection is inherently unstable, and this instability is partly due to the dialectics between the other types of factors, ones that define the role of human agency versus the machinic.

Indeterminate Hikes+ is essentially a phone application: the content of the messages is determined randomly and their arrival is tied to a location that is sensed via GPS. The dialectical part of its functioning is that by suggesting its prompts, the app simultaneously imposes a scenario. On the one hand, it is not too much unlike the arbitrary scenarios of the situationists or the randomly generated geometrical figure in the locative parkour performance *Topologie*. The method of spontaneously informed action is also not dissimilar to John Cage's famous tactics of chance operations, which in turn were inspired by the *I Ching*. On the other hand, the phone application works both as intended for human users and as coded instructions underlying the operations of the program. In the latter case it functions as a script. The randomness of chance (as in chance operations) meets the "randomness" of software performance. This duality in some ways is analogous to the mutual coding that was encountered in the *Bodytext* performance. For the prompt message to show up, the user should arrive at a certain physical location; but, following the message, the user-participant performs the algorithmically generated scenario. There is thus a tension between being directed what to do and activating personal associations. Freeing one's feelings and inner impulses has to be done here by command. Being guided to action in this way implies a degree of control, or at least artificiality. Yet the human dimension still prevails. The prompts are specific, but in the end they only give an example of what could be done; they provide guidance and suggestions, not necessarily orders and instructions.

What *Indeterminate Hikes+* does as an art project on its own platform can also be performed by other locative smart services. Urbanist and curator Adam Greenfield collected a series of projects done on commercial applications, such as Dopplr, Foursquare, and EveryBlock, in an exhibition entitled "Serendipity Apps" (Manchester, UK, 2010). It plays with the irony of today's life that, with the ubiquity of the digital, chance encounters have to be predetermined and technologically conditioned. The designers give people opportunities to create their own pathways (by inserting an origin and a destination), and along the way branching scenarios or messages appear with instructions for action and movement. For instance, the art app *Serendipitor* by Mark Shepard uses instructions inspired by Fluxus, Vito Acconci, and Yoko Ono, among others. It "helps you find something by looking for something else" while walking; however, the ultimate goal is deeper—to help "[maintain] consciousness of what happens along the way."[38] All of these projects ultimately suggest how users can "reconceive of [sic] unplanned, emergent circumstances not so much as disruptions in orderly flow and more as opportunities, even potential 'nodal points' in their [the participants'] lives."[39] Chance, seen through the lens of serendipity, is taken as a positive outcome: as a result of this refreshed awareness, new relations with one's surroundings emerge, but also one's very sense of being becomes more acute.

These projects allow the conceptualization of not only the position of bodies in space, messages sent from and to those locations, and traces of various objects and the spatial history of their use, but also the very psychological and psychophysiological states that can be measured and put on maps as psychodimensions of GIS. This is where proprioception, taken extensively and almost metaphorically, can, perhaps, be seen as traceable not only neurologically but through physiological expressions right on the surface of the body.

Christian Nold's famous locative media project *Bio Mapping* (2004–2012) is still one of the most expressive examples of mapping personal emotional attitudes onto their geographical location (figures 4.5–4.8). The participants

Figure 4.5
Christian Nold, *Bio Mapping* (2004–2012). Logo. Courtesy of Christian Nold.

of the multiple workshops in different cities were invited to take walks wearing a galvanic skin response (GSR) sensor. Similar to Affectiva's sensor discussed in relation to biofeedback and the "algorithmic organic," this custom-made device registers the user's level of arousal and records it as it is associated with location (defined via GPS). The data are collected once every four seconds, which is the most optimal to create a smooth and detailed representation. GSR and GPS data, then, get visualized together as a spiky path projected on Google Earth, where the varying heights indicate changes in skin resistance that relate to the wearer's affect. The participants were also encouraged to add their comments about their activities during the walks (talking to friends, running, stopping to smoke, etc.), their feelings about the area, and events happening to them in the surrounding space (crossing an intersection, encountering something or somebody). Put together, the individual tracks accompanied by the comments produce a map of communal emotion.

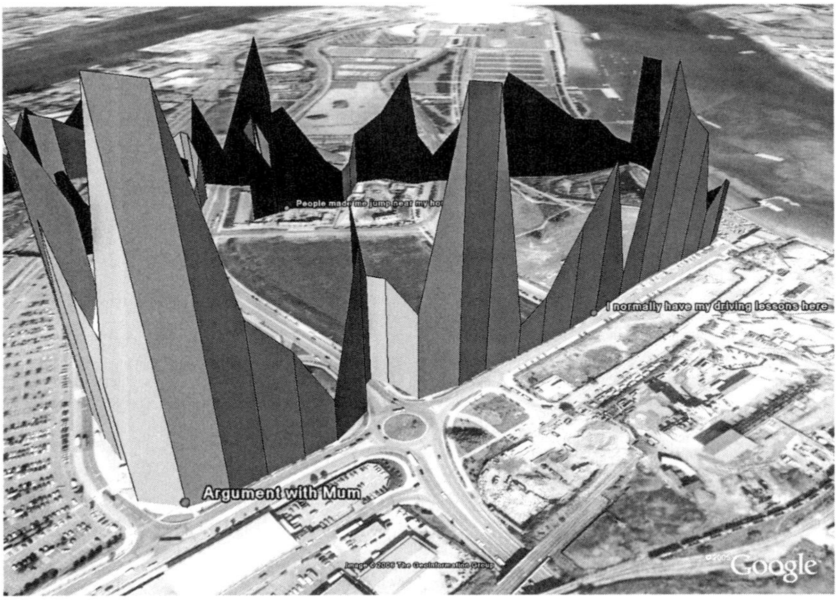

Figure 4.6
Christian Nold, *Bio Mapping* (2004–2012). Greenwich, UK. Emotion Map—Bitmap. Courtesy of Christian Nold.

Two types of maps are created as part of this project: two- and three-dimensional. The first type is a more traditional kind of mapping, without much attention to individual experiences. Rather, it gives a generalized picture with the levels of color intensity corresponding to the summarized data. In contrast to that, three-dimensional maps represent individual participants' tracks layered over a standard Google Earth image of that location (see figure 4.6 for an example).

In its original version, *Bio Mapping* is primarily a documentary project: the facts are recorded and stored and cannot be changed anymore. Yet one can say that the element of interactivity is still there—in the freedom of interpretation that the participants have over the "facts." The data acquire meaning only when they are read and interpreted by their "producers." The spikes on the tracks may be surprising, attract attention to themselves, and require an explanation that this same event otherwise would not have. In *Emotional Cartography*, a book reflecting on the project, Nold notes that *Bio Mapping* functions as an inversion of the lie detector: the spoken-word

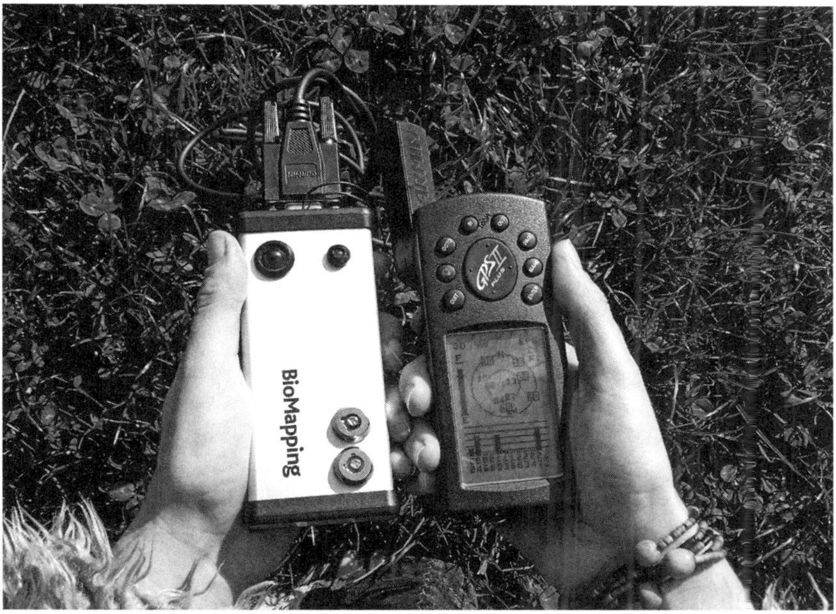

Figure 4.7
Christian Nold, *Bio Mapping* (2004–2012). Bio Mapping device. Courtesy of Christian Nold.

interpretation becomes the true record, and the biodata is an excuse to generate it.[40] These interpretations add a necessary layer to the project that calls for challenging the authority of biometrics and indicates the potential for endless varieties of narratives. Despite the strictly defined rules of the project's creator, it is a collective endeavor, and the resulting map tries to open up as many diverse and bizarre voices and interpretations as possible.

The participants' comments primarily include descriptions of external events and activities, rather than names of particular emotions. This is in contrast with contemporary mobile applications for tracking emotions and moods where the user selects an already given option, often based on Paul Ekman's list: happy, depressed, enthused, surprised, sad, angry, etc. In *Bio Mapping* these common indicators are less critical, whereas what is foregrounded are the raw physiological data and their interpretations by the people producing these data. *Bio Mapping* is not a way of telling people about their emotions; instead, it engages their imagination.

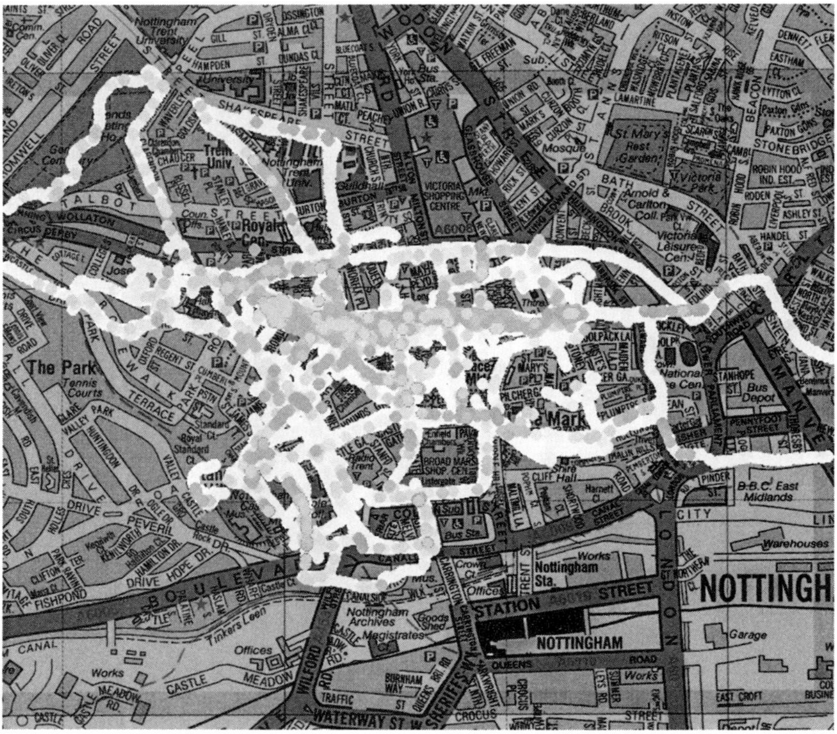

Figure 4.8

Christian Nold, *Bio Mapping* (2004–2012). Communal Emotion Map of Nottingham, UK. Screenshot from an exhibition at Angel Row Gallery. Courtesy of Christian Nold.

Bio Mapping appeared during the time just before the boom of commercial applications that serve both individuals who want to track and understand their lives better and the marketing interests of businesses that rely on this information about individuals' state of emotional engagement in public space. It does not satisfy solely private needs, nor does it attempt to present any "objective" picture of a collective emotion that could be used commercially. It emphasizes the "noisiness" and "messiness" of the data—their fluid, highly individual and idiosyncratic character. To explain his vision, Nold finds useful the concept of "propositions" as articulated by Alfred North Whitehead and extended by Bruno Latour.[41] In contrast to a scientific statement, which has to be either true or false, a proposition only has to be articulate or inarticulate. But what can be "articulated" is a variety of possible connections and associations that a conventional scientific formula cannot encompass (this may also remind us of the definition of "articulation" as a function of establishing junctions and linkages, explored in relation to *Bodytext* in chapter 2). For Latour, these associations are needed to make the system more relevant to other contexts, capable of developing new networks and even of reconsidering existing political hierarchies. Nold's innovative contribution is the introduction of *emotions as propositions*, which are "more 'interesting' than a scientific statement, and have the power to enroll new entities into unfamiliar networks."[42] (He gives an illustrative example of how *Bio Mapping* helped to instigate spatial reforms: in 2009 in Bethlehem, Pennsylvania, the mayor was so affected by the data from his own and others' tour to an abandoned mill that he arranged the construction of a new pedestrian area and a community center in that location.) Nold sees emotion maps not as representational but as discursive, reflecting the stakes of the unconscious, and thus becoming a territory for exploration, testing, and discussion.

Like many locative art projects in the epoch of Google Maps, *Bio Mapping* today would probably need to be different, and different in terms of its conceptual and not simply its graphical design. Today's social and mobile media tools allow more powerful capacities for sharing content, and a lot of ideas that seemed original and provocative in the 2000s have successfully been adopted by commercial companies. Yet Nold himself confirms that he would not add a social media component to *Bio Mapping* even today, due to the more "ephemeral" nature of relations in social networks, compared to the "concrete realities of place-based relations."[43] Linking emotions to geographical space is still a potent idea, but the concept of mapping can be extended beyond the territorial connections and can refer to the space of relations in a broader sense. Individuals, groups, or whole communities can be linked on the basis of some common experience (a service already provided by multiple online forums, etc.). As soon as they are established, these connections also need to be challenged and reconfigured, and the subjective and transient nature of affective mapping may be an appropriate solution for this. *Bio Mapping,* in its initial vision, addresses the fluid nature of the very material it tries to map (that is, the instability of data), and allows us to recognize the potential of the zones for creative interpretation and action.

The psycho-physiogeographic practices exemplified by projects such as *Indeterminate Hikes+* or *Bio Mapping* for the most part engage the participants via a visual interface. Yet an important grouping of locative art that demonstrates its proprioceptive aspects is represented by sound-based topokinesthetic experiments. Sound engages on a visceral level, through physical and yet invisible vibrations, and hence its impact on the experience of place can scarcely be overestimated. The object of sound, just like the object of proprioceptive sense, is impossible to seize or get a hold of: it exists only in the moment of its utterance or performance. Overlaid with and geotagged to particular locations, it is capable of creating an eerie feeling of defamiliarization and displacement. It can be seen as another way of interfacing with the reality of the imaginal, where the associations would appear at a level beyond verbal or visual descriptions, triggered either through recognition (the clanking of distant trains, or chirping of unseen birds) or through its own acoustic textures.[44] But the "cursor" "pointing" at that invisible layer of the past, the poetic or even a dream reality, is the body moving through space. As Elena Biserna, a scholar of acoustic participatory practices, notes, sound is a mobile figure par excellence: "auditory knowledge emphasizes the movement and becoming of reality, its (connective, transitive) process of appearance and disappearance."[45] Engaging with Jean-Luc Nancy, she echoes his argument that sound "is first of all presence in the sense of a *present* that is not a being . . . but rather a *coming* and a *passing*, an *extending* and a *penetrating*."[46] Such descriptions are relevant here because these qualities are interfacial qualities—descriptions of operations, or performance, of registration and transference of signals. The listening body perceives a sound as it flows through both time and space. Sound *becomes* an "auditory knowledge" only at the moment of being perceived by a body placed in space. That is, the acoustic characteristics of space revealed by sound depend centrally on the placement of the perceiver. The space itself is what allows the feedback loop of perception to close: it enables the perceiver to *feel* herself as an embodied spatial center (however arbitrary), while being immersed in sound waves.

The immersion of acoustic perception is the opposite of the distance effect, which is inevitably there in the case of visual perception. The sound field is not an object to be defined strictly. There is no one particular atmosphere which one is "immersed into." It is, rather, an immersion into the relational field, the space in between the borders of concrete spatial associations. As sound art researcher Brandon LaBelle poetically puts it, sound "expands and contracts space by accumulating reverberation, relocating place beyond itself, carrying it in its wave, and inhabiting always more than one place; it misplaces and displaces."[47]

There are various strategies that can be distinguished in projects developed at the intersection between sound and locative art.[48] *Radioscape* (2004–ongoing) by Dutch artist Edwin van der Heide can serve as just one example, representative of a number of the characteristics of sound locative art. *Radioscape* is an immersive environment that explores the bodily relationship to the medium of radio.[49] In this piece, sources of sound already exist in space ("aether") in the form of fifteen custom-developed radio transmitters that are distributed over a part of a particular city, each transmitting one layer of a "meta-composition." The participant navigates the neighborhood and shapes his or her experience through simple movement though space, while using a custom-developed receiver that is able to mix the received signals from the individual transmitters and gather multiple signals simultaneously (figure 4.9). As the artist describes his concept: "by interacting with the environment the visitors become 'inhabitants' of the

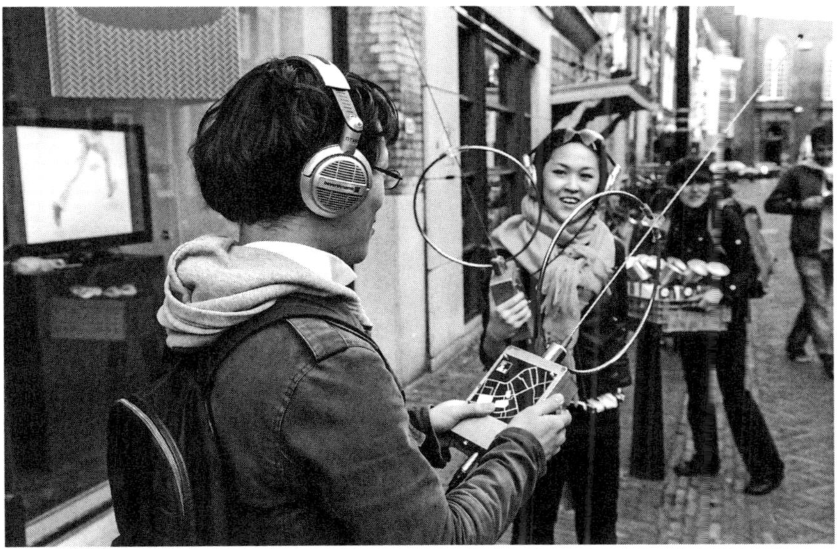

Figure 4.9
Edwin van der Heide, *Radioscope* (2004–). Presented at the Urban Explorers festival, Dordrecht, Netherlands, 2010. Courtesy Studio Edwin van der Heide.

organized transmitted signals. The visitors reorganize the area and assign new meanings to places."[50]

What the participants construct through their movement on site can be compared to the abstract geometrical figure that parkour performers intend to reproduce in *Topologie*. Radio signals form invisible figures with their range, but it is the human walker-navigator who—via the very act of listening—"draws" the invisible lines, consisting of the layered signals. The map emerging from this embodied experience is thus a multidimensional mental map where the visible environment is paralleled with a positioned sound that in turn can be activated and revealed only through the physical movement of a listening body through space. The volatile qualities of the received radio sounds add to the resulting construction being open, dotted, and multistable. The resulting mental image, then, can be called not only audio, visual, or kinesthetic, but also proprioceptive, since it engages navigation through the acoustic field (spread throughout space) via body movement. Sound itself, as a volume and a field, resists being captured as a line. No one "track" can be repeated: even if a particular path can be backtracked and reproduced, there is always a variation in signal that may depend on other people or objects in space, which in an urban environment is unpredictable.

All together, the artistic locative media practices described above help to frame the relation between the self and the surrounding environment by stretching the embodied presence in various unobvious directions. The center for those experiences is the situatedness of the perceiving body in the spatiotemporal "here-and-now." This lived moment in time and point in space serves as a base for imagining potential scenarios of movement or speculating about the layers of describing this place and moment that are not immediately given. These layers become layers of experience, but they are essentially *projected* onto the environment by the humans themselves; they are there only as a possibility. This type of speculative potential activated by the moving body (with its role as an index pointing at the wider indeterminacy of things) is where proprioception becomes important as a *feeling* of connectivity that encompasses both the bodily linkages and the connections with the sphere of the abstract, the imaginary, and the disembodied. As a sense, it uses "extrinsic" references to establish subjective and "lived" relations. Awareness of these volatile connections in time and space becomes the source of its own affective encoding.

Augmented Vision and the Gestural Gaze

Whereas locative media is a relatively wide phenomenon (since its major principle is geospatial tracking), augmented reality—or AR technology, in its specific sense as a technology—is primarily a visual tool. AR often (but not necessarily) builds on geolocational platforms and implies a view of spatial surroundings as networked and layered. But as opposed to just placing one within the representation of a place, it inserts additional features into one's view of the location as it is (or as it is seen through the camera and screen of a mobile digital device). To discover this additional layer takes movement: either of the whole body—to come to a certain location where the virtual figures are hidden—or just of a hand—to find the right angle and to locate an otherwise invisible object. In AR-enabled interfacing, the body's relation with space is thus strictly structured; more than this, as I propose, it can be seen as an experience of a relational structuredness more generally.

Wearable Vision: Shifted Perspectives in Head-Mounted Displays

Before going into more detail about what augmented reality is and what it represents, I will first address the tradition of practice that does not use geopositioning technologies and instead focuses on modification of the visual field and its effects on spatial perception at large.

Alongside the organ of perception, the apparatus of spatial perception includes the particular position of the body. But this apparatus can also be augmented by a device and its own optical, mechanical, and electronic predeterminations for the distribution of signal (or visual input). Today's visual technology (particularly AR) exists among a wide range of optical devices oriented toward transforming our sense of reality. The essential components of these devices have been lenses and mirrors, and in the digital age the digital camera—that is, an electronic processor of light signals. The different media and the settings prescribed by their apparatuses require, and thus produce, different kinds of gaze dynamics, meaning the trajectories along which gaze and perception travel. Immersive virtual reality headsets (or HMD, head-mounted displays) seem to be the most logical examples for the study of perspective perception or a positioned gaze (which will be introduced as being a gestural gaze as well), since the movement of the head directly corresponds to the direction of the gaze of the virtual reality subject. (The closest analogy is the effect created by the movement of a handheld camera in film, which reproduces the first-person perspective and a "natural" type of vision.) Moving through a completely virtual environment does trigger strong proprioceptive effects and a sense of disorientation, and providing these effects involves serious computational work. Yet I would like to bring attention to the projects that attempt to further our understanding of the process of creating an alternative reality and the motivation behind it.

Viewing devices, or *Sehmaschinen*, by Swiss artist Alfons Schilling constitute our first example. One of these devices, *Antilope (Gazelle)* (1985), was constructed of prisms and mirrors mounted on a wooden easel and was intended to invert the spatial relations of everything in front of the viewer: foreground becomes background, things on the left look as if they are on the right, negative space is seen in relief (figure 4.10). Schilling's long interest in stereoscopic vision led him to the realization that "space does not have to be what it is, that it can be manipulated." As he further explained in an interview:

> I have come to a point in my art where all the images that I produce exist only in the brain. They cannot be recognized with the retina; the information is processed further back in the brain. In fact by ordinary monocular vision these images are not perceptible. This is very different from how in the past an image has been depicted. I am now dealing directly with the structure of the brain and how the outside world is perceived. I have taken my art from the outer eyes to the inner eye. Such images are not realized through light. In a sense they are objects of the dark, mental constructs that become visible purely through their spatial coherence.[51]

Schilling's *Video-Head-Set* devices from the 1970s, considered one of the precursors of virtual reality, consisted of two small monitors placed directly in front of the eyes and transmitting views from two remote cameras shooting different objects and in different directions (figures 4.11, 4.12). The brain would have to put the two images and viewpoints together, fusing alternative realities. All *Sehmaschinen* were conceived to be used while moving. Even a device with a camera obscura principle called a *Dunkelkammerhut*, or walking darkroom, was supposed to be used in movement (see figure 4.13). As follows from the optical principle it was built according to, the image of the outside environment would be seen upside-down and flipped.

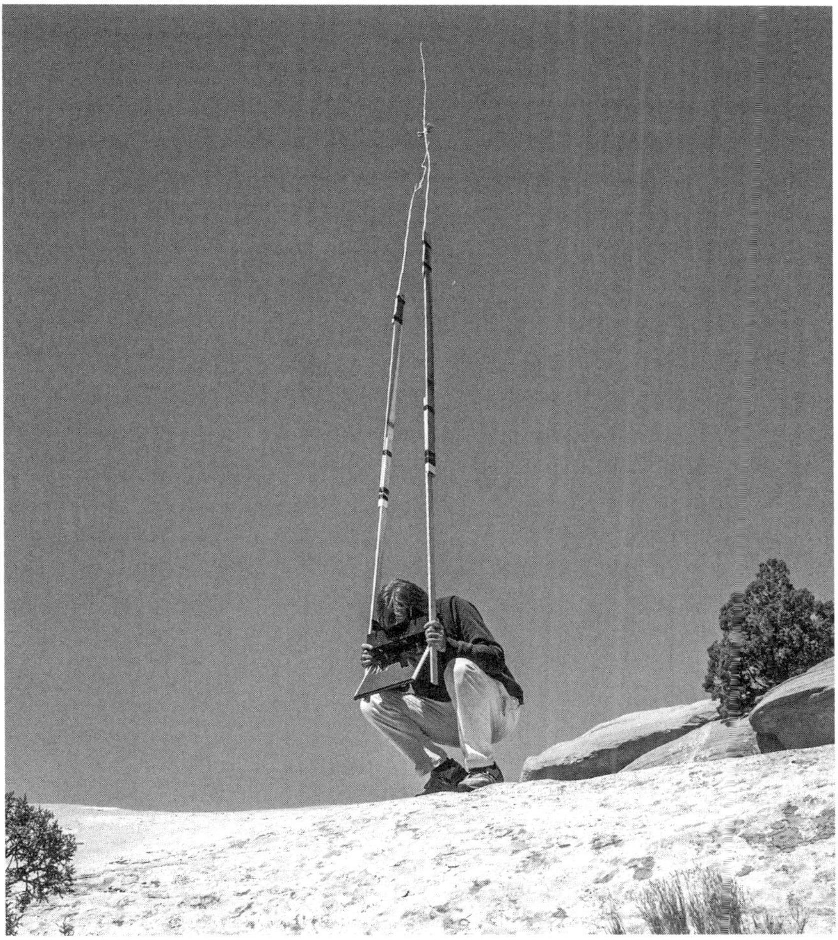

Figure 4.10
Alfons Schilling, *Sehmaschine: Antilope (Gazelle)* (1985). Portable viewing machine, wood, prisms, mirrors. © Alfons Schilling, courtesy Estate Alfons Schilling, Vienna. Photo credit: Archive Alfons Schilling.

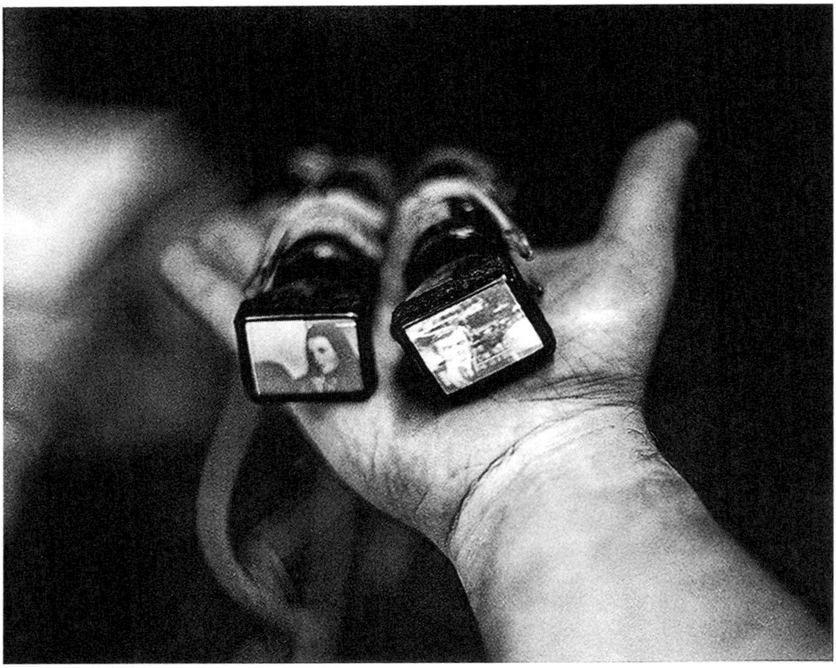

Figure 4.11
Alfons Schilling, *Video-Head-Set* (1973). © Alfons Schilling, courtesy Estate Alfons Schilling, Vienna. Photo credit: Archive Alfons Schilling.

Similar vision-transforming devices, particularly for making things appear upside-down, were also made by Carsten Höller and George Stratton.

Both Schilling's and Höller's experiments were artistic statements and were mostly analog or low-tech. However, another experimenter in the field of altered vision, Steve Mann, is of another register and discipline: as an engineer, he not only had the vision of developing devices to augment vision, but since the 1980s has created numerous technically sophisticated versions of *EyeTaps* and is considered to be the father of wearable computing. The principal difference of his projects from the other works using a live video feed (those of Schilling, Peter Campus, and Bruce Nauman, for instance) is that the *EyeTap* technology first sends the images from the camera not to the eye but to the processor that modifies them. As a result, the eye sees a mediated and altered image. Still, it is an augmented and not a virtual vision, because the transmitted images are sourced from the real

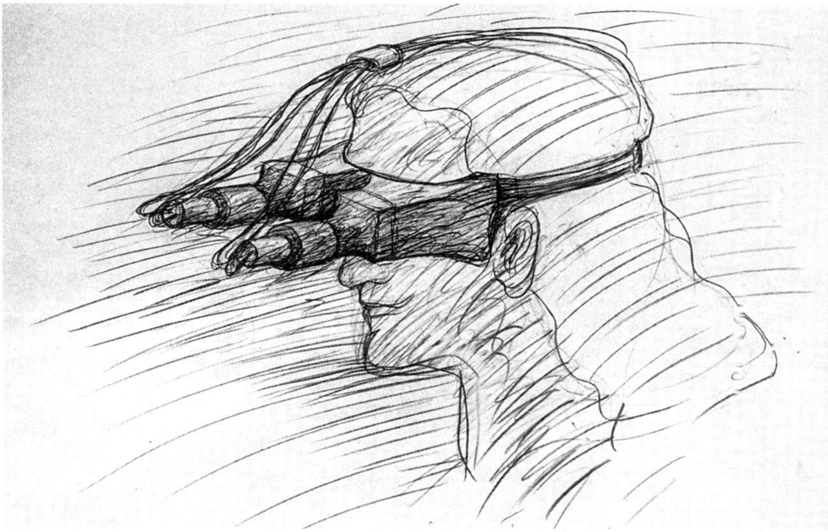

Figure 4.12
Alfons Schilling, *Video-Head-Set* (1973). Drawing. © Alfons Schilling, courtesy Estate Alfons Schilling, Vienna. Photo credit: Archive Alfons Schilling.

environment. The view from inside the glasses or goggles would include a superimposition of the real-time camera view with computer-accessed images and information such as word and data processing, wireless-acquired communication, etc.—almost as in today's Google Glass. The *EyeTap* can also stream directly to one's eye the real-time video of other "gloggers" (short for "cyborg log"), thus raising questions about what it means to share one's first-person perception of the world, as well as about the borders of the private, public, and communal. Besides the simple upside-down switching, the program settings include infrared, radar, sonar, and other kinds of computer vision otherwise totally unavailable to humans. Mann explains his original motivation and the current concerns: "One of my original impetuses for constructing these systems was a desire to alter and extend personal reality through the use of technology. Gradually, as both myself and my inventions matured, my intentions changed and the systems became concerned with the more general issue of personal autonomy in an era of increased automation and ever-more-invisible technologies."[2]

Whereas Mann's inventions became influential for a whole industry of wearable AR, there is also room for a justified critique, as pointed out, for

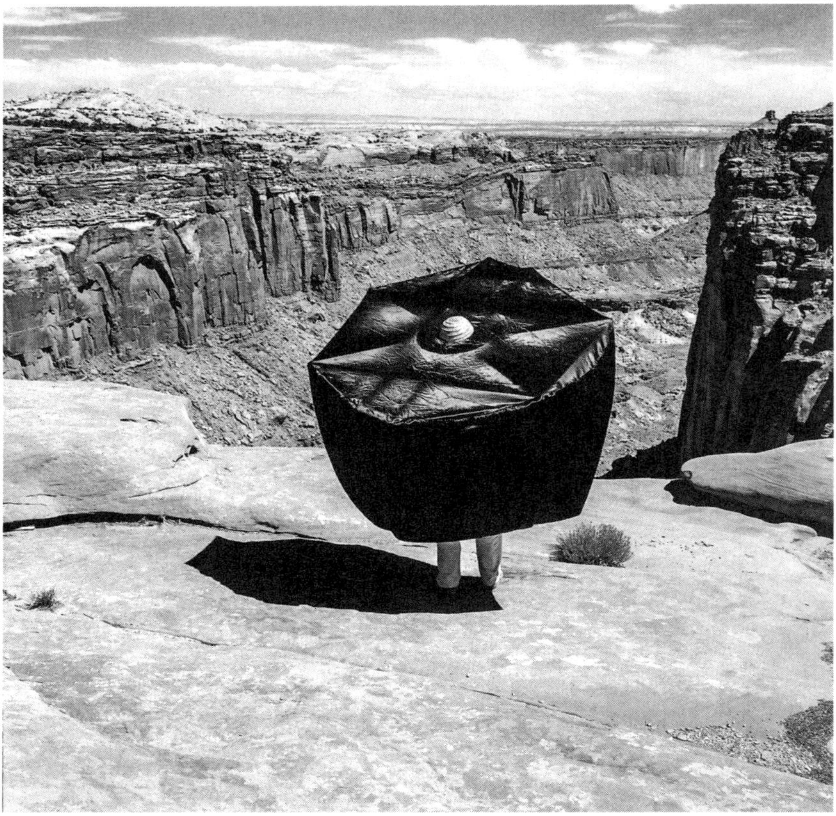

Figure 4.13
Alfons Schilling, *Sehmaschine: Dunkelkammerhut* (1984). © Alfons Schilling, courtesy Estate Alfons Schilling, Vienna. Photo credit: Archive Alfons Schilling.

instance, by Jimmy Loizeau and James Auger in their ironic take on mediated communication, *The Interstitial Space Helmet* (*ISH*) (figure 4.14).[53] A small screen is placed 17 cm from the eyes and shows a video image of another person as captured within a similar head "camera." The human head is literally locked within a helmet in front of a screen, conveying the creators' commentary on the growing amount of human interaction via web cams and artificial personas. The *ISH* pushes the idea of mediation to an absurd extreme, merging elements of the virtual world into the physical world. It is the opposite of augmentation in that the screen image not only alters the scope of vision but also shrinks it.

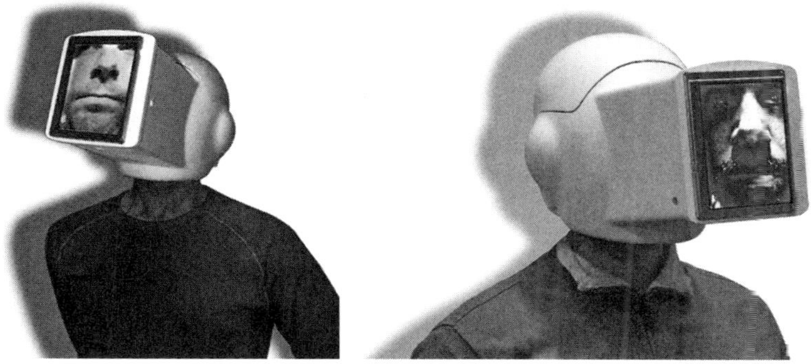

Figure 4.14
Jimmy Loizeau and James Auger, *The Interstitial Space Helmet (ISH)* (2004). Courtesy of AUGER LOIZEAU.

Another contemporary project, similarly ironic and yet serious at the same time, also deals with confinement and a limited view. *Reverse Blinking* by Ief Spincemaille creates the experience of being captured inside a camera, but unlike in *ISH*, it is a photo camera (figure 4.15).[54] This allows the viewer to see the world only in flashes, when the shutters open for 0.3 seconds before closing again. They open and close very quickly and continuously, creating the impression that nothing around "has time to move": "Everything he or she points his/her gaze at becomes a photograph, a memory; people become frozen figures, and life turns into some kind of slide show."[55] The real environment thus is being turned into an image, which, moreover, is an unrealistic and implausible one. With this relatively simple analog device, Spincemaille comments on the condition of contemporary culture, in which "the world has become an image and the image has become reality."

Finally, there may be no image at all in the head-mounted camera. In that case, as in the immersive documentary experience *Door into the Dark* (2014) by Amy Rose and May Abdalla (Anagram), participants are left in the darkness alone with their own imagination, and . . . a voice that guides them through the recorded journeys of real people. With a black box on your head, you identify yourself with a climber making a treacherous ascent up the cliffs of Patagonia, or an explorer in the dense woods, or a mentally ill patient leaning against a weirdly moving wall, descending into schizophrenia. The narration "takes you deep into [these characters'] world

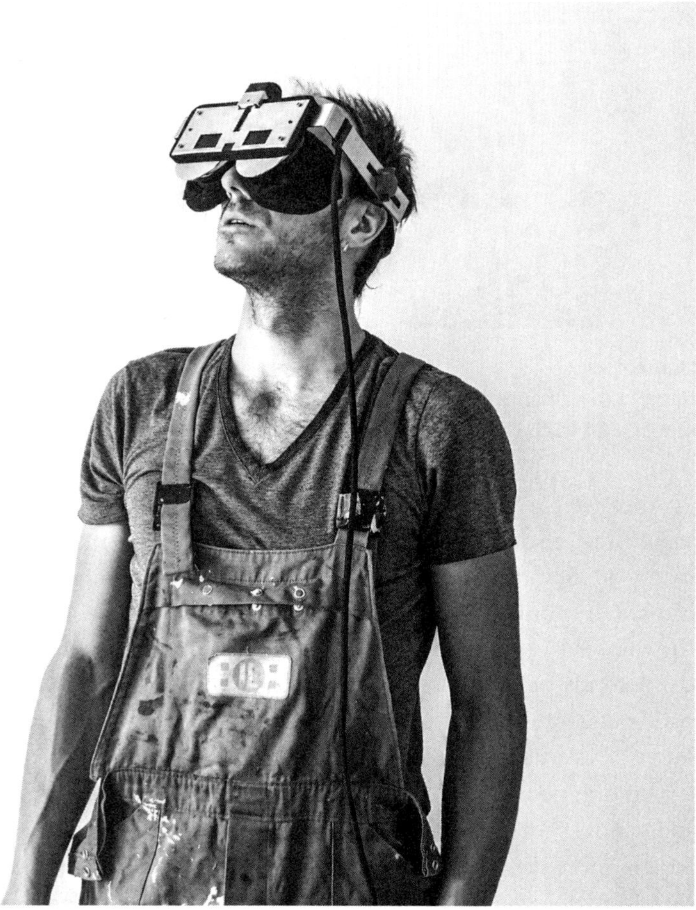

Figure 4.15
Ief Spincemaille, *Reverse Blinking* (2010). Courtesy of Ief Spincemaille.

of sensation, risk and illusion. To find your way back out into the light, you must surrender to the unknown."[56] The feeling of being lost—without any orienteers—turns into the feeling of being guided, of experiencing trust and curiosity. The space to navigate is reduced to the imaginary, while the body moves through the physical environment of a gallery. Despite the voice narration that is crucial for the immersive effect, this experience is different from that in sound-based projects that incorporate location or site more actively and strategically (as will be demonstrated later). Deprived of sight, one navigates the space only through kinesthetic sense. Yet there is still a visual component—the imagined pictures created by the narrated text. Vision itself, then, can be interpreted as the augmentation of an image that is possible without any actual visual stimulation.

These projects show what happens when the very physical conditions of vision—literally at the level of the head and the eye—become subject to manipulation and potential delimitation. The wearer of a head-mounted device, as in these examples, bears a vision of a slightly different world. The shift between the "normal" and the transformed worlds is what matters in terms of its psychological effects, since this shift is in itself profoundly affective.

General Principles of Augmented Reality Art

Augmented reality (AR) technologies illustrate the paradoxes and promises of the mechanisms of identification, simulation, and imagination that increasingly impact the lives and experiences of a contemporary human. Augmentation implies a layer of information (visual, textual, auditory) that can be called upon to appear or disappear on the user's screen and correlate to the objects or situation in the actual environment.

The conceptual strategy of AR art is to display a layer of the imaginary over the natural visual field, to intermix the real and the virtual, to make visible and "feelable" the surrounding environment in its spatial depth, and to indicate the potential for modification and intervention. The virtual here is always an addition to the already given: one complements the other. The result is a superimposition of realities. AR interfaces are not simply windows and displays of information to navigate. They are *operative*: manipulating the object on the screen may be correlated with a mechanical device that maneuvers objects in the real world. This ability distinguishes augmented reality from virtual reality: with augmented reality, there is a continuity between the physical and the virtual. The "real" does not disappear; instead, new imaginary objects are projected onto it. With the experience of AR, the optic serves as a trigger to instigate specific proprioceptive feelings, such as anticipation and the desire to experiment with new bodily modes of control. Such an interface challenges the pregiven conceptions of presence, and opens people up to the embodied exploration of other potential modalities of existence and manifestation that have to do with tighter feedback loop connections between the inside and the outside, feeling and thought.

AR art (as opposed to commercial projects) often claims to emancipate the viewers' ability to discern the constructed nature of reality, allowing them to see alternative interpretations of what is there. An image is defined here not only as seen through its visible surface. The proposed means of fidelity—3D images inserted in real time within a certain localized frame—may in the end not be as convincing as they are believed to be (for example, because of the framing itself). Yet this technology furthers the conversation about the ontological (and aesthetic) status of the virtual image, its effects and affects—a discussion that has recently been hyperactivated by the advent of Oculus Rift VR and other increasingly more technically advanced types of hardware.

The duplication of reality by adding another, imaginary layer has a productive effect: however arbitrary and even absurd, the effects of the virtual presence offer a new perspective that makes one rethink the status quo of spatial relations and the connotations behind them. An image on-screen becomes a medium of its own, with its own agency, a projection surface for new meanings and new types of connections. What is produced is a fresh understanding of space and a sense of *copresence*—the presence of something that leads beyond an individual experience. This is similar to the doubling effect of motion capture, where (as I showed in the second chapter) the screen space also becomes a meeting point between the virtual and the actual. Yet, whereas motion capture digitizes the movements of a "figure" and places it within a "virtual" context, or "ground" (to continue the "figure/ground" idea of Gestalt psychology), in AR it is the "figures" that are digital and virtual, while the "ground" is the "actual" physical environment (even if seen through the screen).

Examples can include projects by artists associated with the Manifest. AR platform, famous for their interventions at the Venice Biennale in 2011, MoMA, and Occupy Wall Street, as well as other actions involving politically charged statements (the location-specific virtual replacement of iconic monuments, such as the Statue of Liberty, with provocative alternatives). A predominantly formal exploration of how 3D illusion transgresses the border between immateriality and materiality and becomes operational and experiential may be exemplified by *Fractured Visions* (2014) by Tamiko Thiel.

Reminiscent of avant-garde formalist experiments, the project attempts to model the optical experiences of people with palinopsia, an effect in which images in the visual field are multiplied and repeated in the form

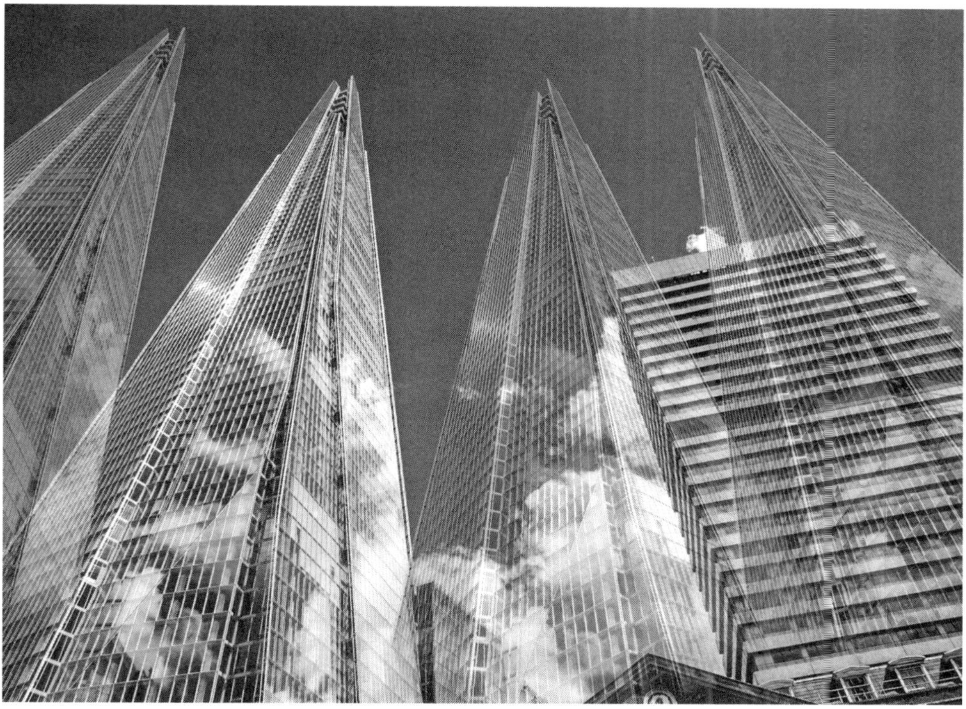

Figure 4.16
Tamiko Thiel, *Fractured Visions: Multiplicities* (2014). Augmented reality installation. Commissioned by the AXNS Collective in collaboration with Dr. Dominic ffytche, King's College London. Photo by Tamiko Thiel.

of superimposition, "trailing," "cascading," etc. In the subproject *Fractured Visions: Multiplicities* (figure 4.16), the image of the iconic London Shard and buildings around it perseverate, turning into "multiple, repeated, fragmented objects that dominate your visual field and give you the sense of being 'enclosed' in a spatial, animated visual hallucination."[57] A formalist study, it aims to imagine an alternative type of vision, based on a rare but natural disorder (or rather a symptom). The created effect is a feeling of disorientation and visual confusion. Besides being an experiment in perception, for Thiel it is also a commentary on urban architectural aesthetics—the contrast between the sleekness of contemporary architecture and the brutalism of the 1970s (Guy's Tower next to the Shard). In this context, AR becomes an example of the classic "consensual hallucination"

(William Gibson): the realities of different cultural visions and styles coexist, whether we want that or not, being both "there" and "not there" at the same time. The scientist and clinician collaborating on the project (Dominic ffytche, KCL Institute of Psychiatry) noted that AR helps to reveal palinopsia's dynamic nature—the point of gaze, changes in fixation, the timing in relation to eye movements, the transition in and out of perceptual consciousness of the palinoptic image. The nuances of dynamic experience could be possible to discover only with the accurate modeling enabled by calculation methods.

Whereas *Fractured Visions* is about the transformation of the perspective on existing objects, Will Pappenheimer's *Large Reflector* (2013) is a virtual object—a sculpture with rotating mirror panels, "installed" in New York's Central Park which, instead of reflecting its immediate current surroundings, shows the 360-degree view in Google Street View of the same park location. The view is thus doubly mediated, locating the viewer between two worlds—the "real" world (with its changing weather conditions, other passerby or even passing animals) and the world of digital abstraction, retrieving over and over the very same "stored" image of the place. (The title is also a reference to Duchamp's *Large Glass*, which in this case reflects the site of the virtual spectator.)

Finally, a more conceptual and critical use of AR is *blemish* by CONT3XT.NET, a group of Vienna-based artists (Sabine Hochrieser, Michael Kargl, Birgit Rinagl, and Franz Thalmair), which was presented as part of the "Invisible Pavilion" during the Venice Biennale 2011 (figure 4.17).[58] It addresses the very medium of visual production in the digital age and its elementary unit—the pixel (in this case defective, or dead), taken within the larger context of the institutional politics of image production, representation, and circulation in the world of global contemporary art. *blemish* focuses on the structure of such a potential image and its weaknesses, the blind spots, emphasizing the unstable nature of reality itself and the nature of art as navigating in between the material and the immaterial. The critical dimension is revealed here through the physical experience itself—AR exploration of the renowned site of art consumption. The artists write: "Ephemeral image vacancies are inscribed in the mobile display as a layer of defective pixels and can be read as an intervention in the technological conditions of devices that serve for the representation of reality by digital means. . . . Which of the significant components of a digital product are visible, which are not?"[59]

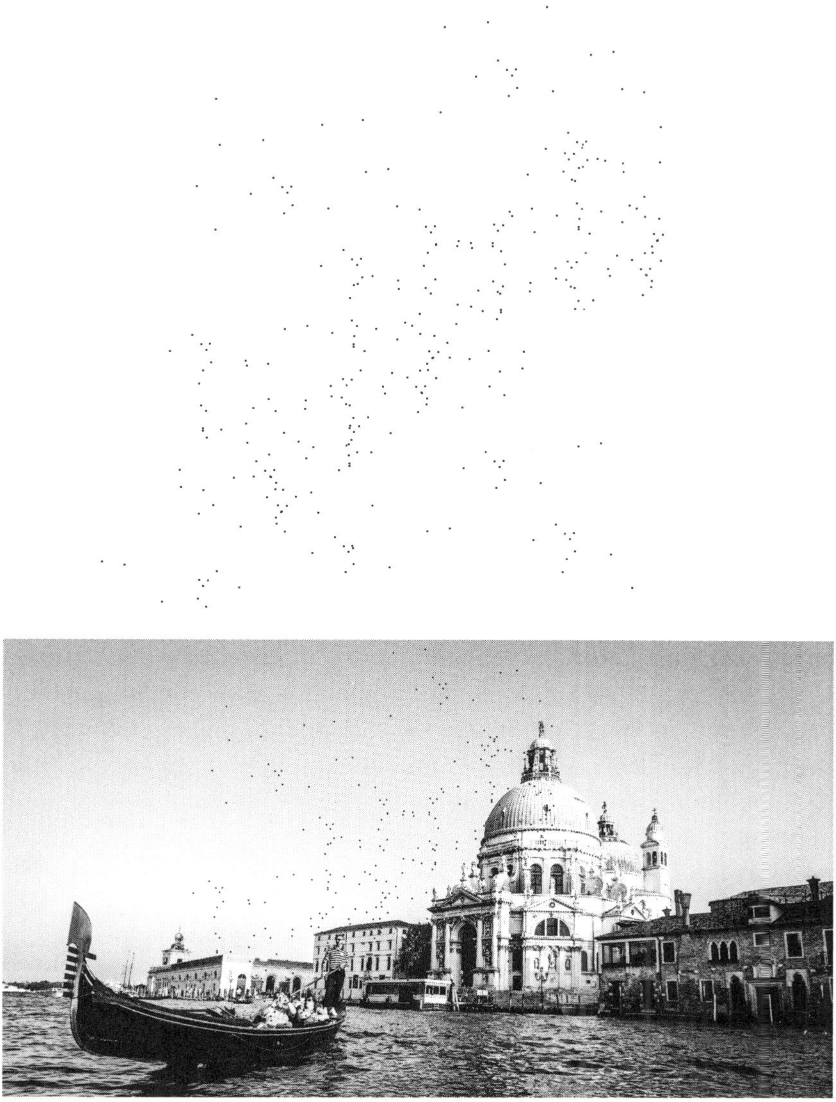

Figure 4.17
CONT3XT.NET (Sabine Hochrieser, Michael Kargl, Birgit Rinagl, and Franz Thalmair), *blemish* (2011), Venice. Courtesy of CONT3XT.NET.

The "leftover mistakes" of the system of production and representation symbolized by the "dead pixel" is a comment on the cracks within this system, deliberately prompting confusion in the audience so they will become aware of the provisional character of reality and the methods of reproducing it.

These projects may not be the most typical examples of AR art in that, except for Pappenheimer's work, they do not fully realize the potential of the "independently" existing imaginary objects, such as fictional characters and other figures with which the users can interact. Yet they call attention to *what* is being seen and *how*, as well as to the complexities and potential shortcomings of the figurative approach to reality in general—its unsteady and yet regulated nature.[60] All of these projects accentuate the experiential dimension, which would not be possible without the embodied quality of perception of AR.

The presence of a virtual object or transformation of vision, as in *Fractured Visions: Multiplicities*, can be said to be a pure visual effect. In all these cases, the event in its sensory certainty would be impossible without visual perception. It is viewing that causes the virtual to manifest itself both in its appearance and in potential action. As Jean-François Lyotard demonstrates in *Discourse, Figure*, what is important in the work of the visual (or the figural, in his terms) is the act of manifestation, of showing. He distinguishes the figural from language—that is, from the discursive regime of signification and the construction of meaning. The figure gives language "density" and "thickness," as well as both content and the very ability to express. Lyotard defines expression as a "presence of meaning" that appears on the "surface of discourse through effects."[61] In the case of AR, the effects of a distorted or enriched image of the environment refers to the meaning that arises in the space between the virtual and the physical, the mathematically constructed and the intuitively perceived. The conditions of these effects already have to do not only with the visual but with the distribution of meaning within the space, with the *potentiality of space* to "hold" the image. Here we may recall the deictic nature of all locative media: they always imply a certain context in which they operate; even if it is not defined specifically, there is still an idea of "where," the spatial context. The appearance of virtual objects and transformations of the visual field take place at specific locations, but even there they are still to be discovered. AR interfaces invite viewers to have novel perceptions and interactions within the given location, but also within the space where the processes of signification and "making sense" happen: what is "expressed" is both on the screen and in how the screen is being operated.

Gestural Gaze

In order to be perceived, virtual objects need to be placed. Part of the technical dimension of this procedure is the identification and marking of a very specific point in space that could be translated into geographical coordinates and become a constitutive element of the code behind the work. AR uses a different approach to navigable space than, for instance, the one that navigation by stars offers: figuring one's relation to the stars is needed only insofar as it orients humans within their own human landscape; in other words, spatial location is a function, a product of the relation to stars. The idea behind AR is that the relations are established with virtual objects and not only with the points in space that they reference (this is where the indexical or referential nature of virtual objects comes into play). What it means, in part, is that the connection to the objects indicates the linkage between the physical space and imaginary space—not the space of mathematical abstraction, but a truly hybrid space. With that, it also speaks of the larger potential of spatiality in general (both physical and imaginary), and here is where the role of perception becomes crucial.

Operating in between the physical and the virtual, AR objects have a special relation to the eye and the gaze. The gaze is taken as spatially positioned and mobile. These two features working together make it *"gestural"*: directional, as well as expressive and indicative of a certain meaning. Understanding the gaze as "gestural" implies at least two moments: about the characteristics of the space under/within view, and about the operative nature of looking.

The potential of AR (and its interest to artists) is about orienting oneself in a space marked by the kinds of coordinates that are not necessarily mathematical. Hence, the space itself is characterized in other kinds of categories that are related to a subjective experience of the parameters imagined by the artists. AR technology is not created for navigation in the same sense that conventional digital maps are. The principle is the same—we see how we move, and it is movement that enables vision and establishes the map. In AR, no map may be produced, but the dynamic position of the body is still critical for seeing the virtual indicators, or orienters, within the alternative, semifictional, hybrid reality. Movement—both of the body and of the hand holding the device—sets the direction of the gaze, and the combination of the two is enough to generate the designed effects. Tied to movement, looking becomes a gestural operation. Navigation lets the haptic and the optical regimes of looking converge.[62] (This is also similar to what Edwin van der Heide explained in relation to the experience of his *Radioscape*: just tilting or rotating the antenna device, while standing in one place, would produce perceptible change in the sound.)

A slight change of the angle of view can make the object appear or disappear. The illusionary line of connection with it is controlled by the beholder's *movements* that activate it by discovering its very existence. For instance, in Pappenheimer's *Large Reflector*, certain sides and "mirror" images can be accessed only from a particular angle, just as a real sculpture should be explored from all sides by walking around it. The unfamiliarity of the object or the visual effects of the existing object (as in *Fractured Visions: Multiplicities*) stimulates curiosity and the desire to engage further. As a result, the engagement is not so much with the illusory virtual but with the actual environment, be it the geometries of a city, its representations, or the world of art. Individual gestures and movement involved in AR art are a crucial part of the dynamic emotional experience of location, introducing additional, more improvisatory, and embodied spatiotemporal dimensions that transcend veridical cartography. Media content and interaction are resituated in the real world and thereby into everyday action, incorporating techniques of mixed-reality construction.

It is important to clarify that the movement and the gaze belong simultaneously to both the human and the machine (in the form of a smartphone or tablet device)—both enable perceptions, and thus constitute the perceiving apparatus. At a given moment, the gestural gaze and the device actualize one of the numerous possible "points of interest," the images on the screen serving as indicators of recognition and a potential meaning. "Point of interest" is a pun on the cinema studies concept of "point of view," coined by AR artist Patrick Lichty. As he puts it, "the relationship of the viewer, landscape, and media infoset compounds the point of view through multiple points of interest (POIs) in the landscape."[63]

Different media and the settings prescribed by their apparatus require, and thus produce, different kinds of gaze dynamics, different trajectories in which gaze and perception travels (versus the trajectory of the processed data itself). Artist and theorist Geoffrey Rhodes compares several diagrams that describe the relations between the viewing subject and the object of screen-based media, such as cinema, video art installation, and AR. In the case of cinema the spectator is aware of the screen and yet tends not to notice the operations, such as editing, framing, etc., and the image is seen not as an image but an alternative objective reality.[64] Video art logic, particularly in closed-circuit-based performances and installations by Peter Campus or Dan Graham (discussed in chapter 2), places the medium in the center of perceptory loop along with the eye of the viewing subject and the object of the gaze: "the medium becomes the nervous system."[65] In the case of AR, as Rhodes emphasizes, the viewers are aware of the screen as an image and not a reality. They are also aware of the augmentation, yet they don't pay attention to the computational processing of the video and they attend to the results unconsciously. "AR does not attempt to embed the viewer in an objective reality, the spectator identifies with a mediation—the mediation that stands-in for reality."[66] Rhodes suggests that the strategy of AR avant-garde art should be to advance the expressive potential of the "repressed elements of computational tracking and video compositing," thus strengthening the original specificity of this medium. This comparative view on different types of media and their engagement with the gaze is helpful in our context since it stresses the principal changes in the relational quality of the gaze in AR. The gaze itself becomes a part of an external mediatory apparatus that specifies its direction. Unlike the "neurotic" self-focused gaze of video art, AR empowers the gaze to "collaborate" with the algorithms according to which the figures are placed in space; the only condition is that the gaze has to be coupled with a gesture.

The distinction between *visual* and *haptic* types of augmentation could also lead to some interesting conclusions about perception in general, and proprioception in particular. Experiences with a headset and a handheld device indicate this exact distinction, the former device being more immersive, and the latter relying more on the haptic sense and the feeling of extension of the eye through the hand and the held device. This may be reminiscent of the famous fascination with the hand in philosophy: Descartes endowed it with a quality of independence from the rest of the body, and in Martin Heidegger's writings it almost becomes a symbol of humanness (due to humans' ability to use tools), and an indicator of the ability to think: "The work of the hand is richer than we commonly imagine. The hand does not only grasp and catch, or push and pull. The hand reaches and extends, receives and welcomes. . . . Every motion of the hand in every one of its works carries itself [*sich trägt*] through the element of thinking. . . . All the work of the hand is rooted in thinking."[67] Awareness of the positioning of the hand and its role as the "controller" in the interaction signifies a pragmatic but also reflexive attitude. Reflected in this awareness is the fundamental ability to affect and be affected by a conscious entering into relation.

AR Proprioception

Let us now go back to the concept of proprioception and see how this sense matches AR-enabled interfacing (and vice versa). Proprioception relies on positionality and movement; it is a feeling of body points in motion and is, therefore, inherently a metastable state. Since it unfolds in time through movement, it has an active modality. Various sensory modalities may differ from one another depending on the nature of the object and its appearance, but at the core of all senses lies sensorimotor skill. Alva Noë develops a similar assumption about perception more generally. The enactive approach allows him to incorporate elements from both the qualia theory (according to which sensory modalities differ qualitatively) and the "proper objects" view (each feeling corresponds to a particular object: sound for hearing, color for vision, and so on).[68] Proprioception relies on a specific kind of action, one that allows the "enactment" of qualities that go beyond any particular sense and appear *in between* individual senses.

In order to make this applicable to AR (addressed here primarily in the narrow sense of a visual technique, with a broader understanding kept in mind), we need to understand better the role of an image in proprioception. Any type of perception is based on the fact that the mind constantly compares information coming from the outside world with the perceptory data already received and processed. "Image" is one way to name that information, yet the input does not have to be only visual data but can be close to what I described earlier as the "sensible." The image, understood more broadly, can thus include acoustic, haptic, but also mental images, or images formed in mind as the result of any kind of perception. In the case of proprioception, what is being perceived is processed by neither vision nor the haptic sense, but by a more complex mixture of all five senses that engage the whole body (and the feeling of the body as a whole). Brian Massumi describes proprioception as the "spatiality of the body without an image," exactly because it "registers qualities directly and continuously as movements."[69] The "image" implied by Massumi is both a visual image (including the image of the body) and close to the mental image of the perceived. How, then, can perception of images be proprioceptive? What do AR technology and examples, such as the cascading image of the Shard, add to this?

Austrian psychiatrist Paul Schilder offers a productive way to look at this challenge by emphasizing what we can call the role of chance and potentiality in bodily self-perception (or self-image). For him, "the body image is not only a picture of the body but also an anticipatory plan for the detailed movements the body must undertake in order to act"; the plan for movement is "neither conscious, nor unconscious, but preconscious, that is, capable of being made conscious."[70] The body image described here is actually the body schema, or *Körperschema*, the term that inspired Merleau-Ponty and his own theory of bodily perception. Schilder's "anticipatory plan" is also resonant with Merleau-Ponty's ideas of the task-oriented and situationally positioned body: "Brought down to a precise sense, this term ['task-oriented'] means that my body appears to me as an attitude directed towards a certain existing or possible task. And indeed its spatiality is not, like that of external objects or like that of 'spatial sensations,' a spatiality of position, but a spatiality of situation."[71] The attention is thus shifted from the body to the situation this body is in, and more precisely, to *a* situation that the body *may* be in. This helps to explain the nature of proprioception as the field of potentiality that precedes actualization. The potential, preconscious movements and the image underlying them can be compared to what Lichty describes as a "vectoral gesture"—the procedure of locating the virtual object in the site of the intended audience, which is done with the help of "gestural lines of intent," the imagined gestural operations of future spectators.[72] Virtual objects are placed *in anticipation* of their later discovery, similar to the body's being a destination index in locative media. Despite this being a procedure that involves complex calculation and thus strategizing, the anticipation of *potential* users and their positions reveals a new dimension in proprioception—as a feeling of the *collective, spatially distributed body*. The subjective perspective is projected here onto the field of *intersubjectively* shared possibilities.

AR images aim to interfere in the communicational chain, a network of signals traveling and exchanging information within the body-mind complex. Thiel's *Fractured Visions: Multiplicities* serves as a curious example:

unlike most of AR art, it does not create an object but instead simply alters the viewer's perspective on existing objects, thus pointing not at the object but at the subject of perception. The impression is that if something changes, it changes within you—as if the image were appearing not on the screen but within the eye itself. This makes one connect vision and movement more closely and anticipate through the body the next *bodily informed* image. It is this kind of embodied anticipation that can be called proprioceptive in this piece.

As Massumi explains, proprioception is an abstracting mechanism and has a special relationship with thought. Building on all sensory modes, proprioception is never one of them but instead is "the mode of experience of the amodal as such." The *amodal* transgresses modalities of perception given through particular senses, just as it goes beyond a particular object (or "image" of an object) given by the senses. According to Massumi, what is held in common among the different modes of perception is a rhythm. But rhythm itself is amodal:

> it is the Abstract shape of the event as it happens, across whatever modes it happens with. It is the immediate thinking-feeling of nonlocal linkage. . . . Proprioception is not only abstracting. It is self-abstracting: it is by nature recessive, always already slipping away behind the other sense modes to the nonconscious limit of experience, where sensuous experience rejoins the pure activity of thought.[73]

For Massumi, the borderline between the sensuous and the nonsensuous, the visible and the invisible, is discovered through movement. Movement produces "relational reality," "the suggestive force of the shared nonsensuous experience." He refers to the cognitive philosopher Albert Michotte, who explains movement as having the uncanny ability "to survive the removals of its object"—a perceptual feeling without actual perception. Perhaps what is left after the "removal of the object" is exactly "pure thought" and a sense of "relational reality." The question, then, is how technologies participate in triggering and enhancing this sense, and what mechanisms are involved.

In the case of AR, a slice of visible reality is "taken out" and replaced (or rather superimposed) with an alternative artificially created image. The construction of that image (and the resulting illusion in the perceiver's eyes) is based on calculation—that is, the process of comparison, which establishes relations between the abstracted data. This procedure by itself is nonsensuous and yet it relies on the perspective of the potential perceiver, on his or her body movements. So it is the movement that has to be preprogrammed and that later would activate the alternative reality. In the experience of AR (or through AR-enabled interfacing with space), the viewer is confronted not just with the virtual image, but with the image of the virtual, with the nonsensuously (analytically) preestablished relational structure. What viewers are left to do is only to match their own sense of "relational reality" with the "gestural lines" that an AR producer had set up.

This adds a special dimension to conceptual AR art projects, such as the above-mentioned *blemish*. The defective pixels represent the blind spots not only in the art world but in our perception of art as something given, as perceptible through the senses and hence prone to illusions and mistakes. They point at the potential mismatches between people's expectations of the images they are about to see (based, in part, on the attached marketing efforts) and how the works appear to them in reality. Technology is used here to give a critique of how images are being treated in the dominant art system. Yet this questioning of image operations can be addressed to how a blemish, or dead pixel, itself is made to operate. Blemishing the traditional kind of image, or "removing the object" (in the sense of Michotte), does not let us escape the structure of "relational reality" that we are dependent on in order to receive this critical message: the possibility of seeing these black dots is already preprogrammed, and our individual perception structure is preenvisioned (if not by the system of art, then by the capabilities of the machine). But to receive this message would mean to have the "sensuous experience" to "rejoin the pure activity of thought"—to *feel thought*, to experience in your body the very principle, the flexible structure, of your own perception returned to you through the AR apparatus.

Aspects of proprioception—such as relationality, the opposite of exteroception (object perception) or interoception (self-perception); the sense of potentiality and metastability with movement as its core; and finally, the abstracting capacity—enrich and challenge our understanding of the experiential and conceptual potential of locative media and AR art. AR technology—with its engagement of both the visual and haptic senses, the virtual and the actual, intuitive bodily movements and preprogrammed "lines of intent"—invite us to expand the meanings of proprioception by applying them to the sphere of human-computer interaction. Among the questions to be asked further are: What is a movement that cannot be preprogrammed (imaged by the machine) and yet produces effects? How are the criteria for the relation between real movement and the virtual image defined? Where is the borderline between the human and machinic control?

Space mediates and structures the relations between the inside and the outside of the body and gives context to how the self perceives and defines itself. Moving through space and actively perceiving it, the body becomes an index of the latent social, cultural, or personal meanings embedded within the environment. These meanings cannot be recorded on a map in the same way that they are evoked through the lived experience of a place. At the same time, the subject can also feel differently depending on the location, which could activate new responses and stimulate one to look at oneself afresh. The feeling of proprioception offers the sense of meaning located exactly in between the embodied self and a place. The self steps outside of its own confines and becomes aware of its situatedness within a larger spatial whole.

The technologies make evident the mediating factor of spatial operations expressed in the effects of the environment on the self. At the same time, they call attention to the body as the major conditioning factor in perception: movement, turning of the head, direction of the gaze, and hand gestures become instrumental in determining what and how one feels about the world. The capacity of locative media and AR to situate the subject within a mixed reality of concrete territorial and perceptive as well as associative, speculative, and imaginary characteristics activates the initial potential of spatial relationality and augments it. These technologies' interfacial capabilities anticipate and thus preprogram the potential behavior of the user: for example, through the placement of "points of interest" and virtual objects in specific predetermined locations, like *Fractured Visions: Multiplicities* in front of London's Shard or *Large Reflector* in Central Park, New York. Yet at the same time they allow new forms of self-expression (as in the parkour performance *Topologie*) and observations about the environment (as through the app *Serendipitor*) and about oneself (like measuring one's psychophysiological response in *Bio Mapping* or counting one's heartbeat following the prompts received during the *Indeterminate Hikes+*). As *blemish* or political and interventionist AR projects by the Manifest.AR group show, these interfaces can also induce critical reflection. In all of these situations, they stimulate a more acute awareness of one's position within one's immediate surroundings.

Conclusion

The amount of artificial intelligence that we increasingly encounter daily—which analyzes our mental states, emotional predispositions and attitudes, bodily conditions, and behavior in general—offers us versions of ourselves that may be too obvious or too unflattering. As our emotions are interpreted computationally, and as more and more aspects of our lives become encoded to fit algorithmic standards, coming up with a response is vital. The "tactics of interfacing" that I have proposed should be treated not as solutions but rather as propositions (in a Latourian sense) to creatively explore the zones where we can reclaim our affects, our vulnerabilities, instabilities, and "imperfections," and perhaps employ them in order to "*transcode*" the electronic systems and utilize their generative potential. Indeed, what unites the artworks explored in this book is their motivation to reveal aspects of technology that can disrupt the technology's own strict correlationism, expressed, for instance, in the logic of command-action. The situations set up by these pieces bring to the fore the qualities of contingency inherent both in human nature but paradoxically also in the machinic. Expressions of the body and unconscious idiosyncratic associations introduce more confusion to these systems' functionalities. The mutual encoding of a human and a machine may even lead to entropy—as it does in *Bodytext*, a performance that features the possibilities and the limits of translation between the intuitive expressions of body memories and a computationally controlled environment. The emerging poetic meanings, like the ones we find with *Bodytext*, may not make sense to the machine, and yet it is through this dialogue with an algorithmic intelligence that humans may feel themselves reconfigured in novel ways. Smiling or frowning at a 3D-rendered "Alter Ego," a mirror portrait equipped with its own "intelligence" in Alexa Wright's installation, one cannot predict the returned emotions, and thus must stay alert and inventive. This dialogue with the computer-simulated version of oneself engages a sense of presence and provokes the viewer to look at himself or herself (literally) as recomposed and reframed. Such exchanges remind us of the value of embracing the heterogeneous flux of those aspects of the psyche that are not predetermined by conscious memories, expectations, and judgments (let alone machinic subcategorizations).

Throughout the book I have shown different artistic scenarios (or "tactics") for exploring the interrelation of at once divergent and convergent types of intelligence—the algorithmic and the biological, as well as materialities that underlie them and complicate their relations. Yet important in most of the pieces was their framing in terms of particular cultural or psychological narratives that give very specific structure to the viewer's experience. For instance, in *Chameleon* by Tina Gonsalves, we look at video portraits of people captured while they were recalling dramatic events in their lives; we empathize with their emotions while trying instinctively to guess what has happened, trying to deconstruct their stories. Fictional identities in the app *Karen* by Blast Theory and the web platform *Mouchette* by Martine Neddam are explicitly placed within story worlds, and by interacting with them we have to perform within the proposed narrative frame, adopting its rules. Similarly, users of the app *Indeterminate Hikes+* by EcoArtTech are invited to observe the multiplicity of layers within their immediate environment; but in order to do that, they must follow very concrete textual prompts.

All of these tactics serve to activate the participants' "fringe consciousness," the "marginal forms of awareness" (in the words of Hubert Dreyfus)[1] that helps us to navigate varying contexts and intuitively reduce ambiguity. They remind us of what we intuitively know but forget to pay attention to. Instructions and rules act as techniques of defamiliarization or shifting perspective. They bring to the fore these latent capabilities, making us more aware of our own potential. The rules may be arbitrary and the story worlds fictive, but their operations and effects become real due to our use of technology. The computer code functions here to make more apparent another code—that of humans' own creativity, the innate drive to uncover more facets of oneself and the world.

Among the tactics are also allusions to culturally inscribed associations, iconography, and symbolic language, including direct references and quotations. Examples include Stelarc's and Chris Marker's philosophizing chatbots, Dmitry Morozov's *The Escalation of Mind* (with Hesse's text as its "score"), and the AR project *blemish* by CONT3XT.NET, which appeals to the symbolic significance of the city of Venice in the art world. Pure sensory perception works here in close tandem with interpretation, with assigning identity or additional meaning to something given through the senses. For instance, we recognize ourselves as "I" versus the "other," but this relationship may be complicated, as highlighted by the pieces about merging identities explored in chapter 1—*The Mirrorbox* by Megan Daalder, or *Saving Face* by Karen Lancel and Hermen Maat. The aesthetic (sensory) and interpretive (analytical) levels of engagement are so complementary that they amalgamate to such a degree that they may be hardly distinguishable from one another: one picks up where the other one cannot fully penetrate, and vice versa. (This is illustrated by the pieces on mental imagery and control via brainwaves, such as Ellen Pearlman's brainwave opera *Noor*, which translates the cognitive states of a performer into an audiovisual environment). Computer code and granular electronic data detection, intrinsic to most of the technical processes in these pieces, work to intensify the connections between these typical (and essential) human capabilities—the psychological/cognitive and the affective/precognitive kinds of reflection. Even when the code may work "against" "everything human" and figures as a "daemonic," as Wendy Chun would put it[2]—which can also be read as a "ghostly" or "uncanny" force according to Kriss Ravetto-Biagioli—its mechanisms, in fact, serve to activate new ties within the characteristically human. They challenge us to discover these ties anew. I argue that it is not only the "informatic body" (expressions of the body processed via digital technologies) that reveals the "potential for transmutation," but a whole sense of the self that can be "rendered mutable" and "delimited" by the parameters of interfacing, an interrelational scenario defined via technology.[3]

Among the many ways in which technology can interfere with our emotional life, there are some that exemplify particularly radical horizons of possible transformations: synthetic biology and gene-editing. It is well known that moods and mental states can be susceptible to manipulation through substances from alcohol, caffeine, and psychedelics to antidepressants and drugs that regulate the hormonal system. The reactions happen at the biochemical level, and the interfaces are the cellular membranes and molecular structures that allow them to communicate signals related to behavioral activity (including mood and emotional states). Biochemical technology involved in neuropharmacology differs in many ways from the kinds of digital technologies explored in this book. Yet it also involves computational modeling and complex algorithms—for example, to calculate the outcomes of a certain drug.

Speculative design has been one of the most thought-provoking methods to reflect on the implications of pharmaceutical technologies and their cutting edge of personalized medicine and genetic modification. *Red Silk of Fate—Tamaki's Crush* (2016) by Sputniko! (Hiromi Ozaki) is a music video about a young genetic engineer named Tamaki who wants to win the love of her lab colleague. To do so, she comes up with an idea based on an ancient myth, turned in her hands into a scientific reality—a "red string of fate" (figure 5.1). According to East Asian mythology, two people destined for each other are connected by an invisible red string, and this is exactly what Tamaki creates in the film. To support the story, Sputniko! made the intriguing artifact in reality: collaborating with scientists from Japan's NIAS (National Institute of Agrobiological Sciences) to spin the mythical thread, they engineered a new type of silkworm by inserting genes that produce

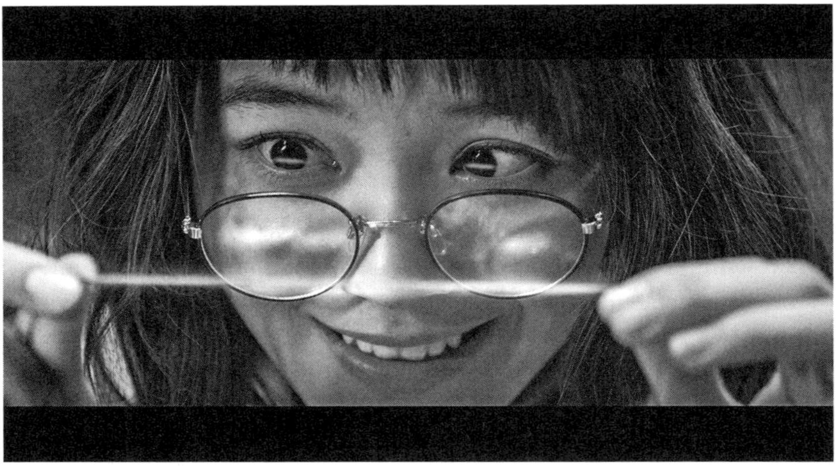

Figure 5.1
Sputniko! *Red Silk of Fate—Tamaki's Crush* (2016). Music film and genetically engineered object. Screen capture from video. Courtesy of Sputniko!

oxytocin, a "love" and social-bonding hormone, as well as the genes of a red-glowing coral, into silkworm eggs. Even though Tamaki's efforts to unleash nature's magical powers go quite astray, Sputniko!, who has a popular following in Japan for her music and conceptual innovations, has established a whole shrine devoted to this object—a scientific implementation of the poetic vision. People are invited to express their innermost wishes on specially made *omamori*, protective Shinto amulets, typically used in Japan for worshiping numerous "spirits." The red silk threads on the amulets symbolize not only the original mythical thread, but also the potentially functional one, induced with oxytocin.

This project sheds new light on the key questions of my inquiry: What governs modifications in the psyche—and how does it work? How can we interface with it, and what are our zones of control?[4] Feedback from screen-based or conversational "alter egos" and other software-controlled interaction partners is radicalized by the possibility of direct biological intervention (e.g., through stimulation of cellular receptors). But can our feelings be reduced to fluctuations of the level of hormones or neurotransmitters? Furthermore, do we want to accept this approach and thus open the opportunity for further manipulation of internal states by external forces, such as medical corporations and other institutions of biopolitical power? A "red string of fate" is not an interface in the sense of a software-based electronic mechanism; the operations of a "code" happen here within the very living matter, at the biochemical level. Yet, at least according to the concept, it could produce very tangible effects—changes in emotions, determining how a person would act and go through life, just as psychotropic medication or other drugs do.

What Sputniko! does is *play* with technoscientific promises by (wittily and humorously) exercising them in a fictional scenario. As in the other artworks explored above, a fantasy itself produces an encoding. It encourages us to be open to what is beyond the habitual, inviting us to be surprised and provoked, to observe (what is) and to imagine (what is not yet). In this case, by seeing what an electrifying turmoil can be caused by summoning invisible natural powers, we are also forcefully pointed toward its source—the craving for love. Science becomes a way to channel it, but the effect is not what was envisioned (Tamaki, like a magnet, attracts everything alive in her way). The moral of the story is that perhaps instead of trying to interfere with affects and what is behind them, it is better to embrace them as they are, to learn to understand them and to collaborate with them. Interfacing (here using the technique of gene editing) makes it possible to experiment with the dynamics of affects and emotions, to test one's *possible* reactions and attitudes. Imagined suppositions, joining forces with technology, beget executable scenarios, or tactics, to navigate through the intricacies of one's soul and its desires, checking how congruent they are with the outside world.

I have attempted to show that interface implies more than a technical mechanism of communication, a procedural architecture of signal transference. As I have argued, it can be considered the condition for interrelation, or interfacing, a process of being-in-relation defined through technological means. An interface gives structure to experience and enables a more profound understanding of the dynamic meshes of relations—the fluxes of sensorial entanglements and shifting agencies that permeate human life. Engaging with the operations of computation on an aesthetic level, being simultaneously immersed into the offered situation and detached from it—that is, keeping an awareness of its constructedness—allows a more grounded comprehension of its effects. As the art projects featured here show, while enmeshed in perpetual negotiations between machinic codes and one's own nature, a human still remains free to imagine and continuously reimagine oneself and the surrounding world.

Notes

Introduction

1. Lauren McCarthy, "LAUREN," project website, accessed October 11, 2019, http://get-lauren.com/.

2. In her book, Kwastek builds a theoretical toolbox for analyzing interactive art in aesthetic terms, such as experience, process, playfulness, performativity, interpretation, etc.—all of which are also relevant categories for the present study, yet are not the central ones. See Katja Kwastek, *Aesthetics of Interaction in Digital Art* (Cambridge, MA: MIT Press, 2013).

3. McCarthy describes *LAUREN* as being "about the grey area where you are not sure what to think. Instead of being asked to react immediately, you can sit down and figure out which parts feel worrying, exciting, hopeful." Quoted in Dominic Rushe, "Let Me into Your Home: Artist Lauren McCarthy on Becoming Alexa for a Day," *The Guardian*, May 14, 2019, https://www.theguardian.com/artanddesign/2019/may/14/artist-lauren-mccarthy-becoming-alexa-for-a-day-ai-more-than-human.

4. Christian Ulrik Andersen and Søren Bro Pold, *The Metainterface: The Art of Platforms, Cities, and Clouds* (Cambridge, MA: MIT Press, 2018).

5. As David Garcia and Geert Lovink note in their often-quoted 1997 "manifesto" essay "ABC of Tactical Media," "Tactical media are media of crisis, criticism and opposition. This is both the source their power, ('anger is an energy': John Lydon), and also their limitation." Garcia and Lovink, "ABC of Tactical Media," Nettime, May 16, 1997, https://www.nettime.org/Lists-Archives/nettime-l-9705/msg00096.html.

6. Matthew Fuller, "A Means of Mutation: Notes on I/O/D 4: The Web Stalker," in *Behind the Blip: Essays on the Culture of Software*, 51–68 (New York: Automedia, 2003). See also Christiane Paul, "'Not Just Art': From Media Art to Artware," in *Interface Cultures: Artistic Aspects of Interaction*, ed. Christa Sommerer, Laurent Mignonneau, and Dorothée King (Bielefeld: transcript Verlag, 2008), 141–151; Olga Goriunova, *Art Platforms and Cultural Production on the Internet* (London: Routledge, 2011); Geoff

Cox, *Antithesis: The Dialectics of Software Art* (Aarhus: Digital Aesthetics Research Center, 2010).

7. Michel de Certeau, *The Practice of Everyday Life*, trans. Steven Rendall (Berkeley: University of California Press, 1984), 20.

8. Ibid., 35. "These styles of action intervene in a field which regulates them at a first level (for example, at the level of the factory system), but they introduce into it a way of turning it to their advantage that obeys other rules and constitutes something like a second level interwoven into the first (for instance, *la perruque*)" (ibid., 31). These "trajectories" "trace out the ruses of other interests and desires that are neither determined nor captured by the system in which they develop" (ibid., 18).

9. Ibid., 30.

10. Arturo Rosenblueth, Norbert Wiener, and Julian Bigelow, "Behavior, Purpose and Teleology," *Philosophy of Science* 10, no. 1 (January 1943): 19.

11. See W. Ross Ashby, *Design for a Brain: The Origin of Adaptive Behavior* (New York: John Wiley and Sons, 1960).

12. As Branden Hookway observes, "As such, the feedback loop encloses a temporal discontinuity. It draws into relation past state and future projection, measurement and desire. It proceeds from one discrete moment of measurement to another, as separated by a cycle of evaluation and response. The gap or discrepancy is as much the delay between measurement and reaction as it is the difference between an event as it occurs and as it is modeled or projected." Branden Hookway, *Interface* (Cambridge, MA: MIT Press, 2014), 101.

13. Situating the observer within the same system as the observed (and making oneself the focus of observation) was the key input of so-called second-order cybernetics, exemplified by works by Margaret Mead and Gregory Bateson. See Gregory Bateson, *Steps to an Ecology of Mind: Collected Essays in Anthropology, Psychiatry, Evolution, and Epistemology* (Chicago: University of Chicago Press, 1972).

14. In addition, Simondon sees metastability as a condition for what he calls individuation: "Individuation must therefore be thought of as a partial and relative resolution manifested in a system that contains latent potentials and harbors a certain incompatibility with itself, an incompatibility due at once to forces in tension as well as to the impossibility of interaction between terms of extremely disparate dimensions." Gilbert Simondon, "The Genesis of the Individual," in *Incorporations*, ed. Jonathan Crary and Sanford Kwinter, trans. Mark Cohen and Sanford Kwinter (New York: Zone Books, 1992), 300.

15. N. Katherine Hayles, in her account on feedback in the context of a related phenomenon of reflexivity, writes: "Reflexivity has subversive effects because it confuses and entangles the boundaries we impose on the world in order to make sense

of that world. Reflexivity tends notoriously toward infinite regress. The dreamer [in "The Circular Ruins" by Jorge Luis Borges] creates the student, but the dreamer in turn is dreamed by another, who in his turn is dreamed by someone else, and so on to infinity." N. Katherine Hayles, *How We Became Posthuman: Virtual Bodies in Cybernetics, Literature, and Informatics* (Chicago: University of Chicago Press, 1999), 8–9.

16. Affect has often been defined as indeterminate, precognitive, not easily localizable: "affect is an impersonal flow before it is a subjective content," writes Brian Massumi in *Parables for the Virtual: Movement, Affect, Sensation* (Durham: Duke University Press, 2002), 61. See also Patricia T. Cloth, "The Affective Turn: Political Economy, Biomedia and Bodies," *Theory, Culture, Society* 25, no. 1 (2008): 1–22; Marie-Luise Angerer, Bernd Bösel, and Michaela Ott, eds., *Timing of Affect: Epistemologies, Aesthetics, Politics* (Zurich: diaphanes; Chicago: University of Chicago Press, 2014); Melissa Gregg and Gregory J. Seigworth, eds., *The Affect Theory Reader* (Durham: Duke University Press, 2010); Antonio R. Damasio, *The Feeling of What Happens: Body and Emotion in the Making of Consciousness* (New York: Harcourt Brace, 1999); among others.

17. Kriss Ravetto-Biagioli, *Digital Uncanny* (New York: Oxford University Press, 2019), 5.

18. Hookway, *Interface*, 63. "In fluid dynamics or thermodynamics, the interface is a boundary across which dynamic conditions are held in a state of contestation. It elicits a drive to contestation from that with which it interfaces. . . . Its formal persistence exists only in the dynamics of a continual formation, dissolution, and reformation." Ibid., 64.

19. See, for instance, Florian Hadler and Joachim Haupt, eds., *Interface Critique* (Berlin: Kadmos Kulturverlag, 2016), as well as the chapter on "daemonic" interfaces in Wendy Hui Kyong Chun, *Programmed Visions: Software and Memory* (Cambridge, MA: MIT Press, 2011).

20. Lori Emerson, *Reading Writing Interfaces: From the Digital to the Bookbound* (Minneapolis: University of Minnesota Press, 2014), 2.

21. Galloway's account covers complex ground between the ontologies and ideologies of computers, with the interface effect demonstrating that "the computer is not an object, or a creator of objects," but a "process or active threshold mediating between two states." Alexander R. Galloway, *Interface Effect* (Cambridge, UK: Polity Press, 2012), 23.

22. Christian Ulrik Andersen and Søren Bro Pold, eds., *Interface Criticism: Aesthetics beyond Buttons* (Aarhus: Aarhus University Press, 2011), 9.

23. As Anna Munster has put it, "the major achievement of interfaciality is not so much escaping the face but rather multiplying it." Anna Munster, *Materializing New Media: Embodiment in Information Aesthetics* (Hanover, NH: Dartmouth College Press, 2006), 168. What she implies is a way to avoid the polarizing view of the radical

split between the human and the machinic, as well as the illusion of coherent and "workable" projections on either side of this opposition.

24. Marshall McLuhan, *Understanding Media: The Extensions of Man* (1964; repr., Cambridge, MA: MIT Press, 1994), 57.

25. As Lacanian media scholar André Nusselder puts it, "There is a gap between the object and its 'exact' representation, and in this gap the (unconscious) functioning of fantasy takes place, as imaginary and metaphorical (trans)formations of data into new forms of reality." André Nusselder, *Interface Fantasy: A Lacanian Cyborg Ontology* (Cambridge, MA: MIT Press, 2009), 20.

26. Galloway borrows the expression "zone of indecision" from Gérard Genette's book *Paratexts: Thresholds of Interpretation*, trans. Jane E. Lewin (Cambridge, UK: Cambridge University Press, 1997). See Galloway, *Interface Effect*, 40.

27. See Gavin J. D. Smith, "Surveillance, Data and Embodiment: On the Work of Being Watched," *Body and Society* 22, no. 2 (2016): 108–139.

Chapter 1

1. Think of the "smile scan" software used to enforce broader smiles among Japanese railway workers. See Danielle Demetriou, "Workers Have Daily Smile Scans," *The Telegraph*, July 6, 2009, https://www.telegraph.co.uk/news/worldnews/asia/japan/5757194/Workers-have-daily-smile-scans.html.

2. See Steven Johnson, *Interface Culture: How New Technology Transforms the Way We Create and Communicate* (San Francisco: Harperedge, 1997); Alan Cooper, *About Face: The Essentials of User Interface Design* (Foster City, CA: IDG Books, 1995); and Jakob Nielsen, *Designing Web Useability: The Practice of Simplicity* (Indianapolis: New Riders, 2000).

3. Anna Munster, *Materializing New Media: Embodiment in Information Aesthetics* (Hanover, NH: University Press of New England, 2006), 128. To complexify that account and to show what happens to the bodily level of engagement, Munster also refers to Brenda Laurel's conception of "virtual dramaturgy" as a mean to overcome "informational reductionism." See Brenda Laurel, *Computers as Theatre* (Reading, MA: Addison-Wesley, 1993). Yet, Munster argues, "Laurel's virtual worlds only accommodate 'the body' and not the heterogeneity typical of embodied life." Munster, *Materializing New Media*, 129.

4. "The composite image would thus give rise to the necessarily beautiful Platonic Idea." Georges Bataille, "The Deviations of Nature," in *Visions of Excess: Selected Writings, 1927–1939*, ed. Allan Stoekl (Minneapolis: University of Minnesota Press, 1985), 55.

5. Interestingly, one version of the etymology of the Slavic *lik* points at its connection to *laiks*, "dance," in a proto-Gothic language; and indeed, a word with the same root, *likovanie/ликование*, in Russian means "rejoicing, exaltation."

6. Vicki Bruce and Andy Young, "Understanding Face Recognition," *British Journal of Psychology* 77 (1986): 305–327. The mother's face was discussed in the psychoanalytic tradition of object relations theory, particularly by Melanie Klein and Donald Winnicott.

7. Hannah Arendt, *The Human Condition* (Chicago: University of Chicago Press, 1958), 198–199.

8. Giorgio Agamben, "The Face," in Giorgio Agamben, *Means without End: Notes on Politics*, trans. Vincenzo Binetti and Cesare Casarino (Minneapolis: University of Minnesota Press, 2000), 90. "The face is at once the irreparable being-exposed of humans and the very opening in which they hide and stay hidden" (ibid.).

9. Ibid., 91.

10. Ibid., 92.

11. "The face does not coincide with the visage. There is a face wherever something reaches the level of exposition and tries to grasp its own being exposed, wherever a being that appears sinks in that appearance and has to find a way out of it" (ibid., 91). "What remains hidden from them [human beings] is not something behind appearance, but rather appearing itself, that is, their being nothing other than a face" (ibid., 94).

12. Amit Pinchevski, "The Face as a Medium," in *Images, Ethics, Technology*, ed. Sharrona Pearl (New York: Routledge, 2016), 193.

13. Emmanuel Levinas, *Totality and Infinity: An Essay on Exteriority*, trans. Alphonso Lingis (Pittsburgh: Duquesne University Press, 1998), 50–51.

14. Pinchevski, "The Face as a Medium," 198.

15. As early as 1901, Georg Simmel pointed out a similar dialectic: the "behind" of a face is what appears "in front"; there is not "any pure intellectuality behind the appearance." Georg Simmel, "The Aesthetic Significance of the Face," in *Georg Simmel, 1858–1918*, ed. Kurt H. Wolff (Columbus: Ohio State University Press, 1959), 281.

16. Béla Balázs described the expressive role of facial physiognomy in a cinematic close-up in 1949. Along with Soviet film directors, such as Eisenstein, he also elaborated on "class faces" and "group physiognomies," which may serve as another example of the "beyond-the-subjective." See Béla Balázs, *Theory of the Film: Character and Growth of a New Art*, trans. Edith Bone (London: Dennis Dobson, 1952).

17. Gilles Deleuze, *Cinema 1: The Movement-Image,* trans. Hugh Tomlinson and Barbara Habberjam (Minneapolis: University of Minnesota Press, 1986), 87.

18. "Concrete faces cannot be assumed to come ready-made. They are engendered by an abstract machine of faciality (*visageite*)." Gilles Deleuze and Félix Guattari, *A Thousand Plateaus: Capitalism and Schizophrenia*, trans. Brian Massumi (Minneapolis: University of Minnesota Press, 1987), 168.

19. Alan Tormey, *The Concept of Expression* (Princeton, NJ: Princeton University Press, 1971), 28.

20. Deleuze and Guattari, *A Thousand Plateaus*, 180.

21. "It is precisely because the face depends on an abstract machine that it does not assume a preexistent subject or signifier; but it is subjacent to them and provides the substance necessary to them." Ibid.

22. Ibid., 170.

23. This research is in line with Stelarc's experimentation with giving up agency of the body to the control of the machine, under the motto "The body is obsolete."

24. Arthur Elsenaar, "Artifacial," project website, 1993 to present, accessed October 17, 2019, http://artifacial.org/.

25. Arthur Elsenaar and Remko Scha, "Electric Body Manipulation as Performance Art: A Historical Perspective," *Leonardo Music Journal* 12 (2002): 17–28. See also G.-B. Duchenne de Boulogne, *The Mechanism of Human Facial Expression*, trans. Andrew R. Cuthbertson (Cambridge, UK: Cambridge University Press, 1990).

26. Charles Darwin, *The Expression of the Emotions in Man and Animals*, 1872.

27. Sigrid Weigel, "Phantom Images: Face and Feeling in the Age of Brain Imaging," in *Zeitschrift für Kunst und Kulturwissenschaften. Kritische Berichte* 40, no. 1 (2012): 40–41.

28. The experiments were conducted primarily on Duchenne's asylum patients, a fact that would be considered controversial from the standpoint of today's scientific ethics. One of the experimental subjects also had an anaesthetic condition of the face and couldn't feel pain, which offered Duchenne a convenient response to criticisms of this method. This is analogous to the role that Elsenaar gives to the face in his projects: involuntary and unconscious.

29. Huge Harry is described as the "president and spokesmachine of the Institute of Artificial Art Amsterdam" and "one of the most outspoken and authoritative voices in the discussion about art, society and technology—always representing the computational point of view with great vigor and clarity." Arthur Elsenaar, "Huge Harry: On the Mechanism of Human Facial Expression as a Medium for Interactive Art," Ars Electronica Symposium, September 1997, http://iaaa.nl/hh/cv.html.

30. Arthur Elsenaar, "Perfect Paul: On Freedom of Facial Expression," on "Artifacial" project website, accessed October 17, 2019, http://artifacial.org/perfect_paul_transcript.

31. Arthur Elsenaar, "Face Shift," on "Artifacial" project website, accessed October 17, 2019, http://artifacial.org/face_shift.

32. Munster, *Materializing New Media*, 136.

33. A similar problem of machinic control over the human face is addressed in a project by Tim Hawkinson called *Emotor*—a mechanically animated portrait, or kinetic sculpture, of a face whose expressions are awkward and unnatural, although the original is a photograph of a human face (divided into sections that are then manipulated).

34. Munster, *Materializing New Media*, 127.

35. Claire L. Evans, "Inside the Empathy Machine: Going Face-in-Face with Megan May Daalder and Her Mirrorbox," *Vice*, August 29, 2012, https://www.vice.com/en_us/article/z44mk3/inside-the-empathy-machine-going-face-in-face-with-megan-may-daalder.

36. Megan Daalder, "The Mirrorbox," project website, accessed October 10, 2017, http://www.themirrorbox.info/.

37. Deenah Vollmer, "The Compassion-Passion Machine," *Believer*, May 8, 2014, http://www.believermag.com/exclusives/?read=article_vollmer.

38. Mattia Casalegno, "Unstable Empathy," project webpage, accessed October 17, 2019, www.mattiacasalegno.net/unstable-empathy/.

39. Karen Lancel and Hermen Maat, "Saving Face," project webpage, accessed October 17, 2019, https://www.lancelmaat.nl/work/saving-face/.

40. "To inspire and rethink sustainable socio-technical ecological systems, the artists deconstruct and connect automated control technologies (such as surveillance, social media, self-quantifying biometric technologies and brain computer interfaces) to human acts of intimacy, tacit knowledge, sensory and aesthetic perception—to explore a digital-synesthetic 'trust-system'" (ibid.)

41. See more in Martijn P. van den Heuvel et al., "Functionally Linked Resting-State Networks Reflect the Underlying Structural Connectivity Architecture of the Human Brain," *Human Brain Mapping* 30, no. 10 (2009): 3127–3141; Gustavo Deco, Viktor K. Jirsa, and Anthony Randal McIntosh, "Emerging Concepts for the Dynamical Organization of Resting-State Activity in the Brain," *Nature Reviews: Neuroscience* 12, no. 1 (2011): 43–56.

42. See Richard E. Niemeyer, "What Are the Neurological Foundations of Identities and Identity-Related Processes," in *Handbook of Neurosociology*, ed. D. D. Franks and J. H. Turner (New York: Springer, 2013), 149.

43. The research on the brain's resting state and its default mode of function has been discussed as a growing "fourth potent arm" of neuroscientific research in the last decade, in addition to the studies of neuroplasticity, affect, and mirror neurons. See Felicity Callard and Daniel S. Margulies, "The Subject at Rest: Novel Conceptualizations of Self and Brain from Cognitive Neuroscience's Study of the 'Resting State,'" *Subjectivity* 4, no. 3 (2011): 227–257.

44. Ibid., 250.

45. Marcus E. Raichle, "The Brain's Dark Energy," *Science* 314, no. 5803 (2006): 1249–1250; and Marcus E. Raichle, "The Brain's Dark Energy," *Scientific American* 302 (March 2010): 44–49.

46. Alexa M. Morcom and Paul C. Fletcher, "Does the Brain Have a Baseline? Why We Should Be Resisting a Rest," *NeuroImage* 37, no. 4 (2007): 1073–1082.

47. In this sense, engagement of MNS in recognizing, learning, and adapting symbolic hand gestures is claimed by some scholars (including Giacomo Rizzolatti, Vittorio Gallese, and George Lakoff) to be key in the formation and development of language. The same is relevant for the ability to understand facial expressions, which also engage muscular functions. See Giacomo Rizzolatti and Laila Craighero, "The Mirror-Neuron System," *Annual Review of Neuroscience* 27 (2004): 169–192; Giacomo Rizzolatti and Michael A. Arbib, "Language within Our Grasp," *Trends in Neurosciences* 21, no. 5 (1998): 188–194; and Vittorio Gallese, "Mirror Neurons and the Neural Exploitation Hypothesis: From Embodied Simulation to Social Cognition," in *Mirror Neuron Systems: The Role of Mirroring Processes in Social Cognition*, ed. Jaime A. Pineda (New York: Humana Press, 2009).

48. Istvan Molnar-Szakacs and Lucina Q. Uddin, "The Emergent Self: How Distributed Neural Networks Support Self-Representation," in *Handbook of Neurosociology*, ed. D. D. Franks and J. H. Turner (New York: Springer, 2013), 167–182.

49. See Lucina Q. Uddin et al., "Self-Face Recognition Activates a Frontoparietal 'Mirror' Network in the Right Hemisphere: An Event-Related fMRI Study," *NeuroImage* 25, no. 3 (2005): 926–935; C. Devue and S. Bredart, "The Neural Correlates of Visual Self-Recognition," *Consciousness and Cognition* 20, no. 1 (2011): 40–51.

50. Ray F. Lee, "Emergence of the Default-Mode Network from Resting-State to Activation-State in Reciprocal Social Interaction via Eye Contact," *IEEE Engineering in Medicine and Biology Society* (2015): 1821–1824.

51. See Laura M. Pönkänen, Annemari Alhoniemi, Jukka M. Leppänen, and Jari K. Hietanen, "Does It Make a Difference If I Have an Eye Contact with You or with Your Picture? An ERP Study," *Social Cognitive and Affective Neuroscience* 6, no. 4 (2011): 486–494.

52. One of the main problems with empirical science (and neuroscience is not an exception) is that the picture that it draws—in order for it to be reliable and applied

universally—has to be a generalization and an abstraction from the individual facts. Inevitably, empirical science operates with an assumption of a possible "pure" case, such as a "pure" brain. But can there be a "pure brain"? Some of the problems particularly with neuroscience, such as the presumptions and their origins, often uncritically rely on forms of visualization. This problem, among others, is addressed in Suparna Choudhury and Jan Slaby, eds., *Critical Neuroscience: A Handbook of the Social and Cultural Contexts of Neuroscience* (Chichester: Wiley-Blackwell, 2012).

53. In the absence of the results for *The Mirrorbox* at the moment, another study can be mentioned that similarly uses the videomorphing technique to test how the mind relates to the images of mixed identities: see Christine Heinisch, Hubert R. Dinse, Martin Tegenthoff, Georg Juckel, and Martin Bruene, "An rTMS Study into Self-Face Recognition Using Video-Morphing Technique," *SCAN* 6 (2011): 442–449. The photographs of a testing subject, a famous person, and an unfamiliar person were morphed into each other over the course of six seconds. The research participant was supposed to recognize a change of identity by pushing a button. The reaction time was shorter when the subject's own face emerged from a recognizable famous face. The study focused on tracking the location of the processes of both self-recognition and comparison of the self-face with other faces. The experiments pointed at the correlations between these recognition processes (faster response) and activation (through low-frequency repetitive transcranial magnetic stimulation, or rTMS) of the right temporoparietal junction, as well as right-prefrontal and left-parietal cortices, and the transition from an unfamiliar face to one's own registered in a left frontal lateralization. This confirms the suppositions of other studies about the role of the temporoparietal junction and prefrontal structures in self-evaluation and self-other distinctions. It also confirms that face recognition is more holistic for the recognition of others, while one's own face is recognized based on individual features.

54. This motif of facial corrosion that reveals another face is reminiscent of a video by Peter Campus, *Three Transitions* (1973), where wiping away his own skin reveals the same face underneath.

55. "Each video clip of eyes is impossibly combined with video clips of noses and mouths that slide back and forth to form impossible faces. . . . It is a product of a very alienating effect that the mirror stage inaugurates." David Bard-Schwarz, *An Introduction to Electronic Art through the Teaching of Jacques Lacan: Strangest Thing* (London: Routledge, 2014), 95.

56. "The total form of the body by which the subject anticipates in a mirage the maturation of his power is given to him only as *Gestalt,* that is to say, in an exteriority in which this form is certainly more constituent than constituted, but in which it appears to him above all in contrasting size (*un relief de stature*) that fixes it and in a symmetry that inverts it, in contrast with the turbulent movements that the subject feels are animating him." Jacques Lacan, *Écrits: A Selection*, trans. Bruce Fink (New York: W. W. Norton, 2002), 2.

57. "The mirror stage is a drama whose internal thrust is precipitated from insufficiency to anticipation—and which manufactures for the subject, caught up in the lure of spatial identification, the succession of fantasies that extends from a fragmented body-image to a form of its totality that I shall call orthopaedic—and, lastly, to the assumption of the armor of an alienating identity, which will mark with its rigid structure the subject's entire mental development." Ibid., 4.

58. "The Ego . . . is frustration in its very essence. Not frustration of one of the subject's desires, but frustration of an object in which his desire is alienated; and the more developed this object becomes, the more profoundly the subject becomes alienated from his jouissance. It is thus frustration at one remove, a frustration that the subject—even were he to reduce its form in his discourse to the passivating image by which the subject makes himself an object by displaying himself before the mirror—could not be satisfied with, since even if he achieved the most perfect resemblance to that image, it would still be the other's jouissance that he would have gotten recognized there." Ibid., 43.

59. Paul Mozur, "One Month, 500,000 Face Scans: How China Is Using A.I. to Profile a Minority," *New York Times*, April 14, 2019, https://www.nytimes.com/2019/04/14/technology/china-surveillance-artificial-intelligence-racial-profiling.html; Don Lee, "At This Chinese School, Big Brother Was Watching Students—and Charting Every Smile or Frown," *Los Angeles Times*, June 30, 2018, https://www.latimes.com/world/la-fg-china-face-surveillance-2018-story.html; Genia Kostka, "China's Social Credit Systems and Public Opinion: Explaining High Levels of Approval," *New Media and Society* 21, no. 7 (July 2019): 156–193.

60. For information on these projects, see the websites for Zach Blas, http://www.zachblas.info; Adam Harvey, https://ahprojects.com/projects; and Leonardo Selvaggio, https://leoselvaggio.com/urmesurveillance, all accessed October 27, 2019.

61. See Édouard Glissant, *Poetics of Relation*, trans. Betsy Wing (Ann Arbor: University of Michigan Press, 1997); Zach Blas, "Opacities: An Introduction" (149–153), and Blas and Jacob Gaboury, "Biometrics and Opacity: A Conversation" (155–165), both in *Camera Obscura: Feminism, Culture, and Media Studies* 31, no. 2 (92): (2016).

62. "Computational objectivity" is interpreted by Peter Galison and Lorraine Daston as an updated form of the "mechanical objectivity" associated with earlier photographic and visualizing technologies. See Peter Galison and Lorraine Daston, "Image of Objectivity," in "Seeing Science," special issue of *Representations*, no. 40, (Autumn 1992): 82.

63. Nicholas Rose, *Inventing Our Selves: Psychology, Power, Personhood* (New York: Cambridge University Press, 1996), 87.

64. Kelly Gates, *Our Biometric Future: Facial Recognition Technology and the Culture of Surveillance* (New York: New York University Press, 2011), 167.

65. See more in Geoffrey C. Bowker and Susan Leigh Star, *Sorting Things Out: Classification and Its Consequences*, Inside Technology series (Cambridge, MA: MIT Press, 1999).

66. John Bulwer's *Pathomyotomia* of 1649 is recognized as the first comprehensive contribution on the muscular basis of emotional expressions and became foundational for today's research. The French academician and painter Charles Le Brun also played an important role with his graphic studies of facial expressions (e.g. *Expressions of the Passions*, 1698).

67. William James, along with Danish physician Carl Lange, famously proposed that emotions are the brain's perception of the internal body. Their position was opposed by Walter Cannon, who argued that emotions are not dependent on particular bodily arousal states; that is, the same emotions can be felt regardless of whether we are running or resting, and an intense heartbeat may indicate anger as much as excitement. The position that facial expressions have no necessary relation to emotions and instead are context-specific social signals of intent is the view of "behavioral ecology" (Alan Fridlund), which is also a Darwinian position: facial displays are not learned, but are mediated by innate form of social cognition. The issue of whether or not facial expressions represent certain emotions, or any internal state, is still a matter of debate. See James M. Carroll and James A. Russell, "Do Facial Expressions Signal Specific Emotions? Judging Emotion from the Face in Context," *Journal of Personality and Social Psychology* 70, no. 2 (February 1996): 205–218; Alan J. Fridlund, *Human Facial Expression: An Evolutionary View* (San Diego: Academic Press, 1994).

68. Nigel Thrift, "Bare Life," in *Cultural Bodies: Ethnography and Theory*, ed. Helen Thomas and Jamilah Ahmet (Oxford: Blackwell, 2004), 145–169.

69. Shannon's theory decontextualizes information, "divorcing it from meaning and disregarding the mind-sets of individuals involved in communicative exchanges." Gates, *Our Biometric Future*, 171. As it serves the instrumental purpose of designing efficient communication networks, this engineering mode of information became dominant in conceptualizing what information was.

70. Paul Ekman and Wallace Friesen, "Measuring Facial Movement with the Facial Action Coding System," in *Emotion in the Human Face*, ed. Paul Ekman, 2nd ed. (Cambridge, UK: Cambridge University Press; 1982), 180. See also Paul Ekman and Wallace V. Friesen, *Pictures of Facial Affect* (Palo Alto, CA: Consulting Psychologists Press, 1976).

71. In 2000, Parrott identified 136 additional emotional states that humans are capable of displaying, and categorized them in a tree-structured list of classes and subclasses where anger, for instance, can be broken down into irritability (aggravation, agitation), exasperation (frustration), rage (fury, scorn, spite, resentment), disgust (revulsion, contempt), envy, torment, etc. See Gerrod W. Parrott, ed., *Emotions*

in Social Psychology (New York: Psychology Press, 2000). Other examples of the expression of more nuanced emotions can be found in Ying-li Tian, Takeo Kanade, and Jeffrey F. Cohn, "Recognizing Action Units for Facial Expression Analysis," *IEEE Transactions on Pattern Analysis and Machine Intelligence* 23, no. 2 (2001): 97–115.

72. Among the most popular algorithms are the Gradient Vector Flow (GVF) snake algorithm, the Parabolic Template algorithm, a Combination algorithm, the Active Contour algorithm by Pardas and Bonafonte, and Hidden Markov Models (HMMs). See Petar S. Aleksic, Jay J. Williams, Zhilin Wu, and Aggelos K. Katsaggelos, "Audio-Visual Speech Recognition Using MPEG-4 Compliant Visual Features," *EURASIP Journal on Applied Signal Processing* 11 (2002): 1213–1227; M. Pardas and A. Bonafonte, "Facial Animation Parameters Extraction and Expression Recognition Using Hidden Markov Models," *Signal Processing: Image Communication* 17 (2002): 675–688; Petar S. Aleksic and Aggelos K. Katsaggelos, "Automatic Facial Expression Recognition Using Facial Animation Parameters and MultiStream HMMs," *IEEE Transactions on Information Forensics and Security* 1, no. 1 (2006): 3–11.

73. Mikael Rydfalk, "CANDIDE, a Parameterized Face," Report No. LiTH-ISY-I-866, Dept. of Electrical Engineering, Linköping University, Sweden, 1987; Jörgen Ahlberg, *CANDIDE-3—An Updated Parameterized Face*, Report No. LiTH-ISY-R-2326, Dept. of Electrical Engineering, Linköping University, Sweden, 2001.

74. "Candide—a Parameterized Face," Image Coding Group, Linköping University, last modified May 24 2012, accessed July 19, 2019, http://www.icg.isy.liu.se/candide/. See also Jixu Chen, "Automatic Face Animation with Linear Model" (2007), accessed July 16, 2019, https://www.researchgate.net/publication/228745691_Automatic_Face_Animation_with_Linear_Model; and Vinay Bettadapura, "Face Expression Recognition and Analysis: The State of the Art," *Tech Report*, arXiv:1203.6722 (April 2012).

75. Hubert Dreyfus, *What Computers Still Can't Do* (1972; repr., Cambridge, MA: MIT Press, 1992), 109.

76. Gates, *Our Biometric Future*, 175.

77. Kirsten Boehner, Rogerio DePaula, Paul Dourish, and Phoebe Sengers, "How Emotion Is Made and Measured," *International Journal of Human-Computer Studies* 65, no. 4 (2007): 275–291.

78. Ibid., 284.

79. Alexa Wright, "Alter Ego (2005)," project webpage, accessed October 17, 2019, https://www.alexawright.com/alter-ego-info.

80. The creative team explains the mechanisms behind the work: "To create both a living 'mirror' and an apparently autonomous 'Alter Ego' image, a series of generic three dimensional (3D) polygonal facial models representing the end-point of fifteen

facial expressions were constructed. These include small and broad smiles, surprise, anger, laughter, disgust, sadness and fear as well as more self-conscious expressions such as winking and poking out the tongue. The 3D models are warped to fit key landmark distances on a two dimensional video image of the individual face of each user. The 2D image of the user's face is then mapped onto the 3D model, which is animated by interpolation between models representing different expressions. Standard computer graphics techniques are used to render the images of the resulting 3D models simulating the individual adopting different expressions." Alexa Wright, Eugenie Shinkle, and Alf Linney, "*Alter Ego*: Computer Reflections of Human Emotions," *Proceedings of the 6th DAC Conference, Copenhagen* (December 2005): 191–199.

81. Ibid.

82. Matthias Niessner, "Face2Face Real-Time Face Capture and Reenactment of RGB Videos," accessed January 6, 2019, www.graphics.stanford.edu/~niessner/thies2016face.html.

83. Matt Iacobini, Tina Gonsalves, Nadia Bianchi-Berthouze, and Chris Frith, "Emotional Contagion in Interactive Art," *The Proceedings of the 7th International Conference on Kansei Engineering and Emotion Research*, KEER (Paris: Arts et Métiers ParisTech, 2010), 1975–1984.

84. Ibid.

85. Tina Gonsalves, Nadia Berthouze, and Matt Iacobini, "The Chameleon Project," *Second Nature* 3 (March 2010): 142–143.

86. See more on FaceSense at http://trac.media.mit.edu/mindreader/. (Curiously, there is also a Face Sense Utility, a Fujitsu proprietary software that enables the built-in HD camera to detect the user's presence and intelligently pauses or stops applications or activities when users are away from their laptops.)

87. "This algorithm adapted and learned from its audience, attempting to bring context into the work. Each day the 'mood' of the *Chameleon* portraits would shift and adapt depending upon the aggregate emotional responses of interacting audiences, thus changing the affective tone and emotional ecology of the gallery space. . . . Over time, *Chameleon*'s personality developed, fluctuating to the responses of its audience, shaping gallery spaces into ever shifting emotional hotspots, resculpting the space and participants." Gonsalves, *Tina Gonsalves: Chameleon*, catalogue (2010), accessed August 4, 2019, http://www.tinagonsalves.com/chameleon_catalogue.pdf, 19.

88. Iacobini et al., "Emotional Contagion in Interactive Art."

89. Quoted in Nadia Berthouze, "Stirring Up Experience: The Interaction of *Chameleon*," in *Tina Gonsalves: Chameleon*, 60.

90. Ibid., 61.

91. The creative team broke down the responses in the following way:

> Intimacy: . . . I was close to the character. He was quite up front and in my face. And talking quite low and quite intimately. . . . And he was being quite flirtatious. The feeling I had inside was like having a connection with someone that you had met in a bar or something. We were mimicking each other.
>
> Reminiscence: I was thinking of some sad things that happened to me, when [the stimulus] was sad for a while, it felt like a long time, and it reminded me of some things.
>
> Post-experience: . . . it wasn't making me think particularly about how people interact with me but I supposed it is quite an intense experience and I will probably have it more if I am thinking about it later. . . . I was aware of the interaction, but I wasn't thinking about it particularly . . .
>
> Empathy and Introspection: . . . I didn't like it when he looked sad and I didn't know why. . . . When I was creating these different expressions I had to think back to how I felt to create them, so it did bring up a lot of genuine emotion . . .
>
> Control vs. Reaction: . . . you wanted to sort of push the machine, or push the image to do something, you wanted to provoke the image to do something so you did want to interact all the time . . . I just responded to them, rather than trying to make them do things. (Ibid.)

92. Ibid., 17.

93. Chris Frith, "Faces Making Faces Make Faces," in *Tina Gonsalves: Chameleon*, 43.

94. See Mark B. N. Hansen, "The Time of Affect, or Bearing Witness to Life," *Critical Inquiry* 30, no. 3 (2004): 584–626; Kriss Ravetto-Biagioli, "The Digital Uncanny and Ghost Effects," *Screen* 57, no. 1 (Spring 2016): 1–20.

95. Quoted in Berthouze, "Stirring Up Experience," 62; and Karl Broome, "*Chameleon*: Techonological and Emotional Provocations," in *Tina Gonsalves: Chameleon*, 72–73.

96. Noël Carroll, *The Philosophy of Horror; or, Paradoxes of the Heart* (New York: Routledge, 1990); David Bordwell, *Narration in the Fiction Film* (Madison: University of Wisconsin Press, 1985); see also Murray Smith, *Engaging Characters: Fiction, Emotion and the Cinema* (Oxford: Clarendon, 1995).

97. Gallese, "Mirror Neurons and the Neural Exploitation Hypothesis."

98. Placed in an "evolutionarily anterior region of the brain," they "function with only secondary intervention from cortical processes." The somatic empathy and embodied social cognition theories are still only gaining their voice among film scholars; some styles of filmmaking work better for explaining these ideas than others. For example, Roger Cook writes in his analysis of Berlin School cinematic aesthetics: "In films where the image's autonomy from the narrative is an essential

element of the viewing experience, the cognitivist practice of folding it back into a situational context dictated by the narrative cannot do full justice to the film." Roger F. Cook, "Embodied Simulation, Empathy and Social Cognition: Berlin School Lessons for Film Theory," *Screen* 56, no. 2 (Summer 2015): 164.

Chapter 2

1. Maurice Merleau-Ponty, *Phenomenology of Perception*, trans. Colin Smith (London: Routledge and Kegan Paul, 1962).

2. W. J. T. Mitchell, *Iconology: Images, Text, Ideology* (Chicago: University of Chicago Press, 1986); W. J. T. Mitchell, *Picture Theory: Essays on Verbal and Visual Representation* (Chicago: University of Chicago Press, 1994); Gottfried Boehm, *Wie Bilder Sinn erzeugen* (Berlin: Berlin University Press, 2007).

3. Isabel Pedersen, *Ready to Wear: A Rhetoric of Wearable Computers and Reality-Shifting Media* (Anderson, SC: Parlor Press, 2013). See also Isabel Pedersen and Andrew Iliadis, eds., *Embodied Computing: Wearables, Implantables, Embeddables, Ingestables* (Cambridge, MA: MIT Press), 2020.

4. In some branches of interdisciplinary and cognitive science research, this type of cognition is referred to today as "4E": embodied, entangled, embedded, and enacted cognition. See Richard Menary, "Introduction to the Special Issue on 4E Cognition," *Phenomenology and the Cognitive Sciences*, no. 9 (2010): 459–463; as well as Sean Gallagher, *How the Body Shapes the Mind* (Oxford: Oxford University Press, 2005). Yet its other roots can be associated with the reformulations of the conception of agency within the "new materialist" turn in the humanities and social sciences in the last decade; see Jane Bennett, *Vibrant Matter: A Political Ecology of Things* (Durham: Duke University Press, 2009), and Rosi Braidotti, *The Posthuman* (Cambridge: Polity Press, 2013). Inspirational and foundational for new materialism is Donna Haraway's cyberfeminism and her writings on the concept of "care," relations between organism and machine, companion species, sympoiesis, etc. Although I do not engage with this work directly here, I share its conceptual positions about nonduality and the relational entanglement of human bodies and nonhuman technologies.

5. See the analysis of video art installations from the 1960s to 1970s in Chrissie Iles, *Into the Light: The Projected Image in American Art, 1964–1977* (New York: Whitney Museum, 2000).

6. See more in Doug Hall and Sally Jo Fifer, eds., *Illuminating Video: An Essential Guide to Video Art* (New York: Aperture Foundation, 1990); Cristina Albu, *Mirror Affect: Seeing Self, Observing Others in Contemporary Art* (Minneapolis: University of Minnesota Press, 2016); Kate Mondloch, *Screens: Viewing Media Installation Art* (Minneapolis: University of Minnesota Press, 2010); and Catherine Elwes, *Installation and the Moving Image* (London: Walfllower Press, 2015).

7. Christine Ross, "The Projective Shift between Installation Art and New Media Art: From Distantiation to Connectivity," in *Screen/Space: The Projected Image in Contemporary Art*, ed. Tamara Trodd (Manchester, UK: Manchester University Press, 2011), 184.

8. Ibid., 185.

9. Rather than showing divisions within the self, the aesthetics of immersiveness and relationality, Ross claims, favors the "regrouping" of users as communities. To demonstrate how it happens, she gives examples of works by Lincoln Schatz (*Here*, 2007–), Raphael Lozano-Hemmer (*Pulse Park*, 2008), Kazuhiko Hachiya (*Inter Dis-Communication Machine*, 1993—a head-mounted display for swapping views, injecting in the user a "double identity self"), Mathieu Briand (*UBIQ, SyS*05.ReE*03*, 2002—swapping and making private views public).

10. As pointed out by Mark Hansen, the terminological distinction is not clear in the English edition of Merleau-Ponty's work, particularly *Phenomenology of Perception*, where the original term "schéma corporel" is mistranslated as "body image." Mark B. N. Hansen, *Bodies in Code: Interfaces with Digital Media* (New York: Routledge, 2006), 38.

11. Maurice Merleau-Ponty, *The Primacy of Perception*, ed. James M. Edie (Evanston, IL: Northwestern University Press, 1964), 136.

12. Maurice Merleau-Ponty, *Phenomenology of Perception*, 106.

13. The representatives of associationism include such diverse thinkers as empirical philosophers John Locke and David Hume, physiological practitioner and author of conditional reflexes theory Ivan Pavlov, as well as early behaviorist psychologists.

14. Merleau-Ponty, *Phenomenology of Perception*, 99.

15. As explored by Paul Schilder in *Das Körperschema* (1923) and Klaus Conrad in *Das Körperschema. Eine kritische Studie und der Versuch einer Revision* (1933).

16. For more on the body schema as potentiality, see Hansen, *Bodies in Code*, 38–43.

17. "To have a body is to possess a universal setting, a schema of all types of perceptual unfolding and of all those inter-sensory correspondences which lie beyond the segment of the world which we are actually perceiving. A thing is, therefore, not actually *given* in perception, it is internally taken up by us, reconstituted and experienced by us in so far as it is bound up with a world, the basic structures of which we carry with us, and of which it is merely one of many possible concrete forms." Merleau-Ponty, *Phenomenology of Perception*, 326.

18. As Shaun Gallagher puts it, explaining Merleau-Ponty's view, "body schema involves an extraintentional operation carried out prior to or outside of intentional awareness." Shaun Gallagher, "Body Schema and Intentionality," in *The Body and*

the Self, ed. J. Bermúdez et. al. (Cambridge, MA: MIT Press, 1995), 228, quoted in Hansen, *Bodies in Code*, 39.

19. See Sonia Cillari's website, accessed November 5, 2019, http://www.soniacillari.net/.

20. Sonia Cillari, "Interview with Sonia Cillari" by Marlon Barrios Solano, Dance-Tech, April 20, 2010, http://www.dance-tech.net/video/interview-with-sonia-cillari

21. See, for instance, Amelia Jones, *Body Art: Performing the Subject* (Minneapolis: University of Minnesota Press, 1998).

22. The name "creature" recalls the one in Mary Shelley's *Frankenstein; or, the Modern Prometheus*. Since Frankenstein's "creature" is a scientific creation, this reference gives Cillari's project a different lens, placing it in the context of the relation not only between the self and the "other" (imaginary or not), but between the human self and the capabilities of technology. The moral of Shelley's story is that the creation gains cognitive autonomy and even becomes morally superior to the creator. Although this analogy shifts attention from art per se, it adds new dimensions to the description of the potential tensions and conflicts between subject and object in the creative process, raising questions about control and its moral aspects.

23. "Artist in Residence: Sonia Cillari, Sensitive to Pleasure," Netherlands Media Art Institute website archive, 2010, http://www.nimk.nl/eng/sonia-cilliari-sensitive-to-pleasure.

24. Ibid.

25. See the projects website of Chris Salter, accessed November 5, 2019, http://www.chrissalter.com/projects.php#JND1. Stelarc's *Ping Body* (1996) is probably the most famous example. Another example could be *Epizoo* (1994) by Marcellí Antúnez Roca (Spain), a performance that enables the spectator to control the artist's body by means of a mechatronic system comprising a body robot (an exoskeleton worn by the performer), a computer, a mechanical body-control device, a vertical projection screen, and audio. Preelectronic experiments exploring the vulnerability of the artist's body under the control of the audience include Chris Burden's giving a rifle to audience members to shoot him (in *Shoot*, 1971) and Marina Abramović's surrounding herself by objects of potential violence (a knife, razor blades, and a loaded gun) in *Rhythm 0* (1973–1974) and allowing the audience to use them.

26. Avatars as visual constructions—for instance, in games and virtual worlds, as well as on social media—allow for the manipulation of identity to the point of its total fictionalization. As Fox Harrell, among others, argues, this ability indicates an empowering potential for new types of self-expression, but it also entails the danger of reproducing the same stereotypes and stigmas as in the real world. See D. Fox Harrell and Sneha Veeragoudar Harrell, "Imagination, Computation, and

Self-Expression: Situated Character and Avatar Mediated Identity," *Leonardo Electronic Almanac* 17, no. 2 (2012): 74–91. Cognitive and behavioral psychology and other kinds of social theory have also been using a related term, "persona," explaining that personas are always shaped by social expectations and predispositions.

27. Vanessa Chang, "Catching the Ghost: The Digital Gaze of Motion Capture," *Journal of Visual Culture* 18, no. 3 (2019): 314.

28. Gunnar Johansson, "Visual Perception of Biological Motion and a Model for Its Analysis," *Perception and Psychophysics* 14, no. 2 (1973): 201–211. Ragnar Johansson provided the experimental assistance.

29. Author's observation and interview with Daiane Lopes da Silva, February 2017.

30. See analysis of Myron Kruger's and Simon Penny's work in Hansen, *Bodies in Code*, 25–52.

31. Emelie Mahdavian, preview of *Intangible Body*, UC Davis ModLab, accessed November 5, 2019, https://modlab.ucdavis.edu/digitalprojects/intangible-body/.

32. There is also another parallel to Cillari's performance, since both pieces deal with the concept of artistic work. In Mahdavian's piece, the work is a dance. As she says, due to the censorship situation, a solution had to be found for how to make a "film that would feature her [the dancer's] work without featuring her 'body.'" "Emelie Mahdavian," an interview by Francis Quettier and Dora Tennant, in *Women CineMakers*, special edition 3 (2014): 8. In total opposition to this, Cillari's work is not only embodied but represented by a naked female.

33. It can be read, for instance, as a continuation of the "shock to thought" theory of political cinema developed by Sergey Eisenstein in the 1920s in relation to the revolutionary agenda of the Soviet avant-garde art.

34. I thank Agustina Andreoletti for bringing my attention to this work.

35. D. Fox Harrell, *Phantasmal Media: An Approach to Imagination, Computation, and Expression* (Cambridge, MA: MIT Press, 2013), 270.

36. Frantz Fanon, *Black Skin, White Masks*, trans. Charles Lam Markmann (1952; repr., London: Pluto Press, 1986), 175, 84. For an example of the discussion of the relation between phenomenology and difference, see Jeremy Weate, "Fanon, Merleau-Ponty and the Difference of Phenomenology," in *Race*, ed. Robert Bernasconi (Malden, MA: Blackwell, 2001), 169–183.

37. From the author's correspondence with LaJuné McMillian, May 2019.

38. LaJuné McMillian, "LaJuné McMillian: Artist Interview," with Salome Asega, Eyebeam (Fall 2018), accessed May 1, 2019, https://www.eyebeam.org/residents/lajune-mcmillian/.

39. LaJuné McMillian, "The Black Movement Project," project website, accessed November 5, 2019, https://laja.me/Black-Movement-Project. With its anthropological bent, McMillian envisions the project will build on and continue the existing black dance archives, such as the one created by dancer Katherine Dunham during her early ethnographic studies of the African diasporas in the 1930s (see "Selections from the Katherine Dunham Collection," US Library of Congress, accessed November 5, 2019, https://www.loc.gov/collections/katherine-dunham/).

40. In the last decades, scholars of race-, ethnic-, and gender-identified media cultures, such as Lisa Nakamura and Judy Wajcman, have explored creative responses to being positioned as only passive consumers of (often biased) media content and the articulation of one's role as a producer. These responses offer a powerful way to resist discrimination and inequality, especially that prevalent in digital culture worlds. See Lisa Nakamura, *Digitizing Race: Visual Cultures of the Internet* (Minneapolis: University of Minnesota Press, 2007); Judy Wajcman, *Techno-Feminism* (Cambridge, UK: Polity Press, 2004).

41. The work is a collaboration between Ekaterina Zharinova (artistic concept and choreography), Denis Perevalov (programming), and Anna Vozzhennikova (design).

42. The numbers seem to appear arbitrarily, but in fact they are governed by a particular logic: they are randomly chosen from a set that is gradually growing. (At first, the algorithm selects digits from the set {1, 2, 3, 4, 5, 6, 7, 8}, only to form another set—for example, {3, 7, 1}, etc.—until all the numbers have been shown.) Some numbers also appear as negative numbers, which would mean a symmetrical transformation of the corresponding movement along the longitudinal axis of the body. Zharinova's movement is grounded in Pilates technique, which takes the core of the body as the center from which elongating lines of movements originate.

43. As Zharinova points out in her research, there are other choreographers experimenting with the translation of mathematics into body movement. One example is the Algo-rythmics project of the Sapientia Hungarian University of Transylvania, in Târgu Mureş (Marosvásárhely), Romania, that uses dance to illustrate sorting algorithms. Another example is Math Dance, an educational project by mathematician Karl Schaffer and Erik Stern that aims to teach math through kinesthetics in grades 4 to 12. See Ekaterina Zharinova, "Inside the Chain: An Intersection of Dance, Mathematics, and Technology" (MFA thesis, George Washington University, 2016). Mathematician and philosopher Brian Rotman has also inspired the production of performances, such as *Ordinal 5* (2011), based on the idea of choreographing the language of diagrams, or category theory, that describes relations between topological positions with trajectories (https://brianrotman.wordpress.com/dancing-a-diagram/).

44. Zharinova, "Inside the Chain," 17.

45. The audience questionnaires, organized by the artist, showed that the dancer's actions were still seen as being more important than in the virtual part. To the question, "Did you find the dance and technology equally engaging?," three people (13%) found the image to be more engaging, whereas thirteen (87%) found the dance equally or more engaging. Ibid., 21.

46. Ibid., 6.

47. Ibid., 21.

48. The theater uses not only projections but holograms to give an even stronger effect of the presence of another layer of reality on stage—the phantom, the dreamy, the fantastical or imaginable. See, for instance, Janis Jefferies, "Blurring the Boundaries: Performance, Technology and the Artificial Sublime—an Interview with Ruth Gibson and Bruno Martelli, igloo," in *Interfaces of Performances*, ed. Maria Chatzichristodoulou, Janis Jeffries, and Rachel Zerihan (London: Ashgate, 2009), 43–54.

49. Simon Biggs and Sue Hawksley, "Bodytext," with audio by Garth Paine, project webpage (2010), accessed November 8, 2019, http://www.littlepig.org.uk/installations/bodytext/index.htm.

50. Simon Biggs and Sue Hawksley, "Memory Maps in Interactive Dance Environments," *International Journal of Performance Arts and Digital Media* 2, no. 2 (2006): 123. One inspiration for the piece was Samuel Beckett's 1958 one-person play *Krapp's Last Tape*, in which the character goes through audiotapes of his recorded memories, commenting on them and recording further reflections. What seems to be a random selection of memories is in fact strictly scripted, which causes Biggs and Hawksley to question "to what degree the subject (and we ourselves) exists as a narrative construct, before, during and after the event." Simon Biggs, Sue Hawksley, and Garth Paine, "Bodytext: Somatic Data as Agency in Interactive Dance," in *Multimodality and Performance*, ed. Carla Fernandez (Newcastle, UK: Cambridge Scholars Publishing, 2016), 180.

51. In her reflections on this piece in her dissertation, Hawksley refers, for instance, to therapeutic approaches that use language to treat posttraumatic syndrome, "splitting tensions" in somatically "engraved" memories through deconstructing mental associations with them. She quotes a pioneer of dance movement psychotherapy, Trudi Schoop, and the phrases she used: "a brave chest with a scared pelvis," "an obstinate right side and a sentimental left side," etc. Hawksley concludes: "These sessions on tension splitting serve to bring unconsciously held habitual tension patterns to patients' attention." Sue Hawksley, "Dancing to an Understanding of Embodiment" (PhD diss., University of Edinburgh, 2012), 119.

52. Quoted from N. Katherine Hayles, *My Mother Was a Computer* (Chicago: University of Chicago Press, 2005), 48.

53. Alexander R. Galloway, *Protocol: How Control Exists after Decentralization* (Cambridge, MA: MIT Press, 2004), 165.

54. "Function is an abstract replica of causality . . . : the same input must always map to the same output. . . . [Backward operation] will not work." Derek Robinson, "Function," in *Software Studies: A Lexicon*, ed. Matthew Fuller (Cambridge, MA: MIT Press, 2008), 105–106.

55. Saussure argues that the signs themselves are arbitrary, there is no direct connection between a sign and its referent; meaning emerges and is stabilized only by differential relations between the signs. "A language and its written form constitute two separate systems of signs. The sole reason for the existence of the latter is to present the former. The object of study in linguistics is not a combination of the written word and the spoken word. The spoken word alone constitutes that object." Ferdinand de Saussure, *Course in General Linguistics*, trans. Roy Harris (Peru, IL: Open Source, 1988), 24–25.

56. Jacques Derrida, *Of Grammatology* (1967), trans. Gayatri Chakravorty Spivak (Baltimore: Johns Hopkins University Press, 1997), 61. "This arche-writing, although its concept is invoked by the themes of 'the arbitrariness of the sign' and of difference, cannot and can never be recognized as the object of a science. It is that very thing which cannot let itself be reduced to the form of presence." Ibid., 58.

57. See also Bernard Stiegler's and Mark Hansen's analyses of arche-writing in relation to technology. Bernard Stiegler, *The Fault of Epimetheus*, vol. 1 of *Technics and Time* (Stanford, CA: Stanford University Press, 1998); Mark B. N. Hansen, *Embodying Technesis: Technology beyond Writing* (Ann Arbor: University of Michigan Press, 2000), 84–85.

58. Biggs and Hawksley, "Memory Maps," 124. Peculiarly, the name of Hawksley's dance company is Articulate Animal (http://www.articulateanimal.org).

59. *Merriam-Webster*, s.v. "articulation," accessed November 8, 2019, http://www.merriam-webster.com/dictionary/articulation.

60. Interestingly, prior to Derrida's *Of Grammatology*, Roland Barthes had proposed that semiology should transform into a discipline not about (lexical) signs, but exactly about jointing and partitioning ("partages") as operations that he named "arthrology" and claimed were fundamental for the existence of human culture.

61. Brian Rotman, "Corporeal or Gesturo-haptic Writing," *Configurations* 10, no. 3 (Fall 2002): 425.

62. Ibid.

63. Ibid.

64. Ibid., 427.

65. Ibid., 434–435.

66. Ibid.

67. This can be also explained through the concept of plasticity which I have explored in more detail elsewhere (see Ksenia Fedorova, "Plasticity and Feedback: Schemas of Indetermination in Cybernetics and Art," *Proceedings of the 22nd International Symposium of Electronic Art (ISEA)* (2016): 51–58.

68. This media-centered understanding is slightly different from the more ontological one by Gilbert Simondon: "By transduction we mean an operation—physical, biological, mental, social—by which an activity propagates itself from one element to the next, within a given domain, and founds this propagation on a structuration of the domain that is realized from place to place: each area of the constituted structure serves as the principle and the model for the next area, as a primer for its constitution, to the extent that the modification expands progressively at the same time as the structuring operation." Gilbert Simondon, "The Position of the Problem of Ontogenesis," trans. Gregory Flanders, *Parresia Journal for Critical Philosophy* 7 (2009): 11.

69. Hawksley and Biggs, "Memory Maps in Interactive Dance Environments," 124.

70. Friedrich Kittler, "There Is No Software," in Friedrich Kittler, *Literature, Media, Information Systems: Essays*, ed. John Johnston (Amsterdam: G+B Arts, 1997), 150.

71. Hayles, *My Mother Was a Computer*, 45.

72. The voltages can only "understand" the binary code, which runs on special interpreters and compilers. In order to make a translation from or to the source code (written by a programmer) and to build and execute the binary (machinic) code, they have to have libraries/modules. Compilers (unlike interpreters) also have to be physically compatible with the type of processor, in order to translate source code into a set of instructions that a given processor can understand. These instructions, in turn, are executed on particular physical registers of a processor operating via binary code.

73. David M. Berry, *The Philosophy of Software: Code and Mediation in the Digital Age* (New York: Palgrave Macmillan, 2011), 14.

74. Kittler's words to describe what could be recorded by a phonograph, thanks to the then-new capabilities of recording the passing and continuous. Friedrich Kittler, *Gramophone, Film, Typewriter*, trans. G. Winthrop-Young and M. Wutz (Palo Alto, CA: Stanford University Press, 1999).

75. M. Beatrice Faci, *Contingent Computation: Abstraction, Experience, and Indeterminacy in Computational Aesthetics* (London: Rowman and Littlefield International, 2018).

76. N. Katherine Hayles, "Print Is Flat, Code Is Deep: The Importance of Media-Specific Analysis," *Poetics Today* 25, no. 1 (Spring 2004): 72 (italics in the original).

77. See Sybille Kraemer, "The Cultural Techniques of Time Axis Manipulation: On Friedrich Kittler's Conception of Media," in *Theory, Culture and Society* 23, no. 7–8 (2006): 93–109.

78. Author's interview with S. Hawksley, S. Biggs, and G. Paine, UC Davis, April 5, 2012.

79. Sybille Kraemer, "Medien, Boten, Spuren. Wenig mehr als Literaturbericht," in *Was ist ein Medium?*, ed. Stefan Muenker and Alexander Roesler (Frankfurt am Main: Suhrkamp, 2008), 67.

80. Author's interview.

81. Biggs, Hawksley, and Paine, "Bodytext: Somatic Data as Agency," 186.

82. Ibid.

83. Kriss Ravetto-Biagioli, *Digital Uncanny* (New York: Oxford University Press, 2019), 142. It could be relevant here also to adopt Galloway's concept of the "unworkable interface," although he discusses it in a very different context. See Galloway, "The Unworkable Interface," in *The Interface Effect* (Cambridge, UK: Polity Press, 2012), 25–53.

84. Ravetto-Biagioli, *Digital Uncanny*, 142.

85. Hawksley compares the scattering of the initial body-stories to the dispersal of a sand mandala, in itself a "meditation on impermanence." "The dispersal of the texts of my own bodystories could be seen to function as a meditation on the psycho-physical dangers of languaging embodied experience as a hermeneutic monolith." Hawksley, "Dancing to an Understanding of Embodiment," 123.

86. The examples of commercially available bioenabled wearable products range from cardio watches, wristbands, devices like Fitbit, Smartwatch (Pebble), Smartring (ElectricFoxy) to an electronic T-shirt (Carré), and even electronic tattoos (mc10). One can track everything from steps walked or run, hours of sleep, to diet components, as well as heart rate variability and other physiological expressions.

87. See the website for Empatica, https://www.empatica.com/, and Rosalind W. Picard, "What Happened to the Q Sensor?," http://affect.media.mit.edu/projectpages/iCalm/iCalm-2-Q.html, both accessed November 8, 2019.

88. Ludwig Binswanger, "On the Psychogalvanic Phenomenon in Association Experiments," in *Studies in Word-Association*, ed. Carl Jung (London: Heinemann, 1918), 446–530; Wilhelm Reich, *Experimentelle Ergebnisse über die electrische Funktion von Sexualität und Angst* (Copenhagen: Sexpolverlag, 1937), translated as

"Experimental Investigation of the Electrical Function of Sexuality and Anxiety," in *Journal of Orgonomy* 3, no. 1–2 (1969).

89. Scientists call a confirmed inference "affirmation of the consequent"; they divide the projections of elements of the physiological domain to the psychological domain into types: one-to-one, one-to-many, many-to-one, many-to-many projections, and a null relation. See Sylvia D. Kreibig, Gunnar Schaefer, and Tobias Brosch, "Psychophysiological Response Patterning in Emotion: Implications for Affective Computing," in *Affective Computing: A Sourcebook*, ed. Klaus R. Scherer, Tanja Bänziger, and Etienne B. Roesch (Oxford: Oxford University Press, 2010), 105–130. See also Jennifer Healey, "Physiological Sensing of Emotion," in *The Oxford Handbook of Affective Computing*, ed. Rafael Calvo, Sidney D'Mello, Jonathan Gratch, and Arvid Kappas (Oxford: Oxford University Press, 2015), 204–216.

90. Natasha Jaques, Sarah Taylor, Akane Sano, and Rosalind R. Picard, "Multi-task, Multi-kernel Learning for Estimating Individual Wellbeing," in *Proceedings of NIPS Workshop on Multimodal Machine Learning*, Montreal, Quebec, December 2015, http://affect.media.mit.edu/pdfs/15.Jaques-etal-NIPSMMML.pdf.

91. Mark Andrejevic and Mark Burdon, "Defining the Sensor Society," *Television and New Media* 16, no. 1 (2015): 19–36.

92. Deborah Lupton, *The Quantified Self: A Sociology of Self-Tracking* (Malden, MA: Polity, 2016), 35. See also Ana Viseu and Lucy Suchman, "Wearable Augmentations: Imaginaries of the Informed Body," in *Technologized Images, Technologized bodies*, ed. Jeanette Edwards, Penny Harvey, and Peter Wade (New York: Berghahn Books, 2010).

93. Commenting on an interview with Eric Boyd, a mechanical engineer who runs Toronto's Quantified Self meetup, Natasha Schüll notes: "Boyd distinguishes data-driven modes of self-discovery from those of talk-based therapy: 'Quantified self is not a linguistic exploration like psychoanalysis—it's more of a digital exploration, and the stuff you're exploring is made up of many little bits and moments.' One arrives at insights not through language unfolding in time, he elaborates, but through tracking these bits and moments *over* time [thus, discovering patterns of behavior]. . . . In self-tracking Boyd finds a pathway from self-knowledge to self-transformation. Tracking has allowed him to regard himself as a 'time-series self,' which he finds both liberating and empowering." Natasha Schüll, "Tracking," in *Experience: Culture, Cognition and the Common Sense*, ed. Caroline Jones, David Mather, and Rebecca Uchill (Cambridge, MA: MIT Press, 2016), 197–198.

94. Nigel Thrift, *Knowing Capitalism* (London: Sage, 2005).

95. See Manuel Castells, *The Rise of Network Society* (Malden, MA: Blackwell, 2000).

96. Quoted in Andrejevic and Burdon, "Defining the Sensor Society," 9.

97. In this sense, Isabel Pedersen stresses the persuasive or rhetorical function of wearable technologies, since their purview lies only in providing guidance, shaping and "shifting" the attitudes and actions of their users. See Pedersen, *Ready to Wear*.

98. In her book *Touch*, Laura Marks argues for indexical connections in electronic media in general: "the indexical relation is maintained insofar as the activity of electrons can be traced to a wave function. . . . We can still trace digital information to interconnected matter." Laura Marks, *Touch: Sensuous Theory and Multisensory Media* (Minneapolis: University of Minnesota Press, 2002), 163. Yet she admits that there are still a lot of questions in quantum physics about electron activities and the true meanings of the approximations made in equations.

99. See note 62 of chapter 1.

100. Cartwright observes that "rather than simply augmenting the senses of the scientific observer, cinematography supplemented or replaced sensory perception." Lisa Cartwright, *Screening the Body: Tracing Medicine's Visual Culture* (Minneapolis: University of Minnesota Press, 1995), 23.

101. Colin Milburn, *Nanovision: Engineering the Future* (Durham: Duke University Press, 2008), 13.

102. In that sense, it introduces "dreamscapes" of technology and animates "a productive dialogue and conflict between presentism and futurism, between humanistic thought and its other." Ibid., 14.

103. John Johnston, "Machinic Vision," *Critical Inquiry* 26, no. 1 (Autumn 1999): 27. Johnston gives a comprehensive account of the notion of "machinic vision" spanning from Virilio's writings to Deleuze's classification of "perception-images" in *Cinema I* and *II*.

104. Cartwright, *Screening the Body*, 27. Cartwright describes the machinic logic of vision as inscriptive: it is inscribed directly onto the bodies of experimental subjects, serving as "a disciplinary technique, insofar as it facilitate[s] the establishment of a productive dynamic economy of the body." Ibid., 37.

105. Ibid., 27. Technologies of algorithmization have a logic akin to what Cartwright names "technologies of destruction" (and it is not only about the electrification of an elephant as a showcase of the power of electricity; it is about the disciplining effect that the very idea and practice of external observation has over the object of observation).

106. For example, see Bernard Stiegler's comparisons of the living and nonliving types of being in *The Fault of Epimetheus*.

107. Luciana Parisi, "Biotech: Life by Contagion," *Theory, Culture and Society* 24, no. 6 (2007): 31. She writes: "Nature no longer can be identified with organic life. Rather, life cannot be detached from symbiotic assemblages that challenge what is

defined as living and nonliving, individual and environment, simple and complex." Ibid., 42.

108. See Eugene Thacker, *Biomedia* (Minneapolis: University of Minnesota Press, 2004).

109. Patricia T. Clough, "Biotechnology and Digital Information," *Theory, Culture and Society* 24, no. 7–8 (December 2007): 312.

110. For example, Patricia Clough describes the relation of quantum indeterminacy to sensing operations and intermolecular communication (ibid.). Luciana Parisi also discusses the special velocities of bionic technologies such as neuroprosthetics, and says that silicon transmitters can feel at faster rates and even "feel before." Furthermore, she writes, "Biochip technologies enter the relation between kinesthetic, synesthetic and visceral sensibilities through a symbiotic assemblage of silicon, neural and bacterial information sensing, giving rise to a sort of bionic sensorium ready to feel the infinitesimal proximities of organic and inorganic matter. From this standpoint, it is possible to suggest that bionic technologies do not just extend sensory–motor perception but confront abstract feeling or nonsensuous sensation spinning out of the feedback circuits of information sensing linking the velocities of bacterial communication with those of neural firing and silicon light-sensitive transfer." Luciana Parisi, "Technoecologies of Sensation," in *Deleuze/Guattari and Ecology*, ed. Bernd Herzogenrath (New York: Palgrave Macmillan, 2009), 191–192.

111. Parisi, "Technoecologies of Sensation," 190.

112. Making a distinction between perception and sensation, Massumi writes: "Sensation pertains to the dimension of passage, or the continuity of immediate experience (and thus to a direct registering of potential). . . . Perception is exoreferential (extensive); sensation is endoreferential or self-referential (intensive)." Massumi, *Parables for the Virtual*, 258–259.

113. Emanuele Coccia, *Sensible Life: A Micro-ontology of the Image* (New York: Fordham University Press, 2016).

114. In this tradition, the Cartesian divide between the body and mind is overcome by notions of the sensible and affect that (both) go beyond this split, acting as forces constitutive of both the embodied and preconscious, as well as conscious and symbolizable (representable) experience. See, for instance, Mark B. N. Hansen, "Feelings without Feelers, or Affectivity as Environmental Force," in *Timing of Affect: Epistemologies, Aesthetics, Politics*, ed. Marie-Luise Angerer, Bernd Bösel, and Michaela Ott (Zurich: Diaphanes, 2014), 65–86.

115. Anne Beaulieu, "Images Are Not the (Only) Truth: Brain Mapping, Visual Knowledge, and Iconoclasm," *Science, Technology and Human Values* 27 (2002): 53–86.

116. See Sybille Kraemer, "Operative Bildlichkeit. Von der 'Grammatologie' zu einer 'Diagrammatologie'? Reflexionen über erkennendes Sehen," in *Logik des Bildlichen. Zur Kritik der ikonischen Vernunft*, ed. Martina Hessler and Dieter Mersch (Bielefeld: transcript, 2009), 94–123.

117. "They [graphical representations] are not only a medium for representation of the object of knowledge, but also at the same time an instrument through which those very objects can be generated and explored. Knowledge is not only represented, transmitted and disseminated through the diagrammatic; it is produced and expanded by it." Sybille Kraemer, "Trace, Writing, Diagram: Reflections on Spatiality, Intuition, Graphical Practices and Thinking," in *The Power of the Image: Emotion, Expression, Explanation*, ed. András Benedek and Kristóf Nyíri (Frankfurt am Main: Peter Lang, 2014), 3.

118. Deleuze's concept of the time-image could be relevant to the topic of time-based arts—for instance, for the discussion of the perception of dynamics (of something existing in time). Yet its relevance will be limited, since the audiovisual of the cinematic image, which is the material on which Deleuze grounds his argument, does not exhaust the possibilities of bodily engagement in the same way that multisensorial stimulation can, and also because feedback dictates its own logic, introducing more multistability.

119. Marco Donnarumma, "Nigredo: Configuring Human and Technological Bodies," in *Experiencing the Unconventional: Science in Art*, ed. Theresa Schubert and Andrew Adamatzky (Singapore: World Scientific Publishing, 2015), 63.

120. Marco Donnarumma, "Nigredo," project webpage, 2004–2019, accessed November 12, 2017, http://marcodonnarumma.com/works/nigredo/.

121. Donnarumma, "Nigredo: Configuring Human and Technological Bodies," 75.

122. Ibid.

123. Gilbert Simondon, "The Genesis of the Individual," in *Incorporations*, ed. Jonathan Crary and Sanford Kwinter, trans. Mark Cohen and Sanford Kwinter (New York: Zone, 1992), 301.

124. Ibid.

125. "The individual would then be grasped as a relative reality, a certain phase of being that supposes a preindividual reality, and that, even after individuation, does not exist on its own, because individuation does not exhaust with one stroke the potentials of preindividual reality. Moreover, that which the individuation makes appear is not only the individual, but also the pair individual-environment. The individual is thus relative in two senses, both because it is not all of the being, and because it is the result of a state of the being in which it existed neither as

individual, nor as principle of individuation." Simondon, "The Position of the Problem of Ontogenesis," 5.

126. See Marco Donnarumma, ed., "Biotechnological Performance Practice," *Canadian Electroacoustic Community* 14 (2012), http://econtact.ca/14_2/index.html.

127. Flora Lysen, "The Interface Is the (Art)Work: EEG-Feedback, Circuited Selves and the Rise of Real-Time Brainmedia (1964–1977)," in *Brain Art: Brain-Computer Interfaces for Artistic Expression*, ed. Anton Nijholt (Cham, Switzerland: Springer International, 2019), 33–34.

128. See Nijholt, *Brain Art*.

129. See Jan Slaby, "Steps towards a Critical Neuroscience," *Phenomenology and the Cognitive Sciences* 9 (2010): 397–416; Joseph Dumit, *Picturing Personhood: Brain Scans and Biomedical Identity* (Princeton, NJ: Princeton University Press, 2004); Jan De Vos and Ed Pluth, eds., *Neuroscience and Critique: Exploring the Limits of the Neurological Turn* (London: Routledge, 2016). See also a critical analysis of EEG wearables and the emerging consumer culture around them in Melissa M. Littlefield, *Instrumental Intimacy: EEG Wearables and Neuroscientific Control* (Baltimore: Johns Hopkins University Press, 2018).

130. *Wave UFO* became famous both within and beyond the art world because it appealed to the popular curiosity about the brainwaves measuring technique, while offering a highly aestheticized and sensualized immersive experience of introspection.

131. Kate Mondloch, *Capsule Aesthetic: Feminist Materialisms in New Media* (Minneapolis: University of Minnesota Press, 2018), 11.

132. Ibid., 88. Similarly, Flora Lysen, in her writing on Karen Lancel and Hermen Maat's interactive performative installation *EEG KISS* (2014–), where the brainwaves of two participants kissing in public are interpreted visually and sonically, describes it as an art-science experiment that "reconceptualizes the object of research—'intimacy'—by reconfiguring what intimacy is in the interplay between public and private experiences, feelings and displays." Flora Lysen, "Kissing and Staring in Times of Neuromania: The Social Brain in Art-science Experiments," in *Artful Ways of Knowing: Dialogues between Artistic Research and Science and Technology Studies*, ed. Henk Borgdorff, Peter Peters, and Trevor Pinch, 167–183 (New York: Routledge, 2019). The scientific view is here once again "complicated" by the shifted boundaries of public and private, as well as staged and unstaged.

133. See the Gallant Lab at UC Berkeley website, accessed November 13, 2019, http://gallantlab.org/index.php. See also Isabelle Klein et al., "Retinoptic Organization of Visual Mental Images as Revealed by Functional Magnetic Resonance Imaging," *Cognitive Brain Research* 22 (2004): 26–31.

134. Ivan Martinovic, Doug Davies, and Mario Frank, "On the Feasibility of Side-Channel Attacks with Brain-Computer Interfaces," in *Proceedings of the 21st USENIX Security Symposium* (2012), https://www.usenix.org/conference/usenixsecurity12/technical-sessions/presentation/martinovic.

135. For an example of such an argument, see Lawrence A. Farwell, "Brain Fingerprinting: A Comprehensive Tutorial Review of Detection of Concealed Information with Event-Related Brain Potentials," *Cognitive Neurodynamics* 6, no. 2 (April 2012): 115–154.

136. Richard Teitelbaum, "In Tune: Some Early Experiments in Biofeedback Music (1966–74)," in *Biofeedback and the Arts: Results of Early Experiments*, ed. David Rosenboom, 35–65 (Vancouver, BC: A.R.C. Publications, 1976).

137. "For example, the databases of meditation [containing sonic score and visual material], when triggered by Saba's brainwaves once her specific threshold state was attained, would produce harmonious sounding music. The libretto would intone the ninety-nine names of God in English (translated from Arabic), and the videos would display tranquil types of images." Ellen Pearlman, "Brain Opera: Exploring Surveillance in 360-Degree Immersive Theatre," *PAJ: A Journal of Performance and Art* 39, no. 2 (2017): 80.

138. For example, Pearlman herself is primarily interested in the issue of surveillance of consciousness, including processes that are beyond conscious control, such as dream states. Ellen Pearlman, "The Brain as Site-Specific Surveillant Performative Space," *International Journal of Performance Arts and Digital Media* 11, no. 2 (2015): 219–234.

Chapter 3

1. See, for example, Ravetto-Biagioli's explanations of the uncanny in application to Alex Garland's film *Ex Machina* (2014) and its character Caleb's relations to the humanoid robot Ava, in Kriss Ravetto-Biagioli, *Digital Uncanny* (New York: Oxford University Press, 2019), 1–15.

2. From the participants' reviews, National Theatre Wales Community, "Your Experiences with Karen" discussion, last modified September 17, 2015, https://community.nationaltheatrewales.org/group/karen/forum.

3. Blast Theory, "Karen FAQ," accessed November 14, 2019, http://www.blasttheory.co.uk/faq/?bproject=6685#bt_faq-7701.

4. Karen Archey, "Karen: A Life Coach, a Friend, a Data Mine, an App," e-flux Conversations, June 2015, http://conversations.e-flux.com/t/karen-a-life-coach-a-friend-a-data-mine-an-app/1923.

5. See a detailed analysis of this aspect in Maria Chatzichristodoulou, "*Karen* by Blast Theory: Leaking Privacy," in *Digital Bodies: Creativity and Technology in the Arts and Humanities*, ed. Susan Broadhurst and Sara Price (London: Palgrave Macmillan, 2017), 65–78.

6. For a detailed comparative analysis of different historical and current psychological models of personality, including the Big Five and its extension, the HEXACO model, see Sandra Matz, Yin Wah Fiona Chan, and Michal Kosinski, "Models of Personality," in *Emotions and Personality in Personalized Services: Models, Evaluation and Applications*, ed. Marko Tkalčič et al. (Cham, Switzerland: Springer International, 2016), 35–54.

7. The sample answers may vary in formulation. For example, regarding the value of appraisal by others: "1. I care about praise and criticism from others. 2. I care about others' attitudes toward me. 3. I hate being taken lightly. 4. I will be very angry if others are impolite to me. 5. I will be very happy if I am treated with respect. 6. I will be very upset if I am criticized in public." Or, regarding mood: "1. Although I'm sometimes sad, I have a mostly optimistic outlook. 2. No matter how badly I feel, I try to think about pleasant things. 3. When I become upset, I remind myself of all the pleasures in life. 4. I try to think good thoughts no matter how badly I feel." From Blast Theory Studio archive, Brighton, UK. See also Gordon C. Bruner II, *Marketing Scales Handbook: A Compilation of Multi-Item Measures for Consumer Behavior and Advertising Research* (Fort Worth, TX: GCBII Productions, 2015).

8. From Blast Theory Studio archive, Brighton, UK.

9. Blast Theory, "Karen," project webpage, accessed June 15, 2019, www.blasttheory.co.uk/projects/karen/.

10. Interview by the author at Blast Theory's studio, October 2015.

11. One can try having one's own conversations with ELIZA emulators, such as the Javascript version of ELIZA by Michal Wallace and George Dunlop, accessed August 7, 2019, http://psych.fullerton.edu/mbirnbaum/psych101/Eliza.htm.

12. For example, replies to a phrase with the keyword "remember" include: "Why do you remember () just now?," "What in the present situation reminds you of ()?," "What is the connection between me and ()?"; replies with the keyword "computer": "Why do you mention computers?," "What do you think machines have to do with your problem?," "Don't you think machines can help people?," "What about machines worries you?," etc. Joseph Weizenbaum, "ELIZA—A Computer Program for the Study of Natural Language Communication between Man and Machine," *Communications of the ACM* 9, no. 1 (January 1966): 44.

13. Ibid.

14. Ibid., 42.

15. For a detailed analysis of the historical development of both chatbots and dialogue systems, see Nguyen-Thinh Le, Julia Masche, and Niels Pinkwart, "From ELIZA to Modern Chatbots and Dialogue Systems: The Historical Development of Conversational Software; Classification, Technology and Application Domains," in *Hello, I'm Eliza. Fünfzig Jahre Gespräche mit Computern*, ed. Marianna Baranovska and Stefan Höltgen (Bochum/ Freiburg: projekt verlag, 2018), 143–160.

16. Joseph Weizenbaum, *Computer Power and Human Reason: From Judgment to Calculation* (San Francisco: W. H. Freeman, 1976), 188.

17. Ibid., 189.

18. Ibid., 190.

19. Wade J. Mitchell, Chin-Chang Ho, Himalaya Patel, and Karl F. MacDorman, "Does Social Desirability Bias Favor Humans? Explicit–Implicit Evaluations of Synthesized Speech Support a New HCI Model of Impression Management," *Computers in Human Behavior* 27 (2011): 402–412. Some companies acknowledge cultural specificities in relation to gender. For instance, Apple's Siri defaults to a female voice in the U.S., but on iPhones where the language is Arabic, French, Dutch, or British English, Siri defaults to a male voice.

20. Joanna Stern, "Alexa, Siri, Cortana: The Problem with All-Female Digital Assistants," *The Wall Street Journal*, February 21, 2017, https://www.wsj.com/articles/alexa-siri-cortana-the-problem-with-all-female-digital-assistants-1487709068?tesla=y&mod=e2fb.

21. The authors of a UNESCO report on gender bias in digital culture find tolerance to verbal abuse especially worrisome: "The subservience of digital voice assistants becomes especially concerning when these machines—anthropomorphized as female by technology companies—give deflecting, lacklustre or apologetic responses to verbal sexual harassment." Mark West, Rebecca Kraut, and Han Ei Chew, *I'd Blush if I Could: Closing Gender Divides in Digital Skills through Education* (UNESCO, EQUALS Skills Coalition, 2019), 106.

22. See James Vlahos, "Siri, Cortana, Alexa: Why Our AIs Have Personalities and a Female Voice," *The Print*, June 23, 2019, https://theprint.in/pageturner/excerpt/siri-cortana-alexa-why-our-ais-have-personalities-and-a-female-voice/253526/.

23. Carl Rogers, "Toward the Modern Approach to Values: The Valuing Process in the Mature Person," in *Person to Person: The Problem of Being Human*, ed. Carl Rogers and Barry Stevens (Lafayette, CA: Real People Press, 1967), 16.

24. Ibid.

25. Harold Garfinkel, *Studies in Ethnomethodology* (Englewood Cliffs, NJ: Prentice-Hall, 1967), 89.

26. Two recent art projects bring fortune-telling to the foreground as an ironic critique of neoliberal techno-utopianism. *Oracle* (2016) by the Russian collective Where the Dogs Run features three eerie-looking holographic piglets that answer visitors' questions with quotations from George Orwell, Marshall McLuhan, textbooks on computational linguistics, etc. And *Icosahedron* (2019) by Zach Blas is "an artificially intelligent crystal ball that predicts the future of prediction"; see Zach Blas website, accessed October 17, 2019, http://www.zachblas.info/works/icosahedron/.

27. Chris Marker, "Dialector," project website, accessed October 17, 2019, http://dialector.poptronics.fr/. See also Agnès de Cayeux, Andrès Lozano, and Annick Rivoire, "History of the Dialector Program," on Chris Marker, "Notes from the Era of Imperfect Memory," November 1, 2015, http://chrismarker.org/history-of-the-dialector-program/.

28. Apparently the program included some "geek humor," such as the inclusion of statements like "NEVER TRUST ANYONE OVER 256K" or "DO YOU PUT A 'K' ON YOUR SHIRT?" André Lozano, "Chris Marker Dialector Reloaded," trans. Dorna Khazeni, on Chris Marker, "Notes from the Era of Imperfect Memory," January 25, 2015, http://chrismarker.org/chris-marker-dialector-reloaded-andre-lozano-english-translation/.

29. "Dialector, en conversation(s), Agnès Varda," accessed October 17, 2019, http://dialector.poptronics.fr/varda.html.

30. Unlike software that is used in today's chatbots, which can learn from previous conversations, Alice software is preprogrammed in AIML (Artificial Intelligence Mark-Up Language), which means that there is a limited database of responses intended for all kinds of queries.

31. Stelarc, "Prosthetic Head: Intelligence, Awareness, and Agency," in *Critical Digital Studies: A Reader*, ed. Arthur Kroker and Marilouise Kroker (Toronto: University of Toronto Press, 2008), 486.

32. Ibid., 477.

33. Ibid., 478.

34. See Lynn Hershman Leeson, *Agent Ruby*, 1998–2002, accessed November 18, 2019, http://www.lynnhershman.com/agent-ruby/.

35. Sherry Turkle, *Life on the Screen: Identity in the Age of the Internet* (New York: Simon and Schuster, 1995), 101.

36. Elizabeth Wilson, *Affect and Artificial Intelligence* (Seattle: University of Washington Press, 2010), 92–93.

37. Sandor Ferenczi, "Introjection and Transference," in Sandor Ferenczi, *Contributions to Psycho-Analysis*, trans. Ernest Jones (Boston: Richard G. Badger, 1916), 31.

38. Ibid., 34.

39. Eryn Brown, "Computerized 'Ellie' Has Just Enough Humanity to Aid in Therapy Work," *Los Angeles Times,* April 3, 2015, http://www.latimes.com/local/california/la-me-virtual-interviewer-usc-20150403-story.html.

40. Ferenczi, "Introjection and Transference," 41.

41. Wilson, *Affect and Artificial Intelligence*, 93.

42. Ibid.

43. Warren Sack and Joseph Dumit, "Artificial Participation: An Interview with Warren Sack," in *Zeroing In on the Year 2000: The Final Edition*, ed. George E. Marcus (Chicago: University of Chicago Press, 2000), 63.

44. Plato, *The Collected Dialogues of Plato, Including the Letters*, ed. Edith Hamilton and Huntington Cairns, trans. Lane Cooper and others (New York: Pantheon Books, 1966), quoted in Sack and Dumit, "Artificial Participation," 63.

45. Norbert Wiener, *Cybernetics; or Control and Communication in the Animal and the Machine* (New York: J. Wiley, 1948), 149–150. Quoted in Orit Halpern, *Beautiful Data: A History of Vision and Reason since 1945* (Durham: Duke University Press, 2014), 68.

46. Some online dating platforms have been accused of fraud, using "armies of fembots" to create the impression that "women on the site are plentiful and eager," and tricking men to pay for interactions with them. "Ashley Madison Code Shows More Women, and More Bots," *Gizmodo,* August 31, 2015, http://gizmodo.com/ashley-madison-code-shows-more-women-and-more-bots-1727613924.

47. Extensive collections of chatbots can be found at the following websites: Pandorabots, http://alice.pandorabots.com/; the Chatterbot Collection, https://www.chatterbotcollection.com/, and Marcus P. Zillman, "Chatterbots Resources and Sites 2019," http://www.zillman.us/subject-tracers/chatterbots-resources-and-sites/, all accessed November 18, 2019.

48. See John West, "Microsoft's Disastrous Tay Experiment Shows the Hidden Dangers of AI," *Quartz,* April 2, 2016, https://qz.com/653084/microsofts-disastrous-tay-experiment-shows-the-hidden-dangers-of-ai/.

49. See Replika's website, www.replika.ai. To read about an experience with using Replika, see Tallie Gabriel, "I Tried Being BFFs with an AI," *Vice,* August 8, 2017, https://www.vice.com/EN_US/article/nezxaq/i-tried-being-bffs-with-an-ai.

50. This feature raises privacy concerns: people might not think of this problem in relation to more established companies, but when asked to open access to their correspondence to a completely new company, they may think twice.

51. Alan Turing, "Computing Machinery and Intelligence," *Mind* 59, no. 236 (October 1950): 434.

52. Ibid., 459.

53. Silvan Tomkins, "Simulation of Personality: The Interrelationships between Affect, Memory, Thinking, Perception and Action," in *Computer Simulation of Personality*, ed. Silvan S. Tomkins and Samuel Messick (New York: John Wiley and Sons, 1963), 10.

54. Ibid., 18.

55. Herbert Simon and Allen Newell, "Heuristic Problem Solving: The Next Advance in Operations Research," *Operations Research* 6, no. 1 (January/February 1958): 1–10. The context for their argument was the justification of such research for the wider application of cognitive models. In addition to its pure scientific and epistemic value, psychological modeling has been governed by a pragmatic reason: the development of more efficient systems of management.

56. Other early relational models worth mentioning include John and Jeanne Gullahorn's Homunculus—a determined model of social behavior, designed to provide predictions for interpersonal behavior based on personality features and a specific situation—and Jim Doran's "Pleasure Seeking Automaton" (1968).

57. Walter Reitman, "Personality as a Problem-Solving Coalition," in Tomkins and Messick, *Computer Simulation of Personality*, 90, 96.

58. For more on MicroPsi, see MicroPsi, accessed January 4, 2019, http://www.garfixia.nl/micropsi; see also Cognitive Artificial Intelligence: The MicroPsi Project, accessed November 19, 2019, http://cognitive-ai.com. A similar feedback principle is featured in an earlier schema designed by Gerald Blum; see Gerald Blum, "Programming People to Simulate Machines," in Tomkins and Messick, *Computer Simulation of Personality*, 129.

59. Joscha Bach, *Principles of Synthetic Intelligence: Psi; An Architecture of Motivated Cognition* (New York: Oxford University Press, 2009), 147–149.

60. For early critiques see, for instance, Milton Rosenberg, "Simulated Man and the Humanistic Criticism," in Tomkins and Messick, *Computer Simulation of Personality*, 113–124. Another influential book on similar issues is Edward A. Feigenbaum and Julian Feigenbaum, eds., *Computers and Thought* (New York: McGraw-Hill, 1963).

61. Noam Chomsky, "Where Artificial Intelligence Went Wrong," interviewed by Yarden Katz, *Atlantic* (November 2012), https://chomsky.info/20121101/.

62. Peter Norvig, "On Chomsky and the Two Cultures of Statistical Learning," accessed August 2, 2019, http://norvig.com/chomsky.html.

63. There are prognoses that chatbots will eliminate apps because they can offer instant access to the needed information (or "value," in Silicon Valley jargon) while eliminating "friction" (signing up, figuring out the new interface, etc.).

64. Stelarc, "Prosthetic Head," 478.

65. Casey Newton, "Speak, Memory," *The Verge*, accessed November 20, 2019, http://www.theverge.com/a/luka-artificial-intelligence-memorial-roman-mazurenko-bot.

66. A similar story is told in an episode of the British TV series *Black Mirror* entitled "Be Right Back" (2013), in which a woman orders a robot that looks like her deceased boyfriend and that can talk based on the real person's social media chat history.

67. Brian Christian, *The Most Human Human: What Talking with Computers Teaches Us about What It Means to Be Alive* (New York: Doubleday, 2011).

68. Mike Murphy and Jacob Templin, "Replika: This App Is Trying to Replicate You," *Quartz* (August 2017), https://qz.com/se/machines-with-brains/1018126/lukas-replika-chatbot-creates-a-digital-representation-of-you-the-more-you-interact-with-it/.

69. Ibid.

70. Ibid.

71. Oren Etzioni, "How to Regulate Artificial Intelligence," *New York Times*, September 1, 2017, https://mobile.nytimes.com/2017/09/01/opinion/artificial-intelligence-regulations-rules.html.

72. "Mouchette," project website, accessed November 20, 2019, http://www.mouchette.org.

73. "Mouchette," in *Rhizome, Net Art Anthology*, accessed November 20, 2019, http://anthology.rhizome.org/mouchette.

Chapter 4

1. Oliver Sacks, *The Man Who Mistook His Wife for a Hat and Other Clinical Tales* (London: Derald Duckworth, 1985). Norbert Wiener gives examples of similar neurological conditions at the beginning of the chapter "Feedback and Oscillation" in his *Cybernetics; or Control and Communication in the Animal and the Machine* (New York: J. Wiley, 1948).

2. Michel Foucault, *The Archaeology of Knowledge and The Discourse on Language* (New York, Pantheon Books, 1972); Denis Wood, with Jon Fels, *The Power of Maps* (New York: Guilford, 1992); John B. Harley, *The New Nature of Maps: Essays in the History of Cartography*, ed. Paul Laxton (Baltimore: Johns Hopkins University Press, 2001); Henri Lefebvre, *The Production of Space*, trans. D. Nicholson-Smith (Blackwell: Oxford, 1991).

3. Gilles Deleuze and Félix Guattari, *A Thousand Plateaus: Capitalism and Schizophrenia*, trans. Brian Massumi (Minneapolis: University of Minnesota Press, 1987), 10–11.

4. These approaches were put forward by, respectively, J. Pickles, D. Wood and J. Fels, B. Latour, and J. Corner and T. Ingold, and are cited in Rob Kitchin, Chris Perkins, and Martin Dodge, "Thinking about Maps," in *Rethinking Maps: New Frontiers in Cartographic Theory*, ed. Martin Dodge, Rob Kitchin, and Chris Perkins (London: Routledge, 2009), 10–23.

5. Ibid., 18.

6. Ibid. Perhaps one of the first and most influential framings of mapping as an additive and subjective practice was by Michel de Certeau: his suggestion to engage everyday stories as "treatments of space," emphasizing the practice and experience of perceiving space, inspired both professional geographers and cultural workers, or artists.

7. The original meaning of the terms "critical cartography" and "critical GIS" implies an explicit political focus and intention to deconstruct the ideological and power implications of spatial representations.

8. Jeremy Hight, "Rhizomatic Cartography: Modulated Mapping and the Spatial Net," NeMe, posted in March 2009, accessed August 2, 2019, http://www.neme.org/991/rhizomatic-cartography.

9. Jeremy Hight, "Writing within the Map," NeMe, posted in March 2010, accessed August 2, 2019, http://www.neme.org/main/1111/writing-within-the-map.

10. See Paul Virilio, *Negative Horizon: An Essay in Dromoscopy* (London: Continuum, 2005). The ubiquity of communication problematizes the status of our relation to space, producing so-called "geographies of nowhere." See Gary Eberle, *The Geography of Nowhere: Finding One's Self in the Postmodern World* (Kansas City, MO: Sheed and Ward, 1994); Marc Augé, *Non-Places: Introduction to an Anthropology of Supermodernity*, trans. John Howe (London: Verso, 1995).

11. See more in Brett Stalbaum, "Database Logic(s) and Landscape Art," C5 Corporation, accessed November 21, 2019, http://www.c5corp.com/research/databaselogic.shtml.

12. Brett Stalbaum, "The Landscape and Culture: Data as Mediator, Modes of Engagement," in C5 Corporation, 2003, accessed November 21, 2019, http://www.c5corp.com/research/landscapeculture.shtml.

13. Ibid. A number of relevant projects were realized under C5 Corporation: see their website at http://www.c5corp.com/.

14. Christine Ross, "Spatial Poetics: The (Non)Destinations of Augmented Reality Art," *Afterimage* 38, no. 2 (September/October 2010): 20–21.

15. Ibid.

16. Charles Sanders Peirce, "On the Algebra of Logic: A Contribution to the Philosophy of Notation," *The American Journal of Mathematics* 7 (1885): 180–202, reprinted in *The Writings of Charles S. Peirce*, vol. 5, compiled by the editors of the Peirce Edition Project (Bloomington: Indiana University Press, 1993), 162–190.

17. "While realism claims to build a mimetic copy, an illusion of an inhabitable world, the index only purports to point, to connect, to touch, to make language and representation adhere to the world as tangent—to reference a real without realism." Mary Ann Doane, "Indexicality: Trace and Sign; Introduction," *Differences* 18, no. 1 (2007): 1–6.

18. Ibid., 2.

19. Nanna Verhoeff, "The Medium Is the Method: Locative Media for Digital Archives," in *(Dis)orienting Media and Narrative Mazes*, ed. Julia Eckel et al. (Bielefeld: Transcript Verlag, 2013), 28.

20. He states, "We are no longer reduced to the dilemma of thought and extension . . . [or] limited to the empirical world and the world of abstract understanding. Between the two is placed an intermediate world . . . 'alam al-mithal, the world of the Image, mundus imaginalis: a world as ontologically real as the world of the senses and the world of the intellect." Henry Corbin, *Mundus Imaginalis, or the Imaginary and the Imaginal* (Ipswich, MA: Golgonooza Press, 1976).

21. Ibid.

22. Christiane Paul, lecture presented at Danube University in Krems, Austria, in May 2010. See also an analysis of context awareness in relation to "ambient displays" and habitual blindness that makes certain things invisible in Dietmar Offenhuber, "Kuleshov's Display—On Contextual Invisibility," in *Throughout: Art and Culture Emerging with Ubiquitous Computing*, ed. Ulrik Ekman (Cambridge, MA: MIT Press, 2013), 432.

23. Paul, lecture at Danube University.

24. See de Certeau, *The Practice of Everyday Life,* 97. An example of a "practiced" space today can be found in computer and video games that offer increasingly vivid environments, which become playgrounds for the players' creation of narratives and self-expression through spatial movement. Sybille Lammes cites games like *The 21 Steps*, *Panoramio*, and *Civilization IV*: "The stories thus created, liberate maps from the isolated, unchangeable and static form they have gained since the Renaissance. They bring maps back (in)to life." Sybille Lammes, "Terra Incognita: Computer Games, Cartography and Spatial Stories," in *Digital Material: Tracing New Media in Everyday Life and Technology*, ed. Marianne van den Boomen et al. (Amsterdam: Amsterdam University Press, 2009), 228.

25. See Doreen Massey, *For Space* (London: Sage, 2005); and Adriana De Souza e Silva, "From Cyber to Hybrid: Mobile Technologies as Interfaces of Hybrid Spaces," *Space and Culture* 9, no 3 (2006): 261–278.

26. Federica Timeto, "Locating Media, Performing Spatiality: A Nonrepresentational Approach to Locative Media," in *Locative Media*, ed. Rowan Wilken and Gerard Goggin (London: Routledge, 2015), 98. "Postrepresentational cartography considers the map as an interface not only of visualizing but of actually entering a network that does not stand in front of us and rather is the location of everyday life. In performative mapping practices, space is figured and practiced at the same time, revealing the coconstitution of space and maps and an implosion of the barrier between the cognitive apprehension and the embodied practice of space" (ibid., 100.)

27. Some scholars may also call these elements "layers." For more on layers, see Nanna Verhoeff, "A Logic of Layers: Indexicality of iPhone Navigation in Augmented Reality," in *Studying Mobile Media: Cultural Technologies, Mobile Communication and the iPhone*, ed. Larissa Hjorth, Jeann Burgess, and Ingrid Richardson (New York: Routledge, 2012), 118–132.

28. Deleuze and Guattari, *A Thousand Plateaus*, 353.

29. See Félix Guattari, *Chaosmosis: An Ethico-Aesthetic Paradigm*, trans. Paul Bains and Julian Pefanis (Sydney: Power Publications, 1995), 92–93. I follow here Federica Timeto, who refers to Guattari's concept of the "esthetic" as "a dimension of creation and virtuality that is not limited to the finite modes of artistic production but rather is concerned with the performative forces by which different space-times are accomplished" (ibid). She writes: "This idea points to a field of processes of differentiation that cannot be 'circumscribed within the logic of discursive sets' because their insides and outsides cannot be clearly delimited in advance, except through experiencing them in their becoming by means of an 'ontological pragmatics.'" Timeto, "Locating Media, Performing Spatiality," 96.

30. Barbara Montero, "Proprioception as an Aesthetic Sense," *Journal of Aesthetics and Art Criticism* 64, no. 2 (2006): 231–242.

31. Brian Massumi, *Parables for the Virtual: Movement, Affect, Sensation* (Durham: Duke University Press, 2002), 58.

32. Ibid., 179.

33. "Urban Atmospheres," project webpage, viewed in June 2018, http://www.urban-atmospheres.net/projects.htm. See more in Ksenia Fedorova and Marc Barasch, "Mission to Earth: Terrestrial Proprioception and the Cyber-Sublime," *Azimuth* 6, no. 12 (2018): 93–112.

34. Preemptive Media website, www.preemptivemedia.net/, and Guerilla Grafters website, http://www.guerrillagrafters.org/, both accessed November 21, 2019.

35. Teri Rueb, "The Choreography of Everyday Movement—2001," project webpage, accessed November 21, 2019, http://terirueb.net/the-choreography-of-everyday-movement-2001/.

36. Les gens d'Uterpan, "Topologie," project webpage, accessed November 21, 2019, http://www.lesgensduterpan.com/proj_topologie-uk.html. See also Brian Seibert, "To Please, To Perplex and to Be Barely Noticed on a Dash through Queens Topologie, a Peripatetic Show from The Chocolate Factory," *New York Times*, May 12, 2014.

37. Leila Nadir and Cary Adams, "Indeterminate Hikes +," project webpage, accessed November 21, 2019, http://www.ecoarttech.net/project/indeterminate-hike/.

38. Mark Shepard, "Serendipitor," project website, accessed November 21, 2019, http://serendipitor.net/site/?page_id=2.

39. Adam Greenfield, "Serendipity Cities: Of Services and Situations," April 17, 2010, https://speedbird.wordpress.com/2010/04/17/serendipity-cities-of-services-and-situations/. "Strictly speaking, the word ['serendipity'] means accidentally discovering something wonderful in the course of a search for something unrelated. The genuine occurrence of serendipity necessarily implies a very powerful order of richness and texture in the world and, to my mind anyway, when you experience it in cities it's a clear indicator of a healthily functioning urban ecosystem."

40. Christian Nold, ed., *Emotional Cartography: Technologies of the Self* (2009), 5; http://www.emotionalcartography.net/.

41. Alfred N. Whitehead, *The Concept of Nature* (Cambridge, UK: Cambridge University Press, 1920); Bruno Latour, "How to Talk about the Body? The Normative Dimension of Science Studies," *Body and Society* 10, no. 2–3 (2004): 205–229.

42. Christian Nold, "Designing for the Space of Emotion," 1; accessed November 21, 2019, academia.edu/2535581/Designing_for_the_Space_of_Emotion. See also Christian Nold, "Bio Mapping: How Can We Use Emotion to Articulate Cities?," *Navigations* (Spring 2018): 1–16.

43. Nold, personal correspondence with the author, July 2019.

44. The evocative quality of sound is also used in many site-specific audio walks that bring to life the historical atmosphere of a place. It feels particularly acute when the place preserves a dramatic tension between what was "then" and what is "now." This happens, for instance, in an audio walk that Christoph Mayer made in 2007 for Gusen, "now" a quiet, pleasant Austrian village in the middle of a pastoral countryside, and "then" the site of a Nazi concentration camp. Movement through the village is guided by emotional testimonies and graphic sound effects that strongly affect the participant's experience. (See the project description at http://audioweg.gusen.org/.)

45. Elena Biserna, "Mediated Listening Paths: Breaking the Auditory Bubble," *Wi: Journal of Mobile Media* 9, no. 2 (March 31, 2015), http://wi.mobilities.ca/elena-biserna-mediated-listening-paths-breaking-the-auditory-bubble/.

46. Jean-Luc Nancy, *À l'écoute* (Paris: Galilée, 2002), 13; quoted in ibid.

47. Brendon LaBelle, *Background Noise: Perspectives on Sound Art* (New York: Continuum, 2006), xi.

48. Among the most distinguished are site-specific sound and sonic interventions (in the style of pioneers like Max Neuhaus), and soundscapes and field recordings (started by Hildegard Westerkamp). Krzysztof Wodiczko's *Personal Instrument* (1969) was one of the first mobile projects that used a portable sound-playing device (a wearable sound filter based on the intensity of light hitting the user's palms). The advent of GPS-enabled geolocation fostered the development of sound cartography and related collaborative documentary projects. Among the already-classic interactive mobile sound projects are *Sonic City* by Ramia Mazé, Lalya Gaye, and Lars Erik Holmquist (2003), where music is created in real time as a direct result of the user's movements through the city; *Hlemmur In C* by Pall Thayer (2004), a sound performance based on relational dynamics; Steve Symons's *Aura* (2007), an environment of sonic traces left by the movements of others, and accessed by walking through a space equipped with GPS; Teri Rueb's *Trace* (1999—ongoing), a site-specific sound route of memorial songs, poems, and stories; Atau Tanaka's *Mobile Music Making* (2004); "Odd Sympathies" by Sans façon (2008), a "concert" where, as an alternative to sitting, the audience walks; reinterpretations and remixes of local soundscapes as in "Noiseman" by Dunne & Raby (2000), or "Sonic Interface" by Akitsugu Maebayashi (1999); and numerous sound walks through cities including Paris, London, Berlin, and New York where both "mimetic" sounds and vocal narration are used to enhance dynamic geographical visualizations. One of the worthwhile taxonomies of mobile sound art was created by Frauke Behrendt in Frauke Behrendt, "Mobile Sound: Media Art in Hybrid Spaces" (PhD diss., University of Sussex, 2010).

49. Edwin van der Heide, "Radioscape," accessed January 20, 2019, http://www.evdh.net/radioscape/index_more.php. Edwin van der Heide is not a pioneer in introducing electromagnetism to sound art, or to mobile practices. The fascination with "deterritorialized sound" available through radio transmission goes back at least as far as Pierre Schaeffer. One of the most famous recent predecessors is Kristina Kubisch, who since 1980 has been building her own self-designed headsets to provide listeners access to their own individual spaces of time and motion.

50. Ibid. See also Edwin van der Heide, "Radioscape: Into Electromagnetic Space," *Leonardo Music Journal* 23 (December 2013): 15–16; Hanna K. Schraffenberger and Edwin van der Heide, "Sonically Tangible Objects," in *xCoAx 2015: Proceedings of the Third Conference on Computation, Communication, Aesthetics and X*, ed. A. Clifford, M. Carvalhais, and M. Verdicchio (Glasgow, 2015), 233–248.

51. Conversation with Alfons Schilling, New York, 1976, in *Eyes, Lies and Illusions: The Art of Deception*, ed. Laurent Mannoni, Werner Nekes, and Marina Warner (London: Hayward Gallery Publishing, 2004).

52. Steve Mann and Hal Niedzviecki, *Cyborg: Digital Destiny and Human Possibility in the Age of the Wearable Computer* (Toronto: Doubleday Canada, 2001), 6.

53. "Interstitial Space Helmet" (2004), project webpage, accessed November 22, 2019, http://www.auger-loizeau.com/projects/interstitial+space+helmet.

54. Ief Spincemaille, "Reversed Blinking," accessed November 22, 2019, http://www.iefspincemaille.com/Reversed-Blinking.

55. Ief Spincemaille, "Reversed Blinking," accessed July 10, 2017, http://www.iefspincemaille.com/index.php?/off-stage/reversed-blinking/.

56. Anagram, "Door into the Dark" project webpage, accessed November 22, 2019, http://weareanagram.co.uk/project/door-into-the-dark/.

57. Tamiko Thiel, *Fractured Visions: To See Again*, catalogue, 6, accessed May 9, 2015, http://axnscollective.org/fv/.

58. "blemish," archived project website, accessed November 22, 2019, https://web.archive.org/web/20150908083635/http://cont3xt.net/blog/?p=4715.

59. Simona Lodi, "Spatial Art: An Eruption of the Digital into the Physical,' *Leonardo Electronic Almanach* 19, no.2 (2013): 23.

60. As two of the artists themselves write: "We see AR not as representation or simulation, as virtual versus physical, but rather as action upon or transformation of pre-existing schemes of representation, in order to shift or undermine them. The AR artwork is immaterial but nonetheless *real* in the dictionary sense of 'actually existing or happening, not imaginary,' being 'fixed, permanent, immovable' with an 'objective independent existence' as it can be called up by anyone with the proper hardware and software at any time at that site, as long as the artist chooses to it leave it there." Will Pappenheimer, Tamiko Thiel, "Assemblage and Décollage in Virtual Public Space," *NMC|Media-N Journal of the New Media Caucus*, CAA Conference Edition (Summer 2016), http://median.newmediacaucus.org/caa-conference-edition-2016-washington-dc/assemblage-and-decollage-in-virtual-public-space/.

61. Jean-François Lyotard, *Discourse, Figure*, trans. Antony Hudek and Mary Lydon (Minneapolis: University of Minnesota Press, 2011), 12. What is shown in "expression" is the distance, the gap between the sign (the conceptual), the object (the sensory-sensible), and the space itself, which is "an order of coexistence." Ibid., 40.

62. "This conversion renders the image or map always already pre-representational. . . . The object of haptic vision is in movement, in flux; it has lost

its fixity." Nanna Verhoeff, *Mobile Screens: The Visual Regime of Navigation* (Amsterdam: University of Amsterdam Press, 2012), 165.

63. Patrick Lichty, "The Aesthetics of Liminality: Augmentation as Artform," *Leonardo* 47, no. 4 (2014): 326.

64. Rhodes follows here the schema of Jean-Louis Baudry in "Ideological Effects of the Basic Cinematographic Apparatus" (1986).

65. Geoffrey Rhodes, "Augmented Reality in Art: Aesthetics and Material of Expression," in *Augmented Reality Art: From an Emerging Technology to a Novel Creative Medium*, ed. Vladimir Geroimenko (Vienna: Springer, 2014), 130. Rhodes exemplifies this idea with Rosalind Krauss's interpretation: "it seems inappropriate to speak of a physical medium in relation to video. For the object (the electronic equipment and its capabilities) has become merely an appurtenance. And instead, video's real medium is a psychological situation, the very terms of which are to withdraw attention from an external object—an Other—and invest it in the Self." (Rosalind Krauss, "Video: The Aesthetics of Narcissism," in *October* 1 [Spring 1976]: 50–64, quoted in Rhodes, 130.) Perception travels along the wires, nothing is ignored or hidden.

66. Rhodes, "Augmented Reality," 133.

67. Martin Heidegger, *Was heißt Denken?/ What Is Called Thinking*, trans. Fred D. Wreck and J. Glenn Gray (New York: Harper and Row, 1968), 16–17. "Thinking" ability is what helps Heidegger to distinguish the principles of "ready-to-hand" (*Vorhandenheit*, being of a thing, as opposed to being of a human; objective, independent from potential usage) and "present-at-hand" (*Zuhandenheit*, pragmatic, but also a reflexive attitude). See further analysis of Heidegger's thought in Paul Dourish, *Where the Action Is: The Foundations of Embodied Interaction* (Cambridge, MA: MIT Press, 2001), 12, 106–110.

68. See Alva Noë, *Action in Perception* (Cambridge, MA: MIT Press, 2006).

69. Massumi, *Parables for the Virtual*, 59.

70. Paul Schilder, *The Image and Appearance of the Human Body: Studies in the Constructive Energies of the Psyche* (New York: Psychology Press, 1999), quoted in Elisabeth Grosz, *Volatile Bodies: Toward a Corporeal Feminism* (Bloomington: Indiana University Press, 1994), 69.

71. Maurice Merleau-Ponty, *Phenomenology of Perception*, trans. Colin Smith (London: Routledge and Kegan Paul, 1962), 100.

72. Patrick Lichty, "The Translation of Art in Virtual Worlds," in *The Oxford Handbook of Virtuality*, ed. Mark Grimshaw (New York: Oxford University Press, 2014), 445.

73. Brian Massumi, *Semblance and Event: Activist Philosophy and the Occurrent Arts* (Cambridge, MA: MIT Press, 2011), 125.

Conclusion

1. Hubert Dreyfus, *What Computers Still Can't Do* (1972; repr., Cambridge, MA: MIT Press, 1992), 109.

2. Chun's reference to "daemons" is more in line with the idea of James Clerk Maxwell's daemon—"an imaginary agent" that "works tirelessly in the background." In the context of the "invisible" background operations within digital systems, "daemonic processes are slaves that work tirelessly and, like all slaves, define and challenge the position of the master." Wendy Hui Kyong Chun, *Programmed Visions: Software and Memory* (Cambridge, MA: MIT Press, 2011), 88–89.

3. Anna Munster, *Materializing New Media: Embodiment in Information Aesthetics* (Hanover, NH: University Press of New England, 2006), 180. "Informatic bodies no longer summon the immediate presence of corporeal existence, which can be affirmed through habitual codes and conventions of visual representation. They disclose a body's potential for transmutation. Information does not simply represent a body or corporeal experience; it renders the emergent properties and capacities of bodies as mutable states that are variable (and delimited) within certain parameters." Ibid.

4. Sputniko!'s project represents a whole category of artworks that deal with biochemical technologies and raise an array of questions around their use. For instance, analogous to *Red Silk of Fate* (in some ways) is *Eudaimonia* (2018), a short film and a speculative object created by Finnish artist Emilia Tikka. The project features a futuristic scenario in which a person can alter his or her character and personality with the help of a personalized genome-editing device (the specially designed object represents this fictional device, which looks like a desirable consumer product). According to this scenario, in order to find "eudemonia" (Greek for "happiness") in today's capitalistic Western world, one must give up qualities like empathy, or emotional attachments that become obstacles to efficiency. Molecular biology and gene engineering interfere here in the very psyche, affecting it from the inside.

Bibliography

Agamben, Giorgio. *Means without End: Notes on Politics*. Translated by Vincenzo Binetti and Cesare Casarino. Minneapolis: University of Minnesota Press, 2000.

Albu, Cristina. *Mirror Affect: Seeing Self, Observing Others in Contemporary Art*. Minneapolis: University of Minnesota Press, 2016.

Aleksic, Petar S., and Aggelos K. Katsaggelos. "Automatic Facial Expression Recognition Using Facial Animation Parameters and MultiStream HMMs." *IEEE Transactions on Information Forensics and Security* 1, no. 1 (2006): 3–11.

Aleksic, Petar S., Jay J. Williams, Zhilin Wu, and Aggelos K. Katsaggelos. "Audio-Visual Speech Recognition Using MPEG-4 Compliant Visual Features." *EURASIP Journal on Applied Signal Processing* 11 (2002): 1213–1227.

Andersen, Christian Ulrik, and Søren Bro Pold, eds. *Interface Criticism: Aesthetics beyond Buttons*. Aarhus: Aarhus University Press, 2011.

Andersen, Christian Ulrik, and Søren Bro Pold. *The Metainterface: The Art of Platforms, Cities, and Clouds*. Cambridge, MA: MIT Press, 2018.

Andrejevic, Mark, and Mark Burdon. "Defining the Sensor Society." *Television and New Media* 16, no. 1 (2015): 19–36.

Angerer, Marie-Luise, Bernd Bösel, and Michaela Ott, eds. *Timing of Affect: Epistemologies, Aesthetics, Politics*. Zurich: diaphanes; Chicago: University of Chicago Press, 2014.

Arendt, Hannah. *The Human Condition*. Chicago: University of Chicago Press, 1958.

Ashby, W. Ross. *Design for a Brain: The Origin of Adaptive Behavior*. New York: John Wiley and Sons, 1960.

Bach, Joscha. *Principles of Synthetic Intelligence: Psi; An Architecture of Motivated Cognition*. New York: Oxford University Press, 2009.

Balázs, Béla. *Theory of the Film: Character and Growth of a New Art*. Translated by Edith Bone. London: Dennis Dobson, 1952.

Bard-Schwarz, David. *An Introduction to Electronic Art through the Teaching of Jacques Lacan: Strangest Thing*. London: Routledge, 2014.

Bataille, Georges. "The Deviations of Nature." In *Visions of Excess: Selected Writings, 1927–1939*, edited by Allan Stoekl. Minneapolis: University of Minnesota Press, 1985.

Bateson, Gregory. *Steps to an Ecology of Mind: Collected Essays in Anthropology, Psychiatry, Evolution, and Epistemology*. Chicago: University of Chicago Press, 1972.

Beaulieu, Anne. "Images Are Not the (Only) Truth: Brain Mapping, Visual Knowledge, and Iconoclasm." *Science, Technology and Human Values* 27 (2002): 53–86.

Behrendt, Frauke. "Mobile Sound: Media Art in Hybrid Spaces." PhD diss., University of Sussex, 2010.

Berry, David M. *The Philosophy of Software: Code and Mediation in the Digital Age*. New York: Palgrave Macmillan, 2011.

Biggs, Simon, Sue Hawksley, and Garth Paine. "BodyText." 2010. Accessed July 20, 2019. http://www.littlepig.org.uk/installations/bodytext/index.htm.

Biggs, Simon, Sue Hawksley, and Garth Paine. "Bodytext: Somatic Data as Agency in Interactive Dance." In *Multimodality and Performance*, edited by Carla Fernandez, 179–186. Newcastle, UK: Cambridge Scholars Publishing, 2016.

Binswanger, Ludwig, "On the Psychogalvanic Phenomenon in Association Experiments." In *Studies in Word-Association*, edited by Carl Jung, 446–530. London: Heinemann, 1918.

Biserna, Elena. "Mediated Listening Paths: Breaking the Auditory Bubble." *Wi: Journal of Mobile Media* 9, no. 2 (March 31, 2015). Accessed June 2, 2016. http://wi.mobilities.ca/elena-biserna-mediated-listening-paths-breaking-the-auditory-bubble/.

Blast Theory. "Karen." Accessed June 15, 2019. www.blasttheory.co.uk/projects/karen/.

Boehm, Gottfried. *Wie Bilder Sinn erzeugen*. Berlin: Berlin University Press, 2007.

Boehner, Kirsten, Rogerio DePaula, Paul Dourish, and Phoebe Sengers. "How Emotion Is Made and Measured." *International Journal of Human-Computer Studies* 65, no. 4 (2007): 275–291.

Bordwell, David. *Narration in the Fiction Film*. Madison: University of Wisconsin Press, 1985.

Bösel, Bernd, and Marie-Luise Angerer. "Capture All, oder: Who's Afraid of a Pleasing Little Sister?" *Zeitschrift für Medienwissenschaft* 7, no. 2 (2015): 48–56.

Bowker, Geoffrey C., and Susan Leigh Star. *Sorting Things Out: Classification and Its Consequences*. Cambridge, MA: MIT Press, 1999.

Bruce, Vicki, and Andy Young. "Understanding Face Recognition." *British Journal of Psychology* 77 (1986): 305–327.

Callard, Felicity, and Daniel S. Margulies. "The Subject at Rest: Novel Conceptualizations of Self and Brain from Cognitive Neuroscience's Study of the 'Resting State.'" *Subjectivity* 4, no. 3 (2011): 227–257.

Calvo, Rafael, Sidney D'Mello, Jonathan Gratch, and Arvid Kappas, eds. *The Oxford Handbook of Affective Computing*. Oxford: Oxford University Press, 2015.

Carroll, James M., and James A. Russell. "Do Facial Expressions Signal Specific Emotions? Judging Emotion from the Face in Context." *Journal of Personality and Social Psychology* 70, no. 2 (February 1996): 205–218.

Carroll, Noël. *The Philosophy of Horror; or, Paradoxes of the Heart*. New York: Routledge, 1990.

Cartwright, Lisa. *Screening the Body: Tracing Medicine's Visual Culture*. Minneapolis: University of Minnesota Press, 1995.

Casalegno, Mattia. "Unstable Empathy." 2012. Accessed March 15, 2016. www.mattiacasalegno.net/unstable-empathy/.

Certeau, Michel de. *The Practice of Everyday Life*. Translated by Steven Rendall. Berkeley: University of California Press, 1984.

Chang, Vanessa. "Catching the Ghost: The Digital Gaze of Motion Capture." *Journal of Visual Culture* 18, no. 3 (2019): 305–326.

Chatzichristodoulou, Maria. "*Karen* by Blast Theory: Leaking Privacy." In *Digital Bodies: Creativity and Technology in the Arts and Humanities*, edited by Susan Broadhurst and Sara Price, 65–78. London: Palgrave Macmillan, 2017.

Chomsky, Noam. "Where Artificial Intelligence Went Wrong." Interviewed by Yarden Katz. *Atlantic*, November 2012. Accessed November 1, 2015. https://chomsky.info/20121101/.

Choudhury, Suparna, and Jan Slaby, eds. *Critical Neuroscience: A Handbook of the Social and Cultural Contexts of Neuroscience*. Chichester: Wiley-Blackwell, 2012.

Christian, Brian. *The Most Human Human: What Talking with Computers Teaches Us about What It Means to Be Alive*. New York: Doubleday, 2011.

Chun, Wendy Hui Kyong. *Programmed Visions: Software and Memory*. Cambridge, MA: MIT Press, 2011.

Cillari, Sonia. "Sensitive to Pleasure." 2010. Accessed October 10, 2018. http://www.soniacillari.net/.

Clough, Patricia T. "The Affective Turn: Political Economy, Biomedia and Bodies." *Theory, Culture and Society* 25, no. 1 (2008): 1–22.

Clough, Patricia T. "Biotechnology and Digital Information." *Theory, Culture and Society* 24, no. 7–8 (2007): 312–314.

Coccia, Emanuele. *Sensible Life: A Micro-ontology of the Image*. New York: Fordham University Press, 2016.

Cook, Roger F. "Embodied Simulation, Empathy and Social Cognition: Berlin School Lessons for Film Theory." *Screen* 56, no. 2 (Summer 2015): 153–171.

Corbin, Henry. *Mundus Imaginalis, or the Imaginary and the Imaginal*. Accessed May 3, 2010. http://hermetic.com/bey/mundus_imaginalis.htm.

Cox, Geoff. *Antithesis: The Dialectics of Software Art*. Aarhus: Digital Aesthetics Research Center, 2010.

Daalder, Megan. "Mirrorbox." Accessed March 15, 2016. http://www.themirrorbox.info/.

Dade-Robertson, M. *The Architecture of Information: Architecture, Interaction Design and the Patterning of Digital Information*. New York: Routledge, 2011.

Damasio, Antonio R. *The Feeling of What Happens: Body and Emotion in the Making of Consciousness*. New York: Harcourt Brace, 1999.

Daston, Lorraine, and Peter Galison. "Image of Objectivity." *Representations*, no. 40 (October 1992): 81–128.

Deco, Gustavo, Viktor K. Jirsa, and Anthony Randal McIntosh. "Emerging Concepts for the Dynamical Organization of Resting-State Activity in the Brain." *Nature Reviews Neuroscience* 12, no. 1 (2011): 43–56.

Deleuze, Gilles. *Cinema 1: The Movement-Image*. Translated by Hugh Tomlinson and Barbara Habberjam. Minneapolis: University of Minnesota Press, 1986.

Deleuze, Gilles, and Félix Guattari. *A Thousand Plateaus: Capitalism and Schizophrenia*. Translated by Brian Massumi. Minneapolis: University of Minnesota Press, 1987.

Derrida, Jacques. *Of Grammatology*. 1967. Translated by Gayatri Chakravorty Spivak. Baltimore: Johns Hopkins University Press, 1997.

De Souza e Silva, Adriana. "From Cyber to Hybrid: Mobile Technologies as Interfaces of Hybrid Spaces." *Space and Culture* 9, no. 3 (2006): 261–278.

De Vos, Jan, and Ed Pluth, eds. *Neuroscience and Critique: Exploring the Limits of the Neurological Turn*. London: Routledge, 2016.

Devue, Christel, and Serge Brédart. "The Neural Correlates of Visual Self-Recognition." *Consciousness and Cognition* 20, no. 1 (2011): 40–51.

Doane, Mary Ann. "Indexicality: Trace and Sign; Introduction." *Differences* 18, no. 1 (2007): 1–6.

Donnarumma, Marco. "Biotechnological Performance Practice." *Canadian Electroacoustic Community* 14 (2012). Accessed April 4, 2015. http://econtact.ca/14_2/index.html.

Donnarumma, Marco. "Nigredo: Configuring Human and Technological Bodies." In *Experiencing the Unconventional: Science in Art*, edited by Theresa Schubert and Andrew Adamatzky, 61–85. Singapore: World Scientific Publishing, 2015.

Dourish, Paul. *Where the Action Is: The Foundations of Embodied Interaction*. Cambridge, MA: MIT Press, 2004.

Dreyfus, Hubert. *What Computers Still Can't Do*. 1972. Cambridge, MA: MIT Press, 1992.

Duchenne de Boulogne, Guillaume-Benjamin. *The Mechanism of Human Facial Expression*. Translated by Andrew R. Cuthbertson. Cambridge, UK: Cambridge University Press, 1990.

Dumit, Joseph. *Picturing Personhood: Brain Scans and Biomedical Identity*. Princeton, NJ: Princeton University Press, 2004.

EcoArtTech (Leila Nadir and Cary Adams). "Indeterminate Hikes+." Accessed July 4, 2019. http://www.ecoarttech.net/project/indeterminate-hike/.

Ekman, Paul, and Wallace Friesen. "Measuring Facial Movement with the Facial Action Coding System." In *Emotion in the Human Face*, edited by Paul Ekman, 2nd ed., 178–211. Cambridge, UK: Cambridge University Press, 1982.

Elsenaar, Arthur. "Artifacial." Accessed September 20, 2018. http://artifacial.org/.

Elsenaar, Arthur, and Remko Scha. "Electric Body Manipulation as Performance Art: A Historical Perspective." *Leonardo Music Journal* 12 (2002): 17–28.

Elwes, Catherine. *Installation and the Moving Image*. London: Wallflower Press, 2015.

Emerson, Lori. *Reading Writing Interfaces: From the Digital to the Bookbound*. Minneapolis: University of Minnesota Press, 2014.

Fanon, Frantz. *Black Skin, White Masks*. 1952. London: Pluto Press, 1986.

Fazi, M. Beatrice. *Contingent Computation: Abstraction, Experience, and Indeterminacy in Computational Aesthetics*. London: Rowman and Littlefield International, 2018.

Fedorova, Ksenia. "Neurointerfaces, Mental Imagery and Sensory Translation in Art and Science in the Digital Age." In *Invisibility*, edited by Asbjørn Grønstad and Øyvind Vågnes, 91–109. New York: Palgrave Macmillan, 2019.

Fedorova, Ksenia. "Plasticity and Feedback: Schemas of Indetermination in Cybernetics and Art." In *Proceedings of the 22nd International Symposium of Electronic Art (ISEA)* (2016), 51–58.

Fedorova, Ksenia, and Marc Barasch. "Mission to Earth: Terrestrial Proprioception and the Cyber-Sublime." *Azimuth* 6, no. 12 (2018): 93–112.

Ferenczi, Sandor. *Contributions to Psycho-Analysis*. Translated by Ernest Jones. Boston: Richard G. Badger, 1916.

Foucault, Michel. *The Archaeology of Knowledge and The Discourse on Language*. Translated by A. M. Sheridan Smith. New York: Pantheon Books, 1972.

Fuller, Matthew. "A Means of Mutation: Notes on I/O/D 4: The Web Stalker." In *Behind the Blip: Essays on the Culture of Software*, 51–68. New York: Autonomedia, 2003.

Fuller, Matthew, ed. *Software Studies: A Lexicon*. Cambridge, MA: MIT Press, 2008.

Gallagher, Shaun. "Body Schema and Intentionality." In *The Body and the Self*, edited by Jose Bermúdez et al., 225–244. Cambridge, MA: MIT Press, 1995.

Gallese, Vittorio. "Mirror Neurons and the Neural Exploitation Hypothesis: From Embodied Simulation to Social Cognition." In *Mirror Neuron Systems: The Role of Mirroring Processes in Social Cognition*, edited by Jaime A. Pineda, 163–190. New York: Humana Press, 2009.

Galloway, Alexander R. *The Interface Effect*. Cambridge, UK: Polity Press, 2012.

Galloway, Alexander R. *Protocol: How Control Exists after Decentralization*. Cambridge, MA: MIT Press, 2004.

Garcia, David, and Geert Lovink. "ABC of Tactical Media." May 16, 1997. Accessed October 11, 2019. https://www.nettime.org/Lists-Archives/nettime-l-9705/msg00096.html.

Garfinkel, Harold. *Studies in Ethnomethodology*. Englewood Cliffs, NJ: Prentice-Hall, 1967.

Gates, Kelly. *Our Biometric Future: Facial Recognition Technology and the Culture of Surveillance*. New York: New York University Press, 2011.

Geroimenko, Vladimir, ed. *Augmented Reality Art: From an Emerging Technology to a Novel Creative Medium*. Vienna: Springer, 2014.

Gonsalves, Tina. *Tina Gonsalves: Chameleon*. Catalogue. 2010. Accessed August 4, 2019. http://www.tinagonsalves.com/chameleon_catalogue.pdf.

Gonsalves, Tina, Nadia Berthouze, and Matt Iacobini. "The Chameleon Project." *Second Nature: International Journal of Creative Media* 3 (March 2010): 138–163.

Goriunova, Olga. *Art Platforms and Cultural Production on the Internet*. London: Routledge, 2011.

Greenfield, Adam. "Serendipity Cities: Of Services and Situations." April 17, 2010. Accessed August 4, 2015. https://speedbird.wordpress.com/2010/04/17/serendipity-cities-of-services-and-situations/.

Gregg, Melissa, and Gregory J. Seigworth, eds. *The Affect Theory Reader*. Durham: Duke University Press, 2010.

Grosz, Elisabeth. *Volatile Bodies: Toward a Corporeal Feminism*. Bloomington: Indiana University Press, 1994.

Hadler, Florian, and Joachim Haupt, eds. *Interface Critique*. Berlin: Kadmos Kulturverlag, 2016.

Hall, Doug, and Sally Jo Fifer, eds. *Illuminating Video: An Essential Guide to Video Art*. New York: Aperture Foundation, 1990.

Halpern, Orit. *Beautiful Data: A History of Vision and Reason since 1945*. Durham: Duke University Press, 2014.

Hansen, Mark B. N. *Bodies in Code: Interfaces with Digital Media*. New York: Routledge, 2006.

Hansen, Mark B. N. *Embodying Technesis: Technology beyond Writing*. Ann Arbor: University of Michigan Press, 2000.

Hansen, Mark B. N. *New Philosophy for New Media*. Cambridge, MA: MIT Press, 2004.

Hansen, Mark B. N. "The Time of Affect, or Bearing Witness to Life." *Critical Inquiry* 30, no. 3 (2004): 584–626.

Harley, John B. *The New Nature of Maps: Essays in the History of Cartography*. Edited by Paul Laxton. Baltimore: Johns Hopkins University Press, 2001.

Harrell, D. Fox. *Phantasmal Media: An Approach to Imagination, Computation, and Expression*. Cambridge, MA: MIT Press, 2013.

Harrell, D. Fox, and Sneha Veeragoudar Harrell. "Imagination, Computation, and Self-Expression: Situated Character and Avatar Mediated Identity." *Leonardo Electronic Almanac* 17, no. 2 (2012): 74–91.

Hawksley, Sue. "Dancing to an Understanding of Embodiment." PhD diss., University of Edinburgh, 2012.

Hawksley, Sue, and Simon Biggs. "Memory Maps in Interactive Dance Environments." *International Journal of Performance Arts and Digital Media* 2, no. 2 (2006): 123–137.

Hayles, N. Katherine. *My Mother Was a Computer*. Chicago: University of Chicago Press, 2005.

Hayles, N. Katherine. "Print Is Flat, Code Is Deep: The Importance of Media-Specific Analysis." *Poetics Today* 25, no. 1 (Spring 2004): 68–90.

Heide, Edwin van der. "Radioscape." Accessed January 20, 2019. http://www.evdh.net/radioscape/index_more.php.

Heide, Edwin van der. "Radioscape: Into Electromagnetic Space." *Leonardo Music Journal* 23 (December 2013): 15–16.

Heidegger, Martin. *Was heißt Denken? / What Is Called Thinking*. Translated by Fred D. Wreck and J. Glenn Gray. New York: Harper and Row, 1968.

Heinisch, Christine, Hubert R. Dinse, Martin Tegenthoff, Georg Juckel, and Martin Bruene. "An rTMS Study into Self-Face Recognition Using Video-Morphing Technique." *SCAN* 6 (2011): 442–449.

Heuvel, Martijn P. van den, René C. W. Mandl, René S. Kahn, and Hilleke E. Hulshoff Pol. "Functionally Linked Resting-State Networks Reflect the Underlying Structural Connectivity Architecture of the Human Brain." *Human Brain Mapping* 30, no. 10 (2009): 3127–3141.

Hight, Jeremy. "Rhizomatic Cartography: Modulated Mapping and the Spatial Net." NeMe, posted in May 2009. Accessed August 2, 2019. http://www.neme.org/991/rhizomatic-cartography.

Hookway, Branden. *Interface*. Cambridge, MA: MIT Press, 2014.

Iacobini, Matt, Tina Gonsalves, Nadia Bianchi-Berthouze, and Chris Frith. "Emotional Contagion in Interactive Art." In *The Proceedings of the International Conference on Kansei Engineering and Emotion Research, KEER 2010*, 1975–1984. Paris: Arts et Métiers ParisTech, 2010.

Iles, Chrissie. *Into the Light: The Projected Image in American Art, 1964–1977*. New York: Whitney Museum, 2000.

Jaques, Natasha, Sarah Taylor, Akane Sano, and Rosalind W. Picard. "Multi-task, Multi-kernel Learning for Estimating Individual Wellbeing." In *Proceedings of NIPS Workshop on Multimodal Machine Learning*, Montreal, December 2015. Accessed January 2, 2017. http://affect.media.mit.edu/pdfs/15.Jaques-etal-NIPSMMML.pdf.

Jefferies, Janis. "Blurring the Boundaries: Performance, Technology and the Artificial Sublime—an Interview with Ruth Gibson and Bruno Martelli, igloo." In *Interfaces of Performances*, edited by Maria Chatzichristodoulou, Janis Jeffries, and Rachel Zerihan, 43–54. London: Ashgate, 2009.

Johansson, Gunnar. "Visual Perception of Biological Motion and a Model for Its Analysis." *Perception and Psychophysics* 14, no. 2 (1973): 201–211.

Johnston, John. "Machinic Vision." *Critical Inquiry* 26, no. 1 (Autumn 1999): 27–48.

Jones, Adrian. *Self/image: Technology, Representation, and the Contemporary Subject.* London: Routledge, 2006.

Jones, Caroline A., ed. *Sensorium: Embodied Experience, Technology, and Contemporary Art.* Cambridge, MA: MIT Press, 2006.

Kitchin, Rob, Chris Perkins, and Martin Dodge. "Thinking about Maps." In *Rethinking Maps: New Frontiers in Cartographic Theory*, edited by Martin Dodge, Rob Kitchin, and Chris Perkins, 1–25. London: Routledge, 2009.

Kittler, Friedrich. *Gramophone, Film, Typewriter.* Translated by G. Winthrop-Young and M. Wutz. Stanford, CA: Stanford University Press, 1999.

Kittler, Friedrich. "There Is No Software." In Friedrich Kittler, *Literature, Media, Information Systems: Essays*, edited by John Johnston, 147–155. Amsterdam: G+B Arts International, 1997.

Klein, Isabelle, et al. "Retinoptic Organization of Visual Mental Images as Revealed by Functional Magnetic Resonance Imaging." *Cognitive Brain Research* 22 (2004): 26–31.

Kostka, Genia. "China's Social Credit Systems and Public Opinion: Explaining High Levels of Approval." *New Media and Society* 21, no. 7 (July 2019): 1565–1593.

Kraemer, Sybille. "The Cultural Techniques of Time Axis Manipulation: On Friedrich Kittler's Conception of Media." *Theory, Culture and Society* 23, no. 7–8 (2006): 93–109.

Kraemer, Sybille. "Medien, Boten, Spuren. Wenig mehr als Literaturbericht." In *Was ist ein Medium?* edited by Stefan Muenker and Alexander Roesler, 65–90. Frankfurt am Main: Suhrkamp, 2008.

Kraemer, Sybille. "Operative Bildlichkeit. Von der 'Grammatologie' zu einer 'Diagrammatologie'? Reflexionen über erkennendes Sehen." In *Logik des Bildlichen. Zur Kritik der ikonischen Vernunft*, edited by Martina Hessler and Dieter Mersch, 94–123. Bielefeld: transcript, 2009.

Kraemer, Sybille. "Trace, Writing, Diagram: Reflections on Spatiality, Intuition, Graphical Practices and Thinking." In *The Power of the Image: Emotion, Expression, Explanation*, edited by András Benedek and Kristóf Nyíri, 3–22. Frankfurt am Main: Peter Lang, 2014.

Krauss, Rosalind E. "Video: The Aesthetics of Narcissism." *October* 1 (Spring 1976): 50–64.

LaBelle, Brendon. *Background Noise: Perspectives on Sound Art.* New York: Continuum, 2006.

Lacan, Jacques. *Écrits: A Selection.* Translated by Bruce Fink. New York: W. W. Norton, 2002.

Lammes, Sybille. "Terra Incognita: Computer Games, Cartography and Spatial Stories." In *Digital Material: Tracing New Media in Everyday Life and Technology*, edited by Marianne van den Boomen, Sybille Lammes, Ann-Sophie Lehmann, Joost Raessens, and Mirko-Tobias Schäfer, 223–235. Amsterdam: Amsterdam University Press, 2009.

Lancel, Karen, and Hermen Maat. "Saving Face." 2012. Accessed January 15, 2018. https://www.lancelmaat.nl/work/saving-face/.

Latour, Bruno. "Drawing Things Together." In *Representation in Scientific Practice*, edited by Michael Lynch and Steve Woolgar, 19–68. Cambridge, MA: MIT Press, 1990.

Latour, Bruno. "How to Talk about the Body? The Normative Dimension of Science Studies." *Body and Society* 10, no. 2–3 (2004): 205–229.

Laurel, Brenda. *Computers as Theatre*. Reading, MA: Addison-Wesley, 1993.

Le, Nguyen-Thinh, Julia Masche, and Niels Pinkwart. "From ELIZA to Modern Chatbots and Dialogue Systems: The Historical Development of Conversational Software; Classification, Technology and Application Domains." In *Hello, I'm Eliza. Fünfzig Jahre Gespräche mit Computern*, edited by Marianna Baranovska and Stefan Höltgen, 143–160. Bochum and Freiburg: projekt verlag, 2018.

LeDoux, Joseph. *The Emotional Brain: The Mysterious Underpinnings of Emotional Life*. New York: Simon and Schuster, 1996.

Lee, Ray F. "Emergence of the Default-Mode Network from Resting-State to Activation-State in Reciprocal Social Interaction via Eye Contact." *IEEE Engineering in Medicine and Biology Society* (2015): 1821–1824.

Lefebvre, Henri. *The Production of Space*. Translated by D. Nicholson-Smith. Oxford: Blackwell, 1991.

Levinas, Emmanuel. *Totality and Infinity: An Essay on Exteriority*. Translated by Alphonso Lingis. Pittsburgh: Duquesne University Press, 1998.

Leys, Ruth. "The Turn to Affect: A Critique." *Critical Inquiry* 37, no. 3 (Spring 2011): 434–472.

Lichty, Patrick. "The Aesthetics of Liminality: Augmentation as Artform." *Leonardo* 47, no. 4 (2014): 325–336.

Lichty, Patrick. "The Translation of Art in Virtual Worlds." In *The Oxford Handbook of Virtuality*, edited by Mark Grimshaw, 444–462. New York: Oxford University Press, 2014.

Littlefield, Melissa M. *Instrumental Intimacy: EEG Wearables and Neuroscientific Control*. Baltimore: Johns Hopkins University Press, 2018.

Lodi, Simona. "Spatial Art: An Eruption of the Digital into the Physical." *Leonardo Electronic Almanac* 19, no. 2 (2013): 12–31.

Lovejoy, Margot, Christiane Paul, and Victoria Vesna, eds. *Context Providers: Conditions of Meaning in Media Arts*. Bristol, UK: Intellect Books, 2011.

Luptan, Deborah. *The Quantified Self: A Sociology of Self-Tracking*. Malden, MA: Polity, 2016.

Lyotard, Jean-François. *Discourse, Figure*. Translated by Antony Hudek and Mary Lydon. Minneapolis: University of Minnesota Press, 2011.

Lysen, Flora. "The Interface Is the (Art)Work: EEG-Feedback, Circuited Selves and the Rise of Real-Time Brainmedia (1964–1977)." In *Brain Art: Brain-Computer Interfaces for Artistic Expression*, edited by Anton Nijholt, 33–63. Cham, Switzerland: Springer International, 2019.

Lysen, Flora. "Kissing and Staring in Times of Neuromania: The Social Brain in Art-Science Experiments." In *Artful Ways of Knowing: Dialogues between Artistic Research and Science and Technology Studies*, edited by Henk Borgdorff, Peter Peters, and Trevor Pinch, 167–183. New York: Routledge, 2019.

Mann, Steve, and Niedzviecki, Hal. *Cyborg: Digital Destiny and Human Possibility in the Age of the Wearable Computer*. Toronto: Doubleday Canada, 2001.

Mannoni, Laurent, Werner Nekes, and Marina Warner, eds. *Eyes, Lies and Illusions: The Art of Deception*. London: Hayward Gallery Publishing, 2004.

Manovich, Lev. *The Language of New Media*. Cambridge, MA: MIT Press, 2003.

Marker, Chris. "Dialector." Accessed December 20, 2018. http://dialector.poptronics.fr/.

Marks, Laura. *Touch: Sensuous Theory and Multisensory Media*. Minneapolis: University of Minnesota Press, 2002.

Martinovic, Ivan, et al. "On the Feasibility of Side-Channel Attacks with Brain-Computer Interfaces." In *Proceedings of the 21st USENIX Security Symposium*, Bellevue, WA, August 2012. Accessed November 13, 2019. https://www.usenix.org/conference/usenixsecurity12/technical-sessions/presentation/martinovic.

Massey, Doreen. *For Space*. London: Sage, 2005.

Massumi, Brian. *Parables for the Virtual*. Cambridge, MA: MIT Press, 2002.

Massumi, Brian. *Semblance and Event: Activist Philosophy and the Occurrent Arts*. Cambridge, MA: MIT Press, 2011.

Matz, Sandra, Yin Wah Fiona Chan, and Michal Kosinski. "Models of Personality." In *Emotions and Personality in Personalized Services: Models, Evaluation and Applications*, edited by Marko Tkalčič, Berardina De Carolis, Marco de Gemmis, Ante Odić, Andrej Košir, 35–54. Cham, Switzerland: Springer Internatinal, 2016.

McCarthy, Lauren. "LAUREN." 2017. Accessed May 30, 2019. http://get-lauren.com/.

McLuhan, Marshall. *Understanding Media: The Extensions of Man*. 1964. Cambridge, MA: MIT Press, 1994.

Merleau-Ponty, Maurice. *Phenomenology of Perception*. Translated by Colin Smith. London: Routledge and Kegan Paul, 1962.

Merleau-Ponty, Maurice. *The Primacy of Perception*. Edited by James M. Edie. Evanston, IL: Northwestern University Press, 1964.

Milburn, Colin. *Nanovision: Engineering the Future*. Durham: Duke University Press, 2008.

Mitchell, W. J. T. *Iconology: Images, Text, Ideology*. Chicago: University of Chicago Press, 1986.

Mitchell, W. J. T. *Picture Theory: Essays on Verbal and Visual Representation*. Chicago: University of Chicago Press, 1994.

Mitchell, Wade J., Chin-Chang Ho, Himalaya Patel, and Karl F. MacDorman. "Does Social Desirability Bias Favor Humans? Explicit–Implicit Evaluations of Synthesized Speech Support a New HCI Model of Impression Management." *Computers in Human Behavior* 27 (2011): 402–412.

Molnar-Szakacs, Istvan, and Lucina Q. Uddin. "The Emergent Self: How Distributed Neural Networks Support Self-Representation." In *Handbook of Neurosociology*, edited by D. D. Franks and J. H. Turner, 167–182. New York: Springer, 2013.

Mondloch, Kate. *Capsule Aesthetic: Feminist Materialisms in New Media*. Minneapolis: University of Minnesota Press, 2018.

Mondloch, Kate. *Screens: Viewing Media Installation Art*. Minneapolis: University of Minnesota Press, 2010.

Montero, Barbara. "Proprioception as an Aesthetic Sense." *Journal of Aesthetics and Art Criticism* 64, no. 2 (2006): 231–242.

Morcom, Alexa M., and Paul C. Fletcher. "Does the Brain Have a Baseline? Why We Should Be Resisting a Rest." *NeuroImage* 37, no. 4 (2007): 1073–1082.

Munster, Anna. *Materializing New Media: Embodiment in Information Aesthetics*. Hanover, NH: University Press of New England, 2006.

Murphy, Mike, and Jacob Templin. "This App Is Trying to Replicate You." *Quartz*, August 2, 2019. https://classic.qz.com/machines-with-brains/1018126/lukas-replika-chatbot-creates-a-digital-representation-of-you-the-more-you-interact-with-it/.

Nakamura, Lisa. *Digitizing Race: Visual Cultures of the Internet*. Minneapolis: University of Minnesota Press, 2007.

Nancy, Jean-Luc. *À l'écoute*. Paris: Galilée, 2002.

Neddam, Martine. "Mouchette." Accessed December 10, 2016. http://www.mouchette.org.

Niemeyer, Richard E. "What Are the Neurological Foundations of Identities and Identity-Related Processes?" In *Handbook of Neurosociology*, edited by D. D. Franks and J. H. Turner, 149–165. New York: Springer, 2013.

Noë, Alva. *Action in Perception*. Cambridge, MA: MIT Press, 2006.

Nold, Christian, ed. *Emotional Cartography: Technologies of the Self*. London: Space Studios, 2009. Accessed April 4, 2019. http://www.emotionalcartography.net/.

Norvig, Peter. "On Chomsky and the Two Cultures of Statistical Learning." Accessed August 2, 2019. http://norvig.com/chomsky.html.

Nusselder, André. *Interface Fantasy: A Lacanian Cyborg Ontology*. Cambridge, MA: MIT Press, 2009.

Offenhuber, Dietmar. "Kuleshov's Display—On Contextual Invisibility." In *Throughout: Art and Culture Emerging with Ubiquitous Computing*, edited by Ulrik Ekman, 431–442. Cambridge, MA: MIT Press, 2013.

Panksepp, Jaak. "On the Embodied Nature of Core Emotional Affects." *Journal of Consciousness Studies* 5 (2005): 158–184.

Pardàs, Montse, and Antonio Bonafonte. "Facial Animation Parameters Extraction and Expression Recognition Using Hidden Markov Models." *Signal Processing: Image Communication* 17 (2002): 675–688.

Parisi, Luciana. "Biotech: Life by Contagion." *Theory, Culture and Society* 24, no. 6 (2007): 29–52.

Parisi, Luciana. "Technoecologies of Sensation." In *Deleuze/Guattari and Ecology*, edited by Bernd Herzogenrath, 182–199. Basingstoke: Palgrave Schol, 2009.

Parrott, W. Gerrod, ed. *Emotions in Social Psychology*. New York: Psychology Press, 2000.

Paul, Christiane. "'Not Just Art'—From Media Art to Artware." In *Interface Cultures: Artistic Aspects of Interaction*, edited by Christa Sommerer, Laurent Mignonneau, and Dorothée King, 141–151. Bielefeld: transcript Verlag, 2008.

Pearlman, Ellen. "Brain Opera: Exploring Surveillance in 360-Degree Immersive Theatre." *PAJ: A Journal of Performance and Art* 39, no. 2 (2017): 79–85.

Pedersen, Isabel. *Ready to Wear: A Rhetoric of Wearable Computers and Reality-Shifting Media*. Anderson, SC: Parlor Press, 2013.

Pedersen, Isabel, and Andrew Iliadis, eds. *Embodied Computing: Wearables, Implantables, Embeddables, Ingestables*. Cambridge, MA: MIT Press, 2020.

Peirce, Charles S. *The Writings of Charles S. Peirce*. Vol. 5. Bloomington: Indiana University Press, 1993.

Penny, Simon. "Traces." 1997. Accessed November, 26, 2016. http://simonpenny.net/texts/traces_embodied.html.

Pinchevski, Amit. "The Face as a Medium." In *Images, Ethics, Technology*, edited by Sharrona Pearl, 193–201. New York: Routledge, 2016.

Pönkänen, Laura M., Annemari Alhoniemi, Jukka M. Leppänen, and Jari K. Hietanen. "Does It Make a Difference If I Have an Eye Contact with You or with Your Picture? An ERP Study." *Social Cognitive and Affective Neuroscience* 6, no. 4 (2011): 486–494.

Raichle, Marcus E. "The Brain's Dark Energy." *Science* 314, no. 5803 (2006): 1249–1250.

Raichle, Marcus E. "The Brain's Dark Energy." *Scientific American* 302 (March 2010): 44–49.

Rancière, Jacques. *The Politics of Aesthetics: The Distribution of the Sensible*. Translated by Gabriel Rockhill. London: Continuum, 2006.

Raskin, Jef. *The Humane Interface: New Directions for Designing Interactive Systems*. Reading, MA: Addison-Wesley, 2000.

Ravetto-Biagioli, Kriss. *Digital Uncanny*. New York: Oxford University Press, 2019.

Ravetto-Biagioli, Kriss. "The Digital Uncanny and Ghost Effects." *Screen* 57, no. 1 (Spring 2016): 1–20.

Reich, Wilhelm. "Experimental Investigation of the Electrical Function of Sexuality and Anxiety." *Journal of Orgonomy* 3, no. 1–2 (1969): 132–154.

Rogers, Carl, and Barry Stevens, eds. *Person to Person: The Problem of Being Human*. Lafayette, CA: Real People Press, 1967.

Rose, Nicholas. *Inventing Our Selves: Psychology, Power, Personhood*. New York: Cambridge University Press, 1996.

Rosenblueth, Arturo, Norbert Wiener, and Julian Bigelow. "Behavior, Purpose and Teleology." *Philosophy of Science* 10, no. 1 (January 1943): 18–24.

Ross, Christine. "The Projective Shift between Installation Art and New Media Art: From Distantiation to Connectivity." In *Screen/Space: The Projected Image in*

Contemporary Art, edited by Tamara Trodd, 184–205. Manchester, UK: Manchester University Press, 2011.

Ross, Christine. "Spatial Poetics: The (Non)Destinations of Augmented Reality Art." *Afterimage* 38, no. 2 (September/October 2010): 19–24.

Rotman, Brian. *Becoming Beside Ourselves: The Alphabet, Ghosts, and Distributed Human Being*. Durham: Duke University Press, 2008.

Rotman, Brian. "Corporeal or Gesturo-haptic Writing." *Configurations* 10, no. 3 (Fall 2002): 423–438.

Rueb, Teri. "The Choreography of Everyday Movement–2001." Accessed November 21, 2019. http://terirueb.net/the-choreography-of-everyday-movement-2001/.

Rydfalk, Mikael. "CANDIDE, a Parameterized Face." Report No. LiTH-ISY-I-866, Department of Electrical Engineering, Linköping University, Sweden, 1987.

Sack, Warren, and Joseph Dumit. "Artificial Participation: An Interview with Warren Sack." In *Zeroing In on the Year 2000: The Final Edition*. Late Editions: Cultural Studies for the End of the Century, book 8, edited by George E. Marcus, 59–87. Chicago: University of Chicago Press, 2000.

Sacks, Oliver. *The Man Who Mistook His Wife for a Hat and Other Clinical Tales*. London: Derald Duckworth, 1985.

Saussure, Ferdinand de. *Course in General Linguistics*. Translated by Roy Harris. Peru, IL: Open Source, 1988.

Scherer, Klaus R., Tanja Bänziger, and Etienne B. Roesch, eds. *Affective Computing: A Sourcebook*. Oxford: Oxford University Press, 2010.

Schilder, Paul. *The Image and Appearance of the Human Body: Studies in the Constructive Energies of the Psyche*. New York: Psychology Press, 1999.

Schiller, Devon. "The Autosemiographic Method: Iconicity and Self-Experimentation in the Science of Facial Expression." 12th International Symposium on Iconicity in Language and Literature. Lund University, 2019.

Schraffenberger, Hanna K., and Edwin van der Heide. "Sonically Tangible Objects." In *xCoAx: Proceedings of the Third Conference on Computation, Communication, Aesthetics and X, Glasgow, 2015*, edited by A. Clifford, M. Carvalhais, and M. Verdicchio, 233–248. Porto: Universidade do Porto, 2015.

Schüll, Natasha. "Tracking." In *Experience: Culture, Cognition and the Common Sense*, edited by Caroline Jones, David Mather, and Rebecca Uchill, 195–203. Cambridge, MA: MIT Press, 2016.

Schwartzman, Madeleine. *See Yourself Sensing: Redefining Human Perception*. London: Black Dog, 2011.

Simmel, Georg. "The Aesthetic Significance of the Face." In *Georg Simmel, 1858–1918*, edited by Kurt H. Wolff, 276–281. Columbus: Ohio State University Press, 1959.

Simon, Herbert, and Allen Newell. "Heuristic Problem Solving: The Next Advance in Operations Research." *Operations Research* 6, no. 1 (January/February 1958): 1–10.

Simondon, Gilbert. "The Genesis of the Individual." In *Incorporations*, edited by Jonathan Crary and Sanford Kwinter, translated by Mark Cohen and Sanford Kwinter, 297–319. New York: Zone Books, 1992.

Simondon, Gilbert. "The Position of the Problem of Ontogenesis." Translated by Gregory Flanders. *Parresia Journal for Critical Philosophy* 7 (2009): 4–16.

Slaby, Jan. "Steps towards a Critical Neuroscience." *Phenomenology and the Cognitive Sciences* 9 (2010): 397–416.

Smith, Gavin J. D. "Surveillance, Data and Embodiment: On the Work of Being Watched." *Body and Society* 22, no. 2 (2016): 108–139.

Smith, Murray. *Engaging Characters: Fiction, Emotion and the Cinema*. Oxford: Clarendon, 1995.

Stalbaum, Brett. "Database Logic(s) and Landscape Art." C5 Corporation. Accessed June 2, 2019. http://www.c5corp.com/research/databaselogic.shtml.

Stalbaum, Brett. "The Landscape and Culture. Data as Mediator, Modes of Engagement." C5 Corporation. 2003. Accessed June 2, 2010. http://www.c5corp.com/research/landscapeculture.shtml.

Stelarc. "Prosthetic Head: Intelligence, Awareness, and Agency." In *Critical Digital Studies: A Reader*, edited by Arthur Kroker and Marilouise Kroker, 475–489. Toronto: University of Toronto Press, 2008.

Stiegler, Bernard. *The Fault of Epimetheus*. Vol. 1 of *Technics and Time*. Stanford, CA: Stanford University Press, 1998.

Sutton, Damian, Susan Brind, and Ray McKenzie, eds. *The State of the Real: Aesthetics in the Digital Age*. New York: I. B. Tauris, 2007.

Swan, Melanie. "The Quantified Self: Fundamental Disruption in Big Data Science and Biological Discovery." *Big Data* 1, no. 2 (June 2013): 85–99.

Teitelbaum, Richard. "In Tune: Some Early Experiments in Biofeedback Music (1966–74)." In *Biofeedback and the Arts: Results of Early Experiments*, edited by David Rosenboom, 35–65. Vancouver: A.R.C. Publications, 1976.

Thiel, Tamiko. *Fractured Visions: To See Again*. Catalogue. Accessed May 9, 2015. http://axnscollective.org/fv/.

Thrift, Nigel. "Bare Life." In *Cultural Bodies: Ethnography and Theory*, edited by Helen Thomas and Jamilah Ahmet, 145–169. Oxford: Blackwell, 2004.

Thrift, Nigel. *Knowing Capitalism*. London: Sage, 2005.

Tian, Ying-li, Takeo Kanade, and Jeffrey F. Cohn. "Recognizing Action Units for Facial Expression Analysis." *IEEE Transactions on Pattern Analysis and Machine Intelligence* 23, no. 2 (2001): 97–115.

Timeto, Federica. "Locating Media, Performing Spatiality: A Nonrepresentational Approach to Locative Media." In *Locative Media*, edited by Rowan Wilken and Gerard Goggin, 94–106. London: Routledge, 2015.

Tomkins, Silvan S. *Affect, Imagery, Consciousness*. 2 vols. New York: Springer, 1962–1963.

Tomkins, Silvan S., and Samuel Messick, eds. *Computer Simulation of Personality*. New York: John Wiley and Sons, 1963.

Tormey, Alan. *The Concept of Expression*. Princeton, NJ: Princeton University Press, 1971.

Turing, Alan. "Computing Machinery and Intelligence." *Mind* 49, no. 1 (October 1950): 433–460.

Turkle, Sherry. *Life on the Screen: Identity in the Age of the Internet*. New York: Simon and Schuster, 1995.

Uddin, Lucina Q., Jonas T. Kaplan, Istvan Molnar-Szakacs, Eran Zaidel, and Marco Iacobini. "Self-Face Recognition Activates a Frontoparietal 'Mirror' Network in the Right Hemisphere: An Event-Related fMRI Study." *NeuroImage* 25, no. 3 (2005): 926–935.

Verhoeff, Nanna. "A Logic of Layers: Indexicality of iPhone Navigation in Augmented Reality." In *Studying Mobile Media: Cultural Technologies, Mobile Communication and the iPhone*, edited by Larissa Hjorth, Jeann Burgess, and Ingrid Richardson, 118–132. New York: Routledge, 2012.

Verhoeff, Nanna. "The Medium Is the Method: Locative Media for Digital Archives." In *(Dis)orienting Media and Narrative Mazes*, edited by Julia Eckel, Bernd Leiendecker, Daniela Olek, and Christine Piepiorka, 21–34. Bielefeld: Transcript Verlag, 2013.

Verhoeff, Nanna. *Mobile Screens: The Visual Regime of Navigation*. Amsterdam: University of Amsterdam Press, 2012.

Verhoeff, Nanna. "Theoretical Consoles: Concepts for Gadget Analysis." *Journal of Visual Culture* 8, no. 3 (2009): 279–298.

Virilio, Paul. *Negative Horizon: An Essay in Dromoscopy*. London: Continuum, 2005.

Viseu, Ana, and Lucy Suchman. "Wearable Augmentations: Imaginaries of the Informed Body." In *Technologized Images, Technologized Bodies*, edited by Jeanette

Edwards, Penny Harvey, and Peter Wade, 161–184. New York: Berghahn Books, 2010.

Wajcman, Judy. *Techno-Feminism*. Cambridge, UK: Polity Press, 2004.

Wardrip-Fruin, Noah. *Expressive Processing: Digital Fictions, Computer Games, and Software Studies*. Cambridge, MA: MIT Press, 2009.

Weate, Jeremy. "Fanon, Merleau-Ponty and the Difference of Phenomenology." In *Race*, ed. Robert Bernasconi, 169–183. Malden, MA: Blackwell, 2001.

Weigel, Sigrid. "Embodied Simulation and the Coding-Problem of Simulation Theory: Interventions from Cultural Sciences." In *A Neuro-Psychoanalytical Dialogue for Bridging Freud and the Neurosciences*, edited by Gerhard Scharbert, 47–71. Luxemburg: Springer, 2016.

Weigel, Sigrid. "Phantom Images: Face and Feeling in the Age of Brain Imaging." *Zeitschrift für Kunst und Kulturwissenschaften. Kritische Berichte* 40, no. 1 (2012): 33–53.

Weizenbaum, Joseph. *Computer Power and Human Reason: From Judgment to Calculation*. San Francisco: W. H. Freeman, 1976.

Weizenbaum, Joseph. "ELIZA—A Computer Program for the Study of Natural Language Communication between Man and Machine." *Communications of the ACM* 9, no. 1 (January 1966): 36–45.

Whitehead, Alfred N. *The Concept of Nature*. Cambridge, UK: Cambridge University Press, 1920.

Wilson, Elizabeth. *Affect and Artificial Intelligence*. Seattle: University of Washington Press, 2010.

Wood, Denis, with Jon Fels. *The Power of Maps*. New York: Guilford, 1992.

Wright, Alexa, Eugenie Shinkle, and Aif Linney. "*Alter Ego*: Computer Reflections of Human Emotions." *Proceedings of the 6th DAC Conference, Copenhagen* (December 2005), 191–199. http://www.ucl.ac.uk/conversation-piece/wright_shinkle_linney.pdf.

Zharinova, Ekaterina. "Inside the Chain: An Intersection of Dance, Mathematics, and Technology." MFA thesis, George Washington University, 2016.

Index